ENCYCLOPEDIA OF DRUG AND DRUG DELIVERY

ENCYCLOPEDIA OF DRUG AND DRUG DELIVERY

Vol. 4

DRUG THERAPY

By

Dr. Kamlesh Dashora
Associate Professor
Institute of Pharmacy
Vikram University
Ujjain (M.P.)
(India)

&

Dr. Darshan Dubey
Assistant Professor
Institute of Pharmacy
Vikram University
Ujjain (M.P.)
(India)

D P H

DISCOVERY PUBLISHING HOUSE PVT. LTD.
NEW DELHI-110 002

Published by:
Tilak Wasan
DISCOVERY PUBLISHING HOUSE PVT. LTD.
4383/4A, Ansari Road, Darya Ganj
New Delhi-110 002 (India)
Phone: +91-11-23279245, 43596064-65
Fax : +91-11-23253475
E-mail: parul.wasan@gmail.com
discoverypublishinghouse@gmail.com
web : www.discoverypublishinggroup.com

***First Edition:* 2012**

ISBN: 978-93-5056-038-9 (Set)
978-93-5056-039-6 (Vol. 1)
978-93-5056-040-2 (Vol. 2)
978-93-5056-041-9 (Vol. 3)
978-93-5056-042-6 (Vol. 4)
978-93-5056-043-3 (Vol. 5)

Encyclopedia of Drug and Drug Delivery

Printed at:
Shree Balaji Art Press
Delhi

Preface

The Present title "**Encyclopedia of Drug and Drug Delivery**" is the amazing advancement of pharmaceutical studies. It provides the various fundamental aspects of drug designing to be understood adequately. It has been compiled for undergraduate, postgraduate students, research scholars, teachers, technicians, industrial microbiologists and scientists involved in pharmaceutical research and development. The basic concepts have been clearly explained and their functions are adequately highlighted. The presentation of text is simple and systematic. The selection of chapters and topics is according to the specified syllabi of several Indian Universities and are so structured as to enable the student to move easily from the fundamental to the complex. It describes the ideas and techniques that are creating or new domain of science and technology. It holds the promise of revolutionizing field ranging from medical science to pharmaceutical industries. It is our earnest hope that this title will be of great value to all our students.

Organizational and economic changes in the pharmaceutical industry have led to a marked reduction in the supply of new chemical entities. This combined with an increase in the expiration of patents and tighter regulatory control has lead many companies to rethink existing product design. The growing trend to improve the efficacy of existing drug products, to both extend parent life and allow competition with opportunistic generic companies, has allowed for a greater understanding of formulation design. This trend is set to continue for some time as economic and political pressure increases.

There have been many improvements in understanding the theoretical and practical aspects of drug discoveries. Numerous therapeutic compounds have been investigated to treat a variety of different illnesses. There are many pharmaceutical methods requiring a variety of formulation options. Advanced formulation design may be thought of as a way to optimize the mode of drug designing to a patient that will ultimately lead to a more effective therapy and approach an optimal therapeutic outcome for a particular drug. This book is designed to take the reader through various cutting-edge areas of advanced formulation design by describing therapeutic categories that may be beneficial. The reader will be introduced to some novel technologies and updated on the state of the art in areas designed to overcome poor efficacy issues.

To make the work more comprehensive and informative, the authors have consulted many authoritative books, research journals, abstracts, monographs etc., so there can be no claim to originality except in the manner of treatment.

The authors express their thanks to their friends and colleagues whose continue inspirations have initiated them to bring out this book.

The authors express their gratitude to Mr. Wasan and staff of M/s Discovery Publishing House Pvt. Ltd., for their whole hearted co-operation in the publications of this book.

The authors will remain sincerely responsible for any shortcomings of the book and be grateful to the readers for their suggestions and constructive criticism for the continuous betterment of the book. They have this opportunity to appeal to the readers to send their suggestions straightaway to this publisher.

–Authors

Contents

1

Introduction

Patients with inflammatory arthritis such as RA can experience a range of physical, psychological, social and/or sexual problems. The role of the nurse is to support, guide, educate and empower the patient and their family so that problems can be identified and care that has meaning and relevance for the patient implemented. This has to be a shared process between the patient and the nurse, as conditions such as RA are not curable and the patient must be committed to a long-term treatment plan. Patients with active disease activity (determined by haematological and biochemical markers, X-ray results and clinical findings) will require a combination of drug therapy to reduce symptoms such as pain and stiffness and also to suppress disease activity, minimizing the potential harm that prolonged inflammation can cause.

Analgesia and non-steroidal anti-inflammatory drugs can provide symptoms relief and the DMARDs can suppress disease activity. The nurse requires in-depth knowledge regarding the condition and treatment options and will need to employ the skills of educator to provide the patient with sufficient information to contribute to shared decision making prior to the commencement of drug therapy.

MEANING OF THERAPEUTIC NURSING

Therapeutic nursing has been defined by Powell (1991) as that practice where the nurse has made a difference to the patient's or client's health state and where he or she is aware of how and why this positive difference has occurred. Levine (1973) distinguishes between nursing which is supportive in nature and seeks to prevent further deterioration and that which is therapeutic, promoting adaptation and contributing to the restoration of well-being. The introduction of drug therapy in conjunction with other treatment interventions – for example, exercise – can be seen as fulfilling both a supportive and a therapeutic role, preventing deterioration and leading to an improvement in symptoms, enabling the patient to participate more fully in meaningful life activities. Caring is the most important value of rheumatology nursing. Although often referred to as a basic requirement there is nothing basic about high quality nursing care, which requires a combination of knowledge, understanding and expertise. Caring involves both an action element in identifying and meeting the needs of the patient and an emotional element, which involves having regard for people as individuals and being concerned about what happens to them.

It would appear that in many situations the action element remains the dominant feature of nursing practice with interactions with patients governed by their physical care needs, resulting in the neglect of the emotional needs of the patient. Yet we know that from the patient's perspective it is how the individual nurse relates to the patient, the emotional style, that determines whether a patient perceives a care episode as satisfactory or not. For care to be effective and holistic in nature the nurse must incorporate both the action and emotional element into care delivery. The adoption of this approach is as important in the instigation of drug therapy as in all treatment interventions, as the patient will have to be satisfied on an emotional level as to the value of drug therapy for compliance to occur.

Interaction between Patient and Nursing

The most important element in rheumatology nursing is the relationship that exists between the patient and the nurse. Such a relationship requires both time and knowledge for the nurse to begin to empower the patient so that an informed decision regarding care management can be reached. The relationship needs to be founded on participation. There is the assumption that by involving the patient in their own care they will ultimately obtain the status of participant but it may well be that the patient is participating from the nurse's frame of reference, rather than from their own view point. Asking the patients to write down their expectations regarding drug therapy may provide the starting point from which the partnership can commence. The factors needed for a therapeutic relationship include:

Encouragement

It is important to encourage the patient to participate in as many treatment decisions as possible. When a patient chooses a course of action such as commencing sulfasalazine they are more likely to adhere even if the beneficial effects take some time to emerge, which in the case of this particular drug could be anything from two to four months. Informing the patient about different drug options and enabling the patient to choose from them should heighten his or her sense of control.

Objectives

Until the nurse has spent time exploring the patient's own beliefs about the purpose and outcome of drug therapy it will not be possible to arrive at shared treatment objectives.

Awareness

1. It is important that the patient has access to a knowledgeable practitioner who is familiar with their care, at any time when there is a perceived change in disease activity or self-management. This can be maintained through a telephone helpline system, which provides the patient with a designated point of contact and helps to reduce anxiety and provide support.
2. It is important that family members are included and assist in care planning so that care options decided can be endorsed and implemented within the patient- supportive framework and within their own social setting.

3. Trying to reach unrealistic goals will only demoralize and adversely affect self- esteem. If the patient is not informed that DMARDs take many months before benefit can be assessed they could become disillusioned with the treatment if symptoms did not improve after the first few weeks and discontinue the regime.
4. It is not helpful to be given generalized information and told what percentage of patients respond well to a given treatment. The individual requires information that is relevant to their own particular situation.

AUDIO TELE CARE

Telephone advice lines have become an integral part of rheumatology care and are traditionally run by clinical nurse specialists or other heath professionals in extended roles. This service provides patients with the means of contacting professionals directly involved in their care management, for advice regarding symptoms management and concerns relating to drug therapy. The telephone advice line also provides direct access for other members of the team involved in the patient's medical and nursing care to seek advice. For example, a general practitioner may ring to question whether gold injections should be discontinued due to the slight presence of haematuria on a urinalysis dipstick test.

The immediate advice by the nurse to continue will prevent treatment being stopped unnecessarily. It will also enable a consistency to care management to develop between members of primary and secondary healthcare. This will reinforce communication links and promote a greater appreciation of the different roles of these two groups.

ARGUMENTS

A philosophy of practice is required to enable nurses within rheumatology to work as a united team with identified shared goals for patient management. A philosophy consists of a system of beliefs from which core practice evolves. The establishment of a philosophy has to include debate and discussion from all members of the nursing team. It is a statement of purpose and will require commitment from all concerned for it to be integrated into practice. If a shared approach to care is not adopted then disunity and fragmentation of care will occur. A rheumatology nursing philosophy of care can be divided into four main areas which are interlinked and complementary to each other. These include:

1. Health.
2. Beliefs relating to the environment.
3. Beliefs relating to the individual patient.
4. Beliefs relating to nursing.

Condition of Patient

Health is achieved when the patient is able to function on a physical, psychological and social level. This is not a reinstatement of complete well-being in all these spheres of life; rather that the patient perceives that they are able to make a useful contribution to all areas of activity. Health requires adaptations and the development of coping strategies to minimize the symptoms of arthritis. Health does not mean the removal of all symptoms; this would be an unrealistic outcome and an unfair burden to place on patients. Health and illness are not static entries and vary depending on:

1. Disease activity;
2. Coping strategies;
3. Available resources;
4. Support.

RA is characterized by flares and remissions of disease activity and the patients may find themselves alternating between health and illness or functioning well on a physical level but not on a psychological level. The patient will need to know how to access a knowledgeable nurse when they experience a stage in their condition that presents different or recurring problems, so that the situation can be reassessed collectively and necessary modifications to the treatment programme made.

Treatment Programme

A patient needs to be involved in the planning of all aspects of care to feel committed and able to implement the agreed treatment programme based on the individual's perceived needs and this applies equally to drug therapy. Neglect of individual concerns can lead to non-adherence with treatment. Nurses need to identify any existing internal and external barriers. A patient may not be able to open their medication container due to reduced manual dexterity or be afraid to continue with treatment due to adverse peer pressure. If the orientation of the consultation between the nurse and the patient does not encourage shared discussion, the patient may not feel part of the treatment regime and discontinue their medications if they are not immediately effective.

Plan at Individual Level

The individual's lay beliefs must be explored before any treatment plan is decided so that the care programme that emerges has both relevance and meaning of the individual concerned. Lay beliefs are usually consistent over time and pertinent to the individual concerned. If a patient has in their lay belief system the thought that drug treatment such as gold has only the potential for damaging side effects to occur and not the potential for improvement they may reject this as a treatment option.

The patient will require explanation on the disease process before they can understand the role of drug therapy. If they do not understand that their pain, stiffness and fatigue is related to disease activity they may not perceive drug treatment as a useful adjunct to other therapy. The individual has a right to be an active not a passive recipient of care so informed decisions can be made and the management of the condition viewed as a shared commitment between the patient and the health professional. Patients may not feel equipped to adapt to this role at first but as the nurse begins to share knowledge with the patient and a therapeutic relationship develops the patient may feel more able to contribute to care decisions. The nurse will be a knowledgeable practitioner using evidence-based practice to underpin holistic care management.

CONDITIONS FOR GOOD NURSING

Knowledgeable Practitioner

Once the clinical decision has been made to commence a patient on DMARD therapy, the patient will require to spend time with a knowledgeable practitioner so that the following aspects can be discussed:

1. The patient's expectations of treatment.
2. Potential side effects.
3. The purpose of the medication.
4. The dosage and time of administration.
5. Monitoring regime.
6. The course to take if a problem arises.

Tones (1991) defines empowerment as the process whereby an individual or community of individuals acquire power – that is, the capacity to control other people and resources. By explaining about the particular drug therapy and the rationale for its introduction the patient will begin to determine its acceptability from their particular view point. Klein (1974) describes five categories of patient involvement:

Basic information

This is the process whereby the health professional gives the information to the patient and the patient passively receives it. This will not be useful for commencing patients on drug therapy as the patient will be the individual responsible for the administration of the medication and to adhere with the proposed regime must be committed to the type of treatment being advocated.

Patient-physician communication may be the single most important variable affecting adherence. For communication to be successful the nurse must present information clearly whilst the patient needs to seek clarification and ensure that their concerns are addressed.

1. The nature of the explanation concerning the diagnosis, and the course and purpose of treatment.
2. When rheumatologists were recorded as having made a clear statement about the purpose of the drug treatment 79% of patients adhered with their therapy compared to only 33% of patients when the explanation offered was not clear.
3. An overuse of medical jargon, which excludes the patient from shared discussion.
4. A shared agreement of the goals for treatment are essential. Arluke (1980) found that some patients with RA stopped their medication at the first sign of improvement as they held the belief that drug efficacy would reduce over time.
5. The need to obtain the patient's concerns regarding drug treatment. Several studies have shown that practitioners misperceive patients' needs regarding the amount, content and preferred method for providing information (e.g. drug information sheets) on arthritis and its treatments.

Ability to satisfy

The health professional may consult the patient and may use the information gained. This will not lead to a shared approach towards treatment management as the practitioner may not have obtained the patient's beliefs, values and expectations, which are all potent factors affecting adherence, not only because of their direct relationship to health behaviour but also because they are subject to misinterpretation unless clarification is sought and will ultimately influence how satisfied the patient is with the consultation.

Decision making

Here equality exists and both parties contribute to the decision making and negotiate the treatment. The nurse would begin by ascertaining the patient's concerns regarding treatment and identify potential compliance difficulties and jointly plan how to overcome them.

Decision process

A patient's values are taken into account and underpin the decision process. A patient may prefer not to commence methotrexate if they drink alcohol regularly and will be more likely to comply with an alternative drug where they do not have to abstain from alcohol. The patient holds the right not to comply with treatment. Liang (1989) describes non-compliance as the ultimate experience of independence. If this occurs it should do so only after full negotiation has occurred and not as a result of ignorance or non-disclosure. The nurse would seek to ascertain the reason why the patient was declining drug therapy. It may be due to a lack of knowledge or apprehensions that have not been discussed. If patients decline therapy they must be given the opportunity to reaccess the service at any time to reassess their condition and other treatment options.

COMMENCEMENT

Patients often commence DMARDs at a time when their arthritis is in an illness phase with evidence of increased disease activity. This will often cause the patient to experience an increase in symptomatology inducing pain, stiffness, fatigue, low mood and systemic manifestations such as anaemia. The majority of DMARDs take many months before their ability to affect the disease process can be assessed. Patients respond at different times but will require education, support and guidance at the commencement of treatment and on an ongoing basis.

The nurse will need, in conjunction with the patient, to plan interventions to enable the patient to cope with increased symptomatology. This may include a combination of exercise, relaxation, pacing activities, review of analgesia or an inpatient stay. The objective of ongoing education and care provision are to enable the patients to participate effectively in their own management, develop coping skills, make informed choices about their treatment and weigh up the consequences of their action or inaction.

In a comprehensive review of patient education studies, Lorig *et al.* (1987) demonstrated positive improvement in patient knowledge, self-care beliefs, medication adjustment and compliance; as well as psychosocial characteristics such as anxiety, depression, self-esteem, locus of control and health status. After the patient has received a full explanation regarding the drug treatment proposed and been given the opportunity to discuss concerns and expressed a commitment to the therapy there may be safety preparation that needs to occur prior to commencement of the named drug - for example, if a patient is to commence methotrexate a chest X-ray will need to be obtained.

Safety Measures

Patients receiving DMARDs require regular safety monitoring of therapy and assessment of disease activity. This is usually carried out within the holistic framework of a nurse led follow up clinic. Running drug monitor clinics is a major component of the nurse specialist role Nurse-led clinics allow partnership, intimacy and reciprocity to evolve between the nurse and the patient and provide the medium for an ongoing therapeutic partnership to develop in which the nurse will be assessing all care needs and identifying problems that may require a referral to other members of the healthcare team such as the chiropodist. The clinic will also provide the forum for education and the development of coping strategies.

The nurse's role evolves from supportive to therapeutic, instead of concentrating exclusively on preventing deteriorating. The nurse will incorporate essential nursing functions to promote adaptation to the condition for the patient and their families. For care to be meaningful within an individual context nurses must understand the impact of illness from the patient's view point. The nurse must work actively with the patient to establish realistic achievable goals which incorporate the patient's lay beliefs.

Cytotoxic agents

Patients receiving DMARDs and cytotoxic agents require regular blood tests and in certain cases urinalysis and blood pressure monitoring. The British Society of Rheumatology and the British Health Professionals in Rheumatology have produced guidelines for clinicians and these are available from either society. Kay and Puller (1992) found marked variations amongst respondents in monitoring schedules and in the interpretation of results. At each visit the patient is seen by the same rheumatology nurse to allow the continuation of the therapeutic relationship. The nurse will:

1. Provide the patient with time to discuss any perceived problems or concerns - for example, is it safe to have vaccinations? Is it normal to experience increased stiffness after the administration of intramuscular gold?
2. Take blood to ascertain any bone marrow suppression leading to a reduction in red blood cells, white blood cells and platelet count; elevation in liver function tests, creatinine levels and urea and electrolytes where indicated. Hill's (1994) research demonstrated that the nurse practitioner is able to identify and instigate appropriate action from blood results.
3. Assess safety in administration - for example, ensuring that methotrexate is taken on a weekly basis.
4. Monitor and observe for any drug reactions that have occurred - for example, gastrointestinal, skin, urinary abnormalities and chest manifestations.
5. Assess for evidence of increased disease activity - for example, synovitis. This may necessitate a different treatment intervention - for example, the administration of intra-articular injections. About 12% of nurse specialists within rheumatology are engaged in the provision of intra-articular injections and this number is rapidly increasing as nurses seek to develop their practice

around patient need and provide a more comprehensive programme of care expansion and is in accordance with principles established in the scope of professional practice. This framework encourages and supports nurses to develop practice which has been shown to directly benefit patient care, as long as it does not fragment existing care or involve inappropriate delegation of duties.

6. It would be inappropriate, indeed unsafe practice, to develop one's role without first obtaining the necessary theoretical knowledge and practical expertise necessary to ensure understanding and safe practice. Benner (1984) states that an expert nurse requires a combination of academic achievement, recognized clinical expertise and practical experience.
7. Access other care needs within a holistic framework and refer where appropriate to other members of the multidisciplinary team.
8. Provide the patient with the telephone helpline – providing a designated point of contact with a knowledgeable practitioner should any problems or concerns arise before the next appointment.

Monitor Programme

The Staffordshire Rheumatology Centre uses a computerized drug monitor program. Kay (1989) states that more use could be made of computer technology in drug monitoring. All relevant information about the patient is recorded. This includes demographic data, blood and urine results, medications taken and reasons for discontinuation where appropriate. The computer is programmed to alert the operator to any results that fall outside the normal parameters such as three consecutive falls in platelets. It is also an invaluable communica-

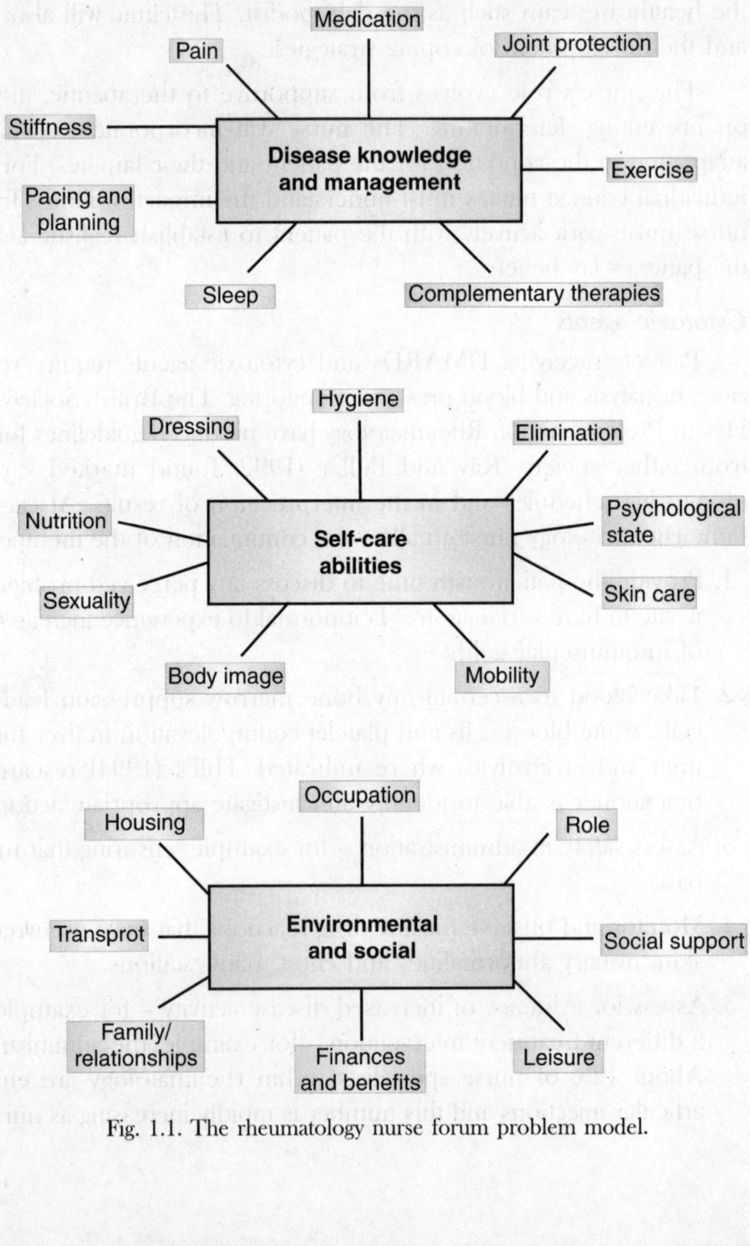

Fig. 1.1. The rheumatology nurse forum problem model.

tion tool as any member of the nursing and medical team can place a message on the patient's computerized record that will be acknowledged by the nurse conducting the clinic. This may include requests for additional investigations to be made while the patient is in clinic so that they do not have the inconvenience of being recalled.

Other units use a metrology graph which includes a visual analogue pain scale, early morning stiffness, articular index, grip strength and the recording of drug regimes. A list of past medication and the reasons for discontinuation is kept on the reverse of the graph. Joint injections, intramuscular steroid injections, infusions, admissions, trauma and even domestic stress are recorded on the chart because they all have a bearing on disease activity. This method of documentation enables the course of the disease to be followed easily.

Care and Safety

The establishment of protocols to guide decision making, ensure continuity of care and safety in practice requires collaboration and commitment from all health professionals involved in the monitoring and assessment of drug therapy. They provide an educational tool for nurses by ensuring that practice is research-based, and act as a dynamic structure updated when new evidence becomes available. The protocols may cover:

1. Urinary abnormalities.
2. Gastrointestinal side effects.
3. Skin manifestations.
4. Chest symptoms.
5. Alteration of drug dosages.

Included below are the protocols used for the daily drug monitor clinic at the Staffordshire Rheumatology Centre. This provides an example of the framework we have adopted enabling these protocols to be a fluid and dynamic tool supporting nurses in the decision making process.

Clinic Protocols

Reactions can occur in many sites. The nurse needs to:

- Be aware of what to look for.
- Know how to respond.

1. Skin manifestations. Changes in skin integrity can be related to:
   (a) The disease process – that is, vasculitis and leg ulcers.
   (b) The drugs used for therapy – that is, gold injections.
   (c) Difficulties faced in maintenance of skin integrity.
      (i) Rash

A rash

- May occur with any medication but is most commonly seen in those patients having gold therapy.
- Can vary from minor eczematous lesions to severe widespread problem.

If a patient presents with a rash:

- Ask if they have recently changed their detergent. Consider sun burn, insect bites.

If the rash is isolated to an odd patch, advise E45 cream. Review in a month's time.

If it is a severe, persistent or generalized rash, this may require a dermatology opinion – arrange referral.

(ii) Itching

If the itching is:

- troublesome – continue with treatment. Advise use of E45 cream or antihistamines. If the scalp is itchy, scaly or dry, advise a coal tar-based solution – for example, Baltar.
- unbearable – that is, cannot sleep with it – stop the medication and refer to the doctor's clinic.

2. Renal manifestations

Renal abnormalities may be due to:

(a) The disease itself – amyloid.

(b) Gold or D-penicillamine and are less likely with sulfasalazine and methotrexate.

(c) Asymptomatic urinary tract infections, which are common in RA patients.

(d) The ultra-sensitivity of lab stick testing.

(i) Proteinuria

- A trace or one + in isolation ignore.
- ++ repeat dipstick testing, if ++ still present then send midstream specimen of urine (MSSU) (patient and GP will be contacted if infection is present).
- If patient is symptomatic send MSSU.
- If MSSU reveals no infection but ++ persists arrange 24-hour urine collection.
- When sending off an MSSU please record on the form whether or not the patient is taking antibiotics.
- If +++ protein stop treatment and collect 24-hour urine, unless the patient is known to have proteinuria.

(ii) Haematuria

- Confirm that the patient is not menstruating.
- A trace or one + in isolation ignore.
- If ++ or +++ MSSU – continue treatment.
- If the patient has observed blood loss, inform doctor.

(iii) Nitrate

- If nitrate positive with protein or blood, send MSSU.
- Nitrate in isolation ignore.

(iv) Glycosuria

- If known diabetic, advise, refer back to diabetic monitoring.
- If not known to be diabetic arrange random blood sugar (if over 11 mmol = diagnostic of diabetes, and will require fasting blood sugar. Fasting blood sugar of over 8 = diagnostic of diabetes.

3. Oral aperture

If the patient presents with a sore mouth:

(a) If dentures are worn ensure these are fitted correctly.

(b) For minor ulcers (i.e. one or two in isolation) continue treatment and suggest: Bonjela, Bioral Gel (apply after meals and bedtime), Bioplex mouth rinse.

(c) For major mouth ulcers – that is, acute severe crop, more than two ulcers: The patient may require Difflan mouth wash and Corlan lozenges or Adcortyl in Orabase. Inform GP.

4. Alimentary canal

(a) Anorexia, dyspepsia and nausea can occur with any drug.

(b) Sulfasalazine - if nausea and/or vomiting occurs. Reduce back to the dose that was not causing this problem and increase as per protocol on fortnightly not weekly basis. If a rash occurs desensitization may be appropriate for early onset (first 14 days/rash - require 2/52 monitor and increase as per protocol).

(c) Leflunomide - if loose stools are experienced patients should be encouraged to remain on the medication and treat the bowel problem.

(d) Penicillamine - taste alteration can occur for up to 12-16 weeks. Support and reassure the patient. May require advice on nutritional supplementation.

(e) Methotrexate - if nausea/vomiting occurs split the dosage over the day, or two days. If nausea/vomiting continues to persist, patient will require domperidone (10 mg three times a day).

(f) Remember gastrointestinal intolerance could be due to NSAIDs - advise over-the-counter antacids and, if it persists, to see doctor.

5. Headaches

(a) Common with sulfasalazine, can occur with leflunomide.

(b) If severe stop drug. Reintroduce after one month at two-weekly intervals.

6. Active joint involvement

(a) Joint stiffness - may follow administration of gold.

7. Breathlessness

   Breathlessness with or without a cough in a patient on gold, leflunomide or methotrexate. Inform doctor.

8. Alopecia

   Can occur whilst taking cytotoxic drugs – that is, cyclophosphamide or methotrexate. If troublesome or severe inform doctor.

9. Muscle

   If the patient notices weakness, double vision, swelling or speech difficulties this raises the possibility of drug-induced myositis (inflammation of muscle) or myasthenia gravis (disorder of the neuromuscular junction) – inform doctor.

Non-Drug-Related Problems

1. Septic joint " if patient has a pyrexia and a hot red joint think of sepsis and inform doctor.
2. Neck instability " sudden onset of weakness in hands and legs. Recent onset of bladder and/or bowel incontinence " inform doctor.
3. Early herpes zoster " if patient on immunosuppressants they may need anti-viral therapy. Stop cytotoxic medication temporarily until herpes zoster has cleared.

Patients who require blood pressure recording.

1. Patients receiving ciclosporin or leflunomide or present with significant protein- urea or haematuria – if the diastolic is greater than 95 Hg discuss with the doctor

Having such a framework enables the nurse to take immediate action when faced with side effects. Ryan (1997) states that during the first year of using such protocols over 800 patients were reviewed in the clinic with only 2% requiring medical referral. Before the instigation of the protocols 16% of patients were referred annually for a medical consultation primarily as a result of limitations being placed on nurse decision making skills. The protocols can also ensure continuity of care when the patient moves back in to the community setting.

CLINICAL FINDINGS

Haematological Findings

The most frequently requested laboratory test is the full blood count. This test examines three groups of cells with different functions: platelets, red cells (erythrocytes) and white cells (leucocytes). Around 2-2.5 mls of venous blood is required in a tube containing an anticoagulant.

Erythrocytes or red blood cells

RBCs are biconcave disks. Their main function is to transport haemoglobin around the body and supply oxygen to the tissues. The average number of RBCs per cubic millimetre of blood is about 5 million. The number of RBCs available in the circulation is regulated so that there is an adequate source to provide oxygenation of the tissues without stopping blood flow. The RBC has a very flexible structure which allows it to alter shape as it passes through the capillaries without damage. Altitude and the degree of exercise undertaken can affect the number of circulating RBCs. RBCs derive from a cell known as the haemocytoblast and are produced in the bone marrow.

Vitamin B_{12} (cyanocobalamin)

Vitamin B_{12} is a nutrient that is required for all cells of the body. An absence or reduced supply of vitamin B_{12} will severely hinder the rate of RBC production, the problem often being the failure to absorb vitamin B_{12} from the gastrointestinal tract, as occurs in pernicious anaemia. Gastric secretion contains a substance called the intrinsic factor which combines with vitamin B_{12} in food, so preparing it for absorption by the gut; but in pernicious anaemia the gastric mucosa does not secrete the intrinsic factor and therefore absorption cannot occur. In the normal process vitamin B_{12} is stored in large quantities in the liver after being absorbed from the gastric system. The total amount of vitamin B_{12} required every day to assist in RBC production is less than 1 microgram and the store in the liver is a thousand times that amount.

Folic acid

Folic acid is part of the vitamin B complex and deficiency can also hinder the production of RBCs.

Anaemia

The typical anaemia of RA is a normocytic normochromic anaemia; the haemoglobin being around 10-12 g/dl. There appears to be a correlation between disease activity and the severity of anaemia. Some disease modifying anti-rheumatic drugs (e.g. gold) may induce anaemia by attacking erythyroid progenitor cells but in other situations studies show that these therapies may normalize erythropiesis. The cause of anaemia in RA is multifactorial. Iron utilization is impaired as indicted by reduced serum iron and transferritin concentration.

As with other forms of chronic inflammation there is an increased synthesis of ferritin and haemosiderin, abnormal retention of iron from FBCs by the reticulo-endothelial system and an increase of lactoferritin, which contributes to the binding and lowering of serum iron. Anaemia means a deficiency of RBCs with haematocritis sometimes as low as 10%. The anaemia of RA can be complicated by:

1. Blood loss – NSAIDs can cause intestinal blood loss;
2. Poor nutrition;
3. Intercurrent infections;
4. Autoimmune haemolytic anaemia;
5. Bone marrow suppression secondary to drug treatment – for example, gold.

A macrocytic macrochromic anaemia may be due to vitamin B_{12} or folate deficiency, alcohol intake or the mechanisms of sulfasalazine or methotrexate.

Leucocytes or white blood corpuscles

When stained and viewed under the microscope it is possible to identify five types of leucocyte. A full blood count measures both their total number and counts the number of each of the five types. This is the differential count.

The white blood cells (WBCs) are the mobile units of the body's protective system. They are formed in the bone marrow and the lymph nodes prior to being transported in the blood to the different parts of the body where they are used. All WBCs have a limited life span and adequate numbers are maintained through marrow production. WBCs play a central role in the process of inflammation. The purpose of the inflammatory response is a complex integration of cellular function and although each type of leucocyte has a different and well-defined function they operate in conjunction communicating via a range of chemical messengers called cytokines.

Table 1.1 White blood cells

Total white blood cell count	4-11 × 10^9 per L
Neutrophils (40-75% to total white cells)	2.5-7.5 × 10^9 per L
Lymphocytes (20-40% of total white cells)	1.5-4.0 × 10^9 per L
Monocytes (2-10% of total white cells)	0.2-0.8 × 10^9 per L
Eosinophils (1-6% of total white cells)	0.04-0.4 × 10^9 per L
Basophils (less than 1% of total white cells)	0.01-0.1 × 10^9 per L

Leucocytosis (a raised WBC count) occasionally occurs in RA but it is not a typical feature. If marked it should raise suspicions of a superseded infection (either systemic or a septic arthritis), a post-infected arthritis or one of the varieties of polyarteritis. Leucopaenia (a reduced WBC count) can occur in systemic lupus erythematosus and in Felty's syndrome but the commonest cause is in the use of DMARDs.

Neutrophils

From 40-75% of circulating WBCs are neutrophils, the first line of defence against bacterial invasions. They are present in large numbers at the place of invasion where they ingest and kill bacteria by a process of phagocytosis.

Monocytes

These are mobile WBCs that filter out of the blood and into the tissues, where they are referred to as macrophages. The key role is probably as an antigen-presenting cell. Lymphocytes cannot respond to naked antigen and it must first be processed by and presented on the surface of an antigen-presenting cell. Macrophages become attached to tissues and remain attached for months, even years. They are involved in the primary clearance of highly resistant bacteria such as mycobacteria.

Monocytes manufacture a range of cytokines that attract or activate other cells involved in the inflammatory response – for example, macrophages produce interleukin 1 which activates a type of lymphocytes to kill viruses.

Basophils

These are seen only rarely in peripheral blood. They mature in the tissues, becoming mast cells. These cells liberate heparin into the blood to prevent blood coagulation and to assist in tissue repair. The mast cells also release histamine and small quantities of bradykinin and serotonin during inflammation.

Eosinophils

These cells are natural phagocytes. They collect at the site of antigen antibody reaction in the tissues and 'digest' the combined antigen antibody complex after the immune process has performed its function. Eosinophils are involved in the pathogenesis of hypersensitivity (allergic) reactions. Their number is greatly increased in the circulating blood during allergic reactions. Eosinophils accompanying RA is sometimes associated with extra-articular manifestations. Although the pathogenesis is not known, immune complexes may be chemotaxis for eosinophils (e.g. attracted towards the source of the chemical). Eosinophilia has also been associated with high titre of rheumatoid factor, elevation of serum gammaglobulins and diminished serum complement levels. Pulmonary complications may be associated with eosinophili or it may be associated with drug therapy such as gold.

Lymphocytes

From 20-40% of circulating WBCs are lymphocytes, which are the cells of the acquired immune system although a routine blood count cannot differentiate between them. There are two types - B lymphocytes and T lymphocytes. B lymphocytes produce antibodies and T lymphocytes are responsible for the elimination of viruses and other micro-organisms that infect the cell of the host and are not 'visible' for antibody attack. Both types of lymphocyte function are independent and a normal immune response requires adequate numbers of both B and T lymphocytes.

Leucocytopenia

A low WBC count is significant because it reduces the body's protection against infection. If the neutrophil count falls below 0.5×10^9 per litre patients are likely to become susceptible to frequently recurring bacterial infections. If levels drop further life is threatened by the risk of overwhelming infection. Causes of leucopaenia include:

1. Leukaemia is associated with reduced neutrophil count as bone marrow production of large numbers of abnormal immature white cells continues at the expense of normal white cell production.
2. Hodgkin's disease.
3. Felty's syndrome.
4. Aplastic anaemia, which is a pancytopenia where reduced production of bone marrow stem cells results in reduced numbers of red cells, white cells and platelets. Aplastic anaemia can be inherited as can pancytopenia although most causes of the latter arise as side effects of cytotoxic drugs.
5. Systemic lupus erythematosus.
6. Side effect of DMARDs.
7. Acquired immune deficiency syndrome (AIDS).

Agranulocytosis

The bone marrow stops producing WBCs, leaving the body unprotected against bacteria and other agents that might invade the tissues. Within two days of production ceasing, ulcers appear in the mouth and colon and severe respiratory infection develops. Bacteria then rapidly invade the surrounding tissues and the blood; without treatment death usually occurs within a week.

Platelets

The number of platelets in each cubic millimetre of blood is normally around 300 000. An increased amount of platelets (thrombocytosis) is a frequent finding in active RA and may well correlate with the number of joints involved, the amount of active synovitis and the presence of extra-articular features.

The mechanism of thrombocytosis is uncertain; an increase in the intravascular coagulation with a compensatory increase in platelet production has been suggested. The thrombocytosis does not predispose to an increase in thrombotic events and is not correlated with bone marrow neoplastic changes. A reduced platelet count (thrombocytopenia) is rare in RA except when related to drug treatment or Felty's syndrome.

Pharmaceutical Study

Liver function

Liver function abnormalities may reflect the anaemia, thrombocytosis and increased erythrocyte sedimentation rate of active inflammatory disease such as RA. Examination of liver histology at this time reveals only minimal non-specific change and some perioportal mononuclear cell infiltration. Active RA may be associated with an increase in liver enzymes, especially serum glutamic oxaloacetic transaminase (SGOT) and alkaline phosphatase. Total protein, total albumin and total globulin estimates are also taken.

These reflect hepatic function (particularly albumin which is made in the liver) as well as absorption and inflammation. Apart from in lupoid hepatitis, which can mimic RA, bilirubin and SGOT are likely to be normal. NSAIDs may induce liver enzyme abnormalities and it may be difficult to differentiate between drug effects and disease activity without discontinuation of NSAID therapy. NSAIDs seldom cause serious liver deterioration. Liver involvement may be present in up to 65% of patients with Felty's syndrome.

Bone metabolism

Calcium, phosphate and alkaline phosphatase are all measured. Calcium is bound to protein and this may need a correction factor if protein levels are abnormal as can occur in RA. A slightly raised calcium and raised alkaline phophatase level if of bony origin should alert the clinician to the possibility of osteomalacia. In Paget's disease there is a markedly raised alkaline phosphatase of bony origin. X-ray confirmation of Paget' s disease should be sought since secondary cancer in bone can also cause a raised alkaline phosphatase.

Urinary function

Kidney involvement is usually sparse in RA although a low grade nephropathy, glomerulitis vasculitis and secondary amyloidosis have all been described. More commonly renal abnormalities result from agents used in the treatment of RA, notably gold, D-penicillamine, ciclosporin and the NSAIDs. The renal involvement caused by nephrotoxic drugs is more likely to be seen first on routine urine testing. Electrolytes (sodium, potassium, chloride and bicarbonate) together with urea and creatinine are usually measured. The chronic renal failure of connective tissue disorders may produce a raised potassium and a low bicarbonate level. The renal function in a normal population will deteriorate throughout adult life. Urea may be raised because of dehydration but creatinine is less likely to do so under those circumstances. Creatinine is the most sensitive simple estimation of renal damage and a raised level may indicate amyloidosis or chronic renal failure.

Amyloidosis

This may develop in patients with RA as a result of long-standing active inflammation and cause proteinuria. It is not specific to the kidneys and may affect other organs including the heart, liver, spleen, intestines and skin. The diagnosis of amyloidosis is confirmed by biopsy of the involved tissue. Unfortunately the presence of secondary amyloidosis in patients with RA heralds a poor outcome.

Rheumatic Disease

Rheumatoid factor

This is a commonly requested immunological investigation. Rheumatoid factor (RF) is found in the blood of 75% of patients with RA of more than 12 months duration but in only 40-50% of those with early disease. Therefore the absence of RF does not preclude a diagnosis of RA; indeed the diagnosis of RA is not made on the presence of RF alone but includes a host of other factors. If the RF is absent and the patient has RA the arthritis is said to be sero-negative. The RF, an IgM/IgG complex of two immunoglobulins, can also be found in other conditions including liver disease, sarcoidosis, subacute bacterial endocarditis and in 4% of healthy adults. The RF is detected by one or more of the following tests:

1. Sheep cell agglutination (Rose Waaler) test;
2. Latex agglutination test.

The Rose Waaler test is given as a titre; below 1:32 is not significant, 1:64 is positive and values above 1:500 usually indicate severe rheumatoid disease.

Anti-nuclear antibody

The ANA is regarded as an initial screening test for systemic lupus erythematosus but is not specific for this condition and if positive the clinician will proceed to request DNA binding from the same patient before confirming a diagnosis of lupus. ANA can occur in other connective tissue disorders such as systemic sclerosis and occasionally in low titre in RA. The ANA titre is important and titres of 1:200 or more in the presence of anti-DNA antibodies are likely to confirm the diagnosis of lupus.

Extractable nuclear antigen

The ENA is helpful for the diagnosis of mixed connective tissue disorder.

Immunoglobulins

These can provide useful information when the ESR is raised for no obvious reason. In sero-positive RA serum IgM is usually though not always raised representing RF in the blood. Equally there may be a polycolonal increase of all immunoglobulins such as IgA, IgG and IgM. IgA synthesized in mucous membranes may be raised in Sjogren's syndrome.

Plasma proteins and electrophoresis

The alpha 2 globulin is an acute phase reactant which when elevated indicates tissue destruction in the early phases of rheumatic disease. In the later stages a raised gammaglobulin may indicate antibody formation and very high levels can occur in connective tissue diseases, sarcoidosis and myelomatosis.

Erythrocyte sedimentation rate

The erythrocyte sedimentation rate (ESR) is used as the standard test in many hospitals for evaluating disease activity in rheumatic disease. If inflammation is present in the body the ESR increases. This is due to the changes in the patient's blood. A quantity (2 mls) of anticoagulated blood is sucked into a capillary tube and the speed with which red and white cells settle over a period of one hour is observed. Normal values vary with age and sex. ESR is of value in distinguishing inflammatory polyarthritis from the degenerative conditions. ESR is elevated in RA, AS, acute gout, PMR, systemic connective tissue disorders, reactive and infective arthritis.

A persistently raised ESR should alert the clinician to the possibility of myeloma (a malignant condition of the plasma cells). If suspected, urine should be collected to test for Bence Jones protein, a characteristic protein produced by the cell. The ESR is not normally elevated in degenerative metabolic joint disease or in soft tissue rheumatism. The Westergren method of establishing the ESR is now universally used. ESRs of up to 20 mm/hr in males and 25 mm/hr in females are accepted as being within normal limits in rheumatological practice.

Plasma viscosity

The plasma viscosity (PV) test mimics ESR evaluation but studies the movement of the plasma in the horizontal rather than the vertical plane. PV eliminates the variation caused by the anaemia of RA which can influence the ESR. High values are found in active RA.

C-reactive protein

C-reactive protein (CRP) is an acute inflammatory protein produced by the liver in response to infections, inflammation or acute injury. It rises in active RA falling back to normal levels once the disease has come under control. Changes in the CRP occur faster than in other biochemical assessments. Levels remain normal in systemic lupus erythematosus.

Other acute phase reactants

Fibrinogen, hepatoglobin and caeruloplasmin all behave as acute phase reactants with high levels found in active disease and low values apparent as the conduction comes under control.

URINALYSIS

Urinalysis, the examination of urine, is a valuable tool for the diagnosis of and screening for several conditions. It also plays an important role in the surveillance of drug therapy especially in terms of toxicity. This will be discussed in further depth later on in the chapter. A patient who is asked to supply a specimen of urine for testing requires the following:

1. An explanation of the reason for testing.
2. Instruction in the method of collecting the specimen.
3. The provision of suitable equipment, including a container for the specimen and washing facilities.
4. An appropriate environment including wherever possible visual and auditory privacy and adequate time.

Physical characteristics

The appearance of the urine should be noted for colour and clarity. Colour changes may be due to endogenous pigments such as haemoglobin (red/brown colour), bilirubin (yellow) or intact red cells (smoky red). Exogenous pigments such as contamination with menstrual blood may also cause colour changes. The administration of sulfasalazine can cause orange discoloration.

Cloudiness is caused by suspended particles which will settle on standing to leave a deposit. Normal urine may contain some renal tubular cells and a few white and red blood cells; these will be prevalent in certain diseases contributing to a cloudy appearance.

Smell

Normal, freshly voided urine has very little odour but will develop an ammoniacal smell if left for any length of time. Infected urine has a characteristic fishy smell. The urine of patients with anorexia, or a ketoacidosis diabetic person will have a sweet smell. The administration of sulfasalazine and/or D-penicillamine can also cause a characteristic odour.

Measurement of Specific Gravity

This can be measured using either reagent strips or a specially calibrated hydrometer. Specific gravity of the urine measures the ability of the kidneys to concentrate or dilute urine. A low specific gravity can occur if a patient has a high intake of fluid or it may indicate renal abnormalities or the presence of diabetes insipidious. A raised specific gravity may indicate dehydration. Routine testing of urine may detect:

1. Blood. Often found to be of no significance but should be investigated if persistent. It may be due to trauma, infection, tumour or stones. Haematuria is an early sign of polyarteritis and 50% of patients with lupus will have small amounts of blood and protein in their urine.
2. Ketones. These are not normal constituents of urine; they are the produces of the breakdown of fatty acids. Their presence may indicate starvation, excessive dieting or uncontrolled diabetes.
3. Bilirubin in the urine may be indicative of hepatic or biliary disease. It may also be present if the patient has been prescribed phenothiazides or chlorpromazine leading to a false positive result occurring.
4. *Glucose.* This is not usually found in urine. Its presence may be due to raised blood glucose levels or to reduced renal absorption. It can be associated with conditions such as diabetes mellitus, stress, Cushing's syndrome and acute pancreatitis. Glycosuria occurring in the urine of a patient with rheumatic disease alerts the clinician to the possibility of corticosteroid-induced diabetes mellitus.

5. *Leucocytes.* The presence of leucocytes in the urine is an indication of bladder or renal infection but follow-up testing such as urine culture is required. White blood cells should be sought in the urine when infections such as Reiter's disease are suspected. Early morning specimens are more likely to contain tubercule bacilli than those taken later in the day if the condition is suspected.
6. *Nitrate.* This is not normally present in urine. It is produced when gram-negative bacteria such as *Escherichia coli* convert dietary nitrates (e.g. from the preservative in meat products and cheese and in smoked foods) to nitrites. As *E. coli* is responsible for 80% of urine infections, the presence of nitrites is strongly suggestive of urinary tract infection. The specimen for testing should have been in the bladder for at least four hours before voiding to allow sufficient time for the nitrate/nitrite conversion.
7. *Urobilinogen.* This is normally found in urine but increased levels may indicate liver abnormalities or excessive destruction of red blood cells such as in haemolytic anaemia.
8. Protein can indicate a range of conditions including renal disease, urinary tract infection, hypertension, pre-eclampsia or congestive heart failure. Transient positive tests are not always significant, and normal urine contains small amounts of albumin and globulin although not usually at a level that would be detected positively on a reagent strip. When testing for urinary protein a morning specimen of urine is recommended to ensure sufficient concentration. Proteinuria may be the first sign of a collagen disease such as polyarteritis.
9. In rheumatoid disease it may indicate urinary infections, nephropathy due to gold or D-penicillamine or be secondary to amyloid diseases. If proteinuria persists a 24-hour urine specimen should be requested; more than 0.15 g protein/24 hours is abnormal. Over 1 g/24 hours is found when there is renal tubular damage and in nephrosis well over 5 g protein/24 hours can occur. The presence of albuminuria on routine testing often indicates a simple infection that will require treatment to prevent aggravation of a flare of RA but it may also indicate glomerular dysfunction in conditions such as lupus.

Cataloging

The results of the urinalysis should be recorded in the patient's records as soon as possible after testing. It is worth noting that a negative result on a particular date may acquire significance in retrospect so all results should be recorded in patient's records at the time of testing (Cook, 1996).

DRUGS FOR LONG TERM TREATMENT

Gold Therapy

Gold is used as long-term treatment for inflammatory joint disease (most commonly RA). Initially most patients are given a test dose of 10 mg intramuscularly (IM) and if no adverse effects are experienced progress to 50 mg IM weekly reducing to 50 mg fortnightly as the patient responds, normally around six months. With adequate control, frequency is reduced to monthly (around 12 months) and continued indefinitely until side effects occur. It may take three to six months for patients to respond to treatment with gold. There are vast variations in the implementation of gold therapy. Kay and Puller (1992) found that out of a survey of 100 rheumatologists 89% used a test dose of intramuscular gold but of those who gave a test dose 64% gave a single 10 mg dose, 9% gave 10 mg followed by 20 mg, 4.5% gave 5 mg followed by 10 mg and the remaining rheumatologists gave various slowly increasing incremental doses. Toxicity with injectable gold is common occurring in 30-40% of patients.

Alertness

Elderly, renal or hepatic impairment, urticaria, eczema or inflammatory bowel disease.

Symptoms

1. Severe renal or hepatic impairment
2. Blood disorders
3. Systemic lupus erythematosus
4. Significant pulmonary fibrosis
5. Porphyria
6. Pregnancy and lactation.

Nephrotoxicity. It is the responsibility of the nurse administering the intramuscular injection of gold to test the urine prior to giving the injection. Minor transient proteinuria is common whilst on this therapy. A meta-analysis of clinical trials showed significant proteinuria in patients receiving gold. A trace of proteinuria can be ignored but increasing proteinuria is an indication to stop gold. Treatment should be discontinued if 24-hour protein urinalysis exceeds over 1 g. Recovery is usual when gold is stopped. Minor renal effects in the absence of proteinuria has been noted with the elevation of enzymes and tubular cell secretion, but this increased turnover of tubular cells has not been related to decreased tubular function later in life. Isolated microscopic haematuria is not usually attributed to gold.

Dermatitis

Dermatitis and oral reactions account for 60-80% of all reactions to injectable gold complexes. Oral mucosal lesions occur less often than rashes. Minor ulcers can be treated symptomatically. Check that ill-fitting dentures are not causing friction. Most rashes are usually erythematous and macular but rarely can exfoliate; 85% of rashes are severely pruritic. Raised eosinophils may be the forerunner to a rash, and the rash may be associated with a metallic taste and organ toxicity such as proteinuria. Minor rashes and/or pruritis can be treated symptomatically with aqueous cream or 0.5% hydrocortisone cream. The nurse will also want to establish that the cause of the rash is related to drug therapy so that the treatment is not discontinued unnecessarily.

The nurse will check that the patient has not recently changed their soap, washing powder or commenced new drug therapy. In some units close cooperation with the dermatologist can establish the cause of the rash and/or irritation and thus may provide valuable information in planning long-term care. Gold should be withheld if there is a widespread pruritis, severe or rapidly progressive rash and/or mouths ulcers/stomatitis as these can be associated with exfoliative dermatitis. Once moderate rashes and/or pruritis has settled it may be possible to reintroduce gold at a smaller dose.

Blood diseases

Neutropenia

It is important to consider whether a decrease in WBCs is due to the disease or related condition (e.g. Felty's syndrome) or to the gold treatment. Careful attention needs to be paid to a rapid and/or progressive downward trend in neutrophils accompanied by a fall in platelets as this is often caused by bone marrow suppression as a direct result of gold treatment. (You can also get neutropenia related to gold without an accompanying fall of other blood cells). Increasing neutropenia will require increased surveillance, possible dosage reduction or cessation of gold.

Eosinophilia

This will occur in up to 50% of patients receiving gold at some stage of their treatment and although it does not generally correlate highly with toxicity it should not be dismissed as in some cases it may herald a toxic reaction to treatment.

Thrombocytopenia

This occurs in 1-3% of patients receiving intramuscular gold. It is usually minor but can be serious. It usually responds well to corticosteroids.

Aplastic anaemia

This is rare but has a high mortality rate, although the rate for survival is improving.

Inflammation lung reactions

Hypersensitivity pneumoconitis, distinguishable from rheumatoid lung by its acute or subacute onset, can occur and will necessitate cessation of gold. Obliterative bronchiolitis has been reported in patients with RA but it remains unclear whether this is a result of gold therapy or the condition itself.

Pulmonary involvement in patients with RA

Pulmonary involvement in patients with RA is fairly common although the clinical features may be subtle. Pleurisy and pleural effusions may improve spontaneously or require treatment. The occurrence of persistent pleural effusions will lead to fibrosis. Sero-positive RA patients may develop parenchymal pulmonary nodules which, although usually asymptomatic can cause pleural effusions. Caplan's syndrome (pneumoconiosis) is characterized by large multiple pulmonary nodules and is seen in individuals who have experienced prolonged exposure to coal dust. Isolated pulmonary arteritis is a rare complication of RA and is frequently associated with interstitial fibrosis and nodulosis.

Neurotoxicity due to gold

Neurotoxic reactions are uncommon but the following diverse reactions have been noted, including peripheral neuropathy, Guillan Barre syndrome, cranial nerve palsies and encephalopathy.

Vasomotor reactions

These are often referred to as 'nitroid' reactions and may occur within minutes of administration of the gold injection and necessitate that the patient remains in clinic for a period of time following the injection. Reactions are characterized by weakness, dizziness, nausea, sweating, facial flushing, erythema and hypotension.

Post-injection reactions

Apart from the possibility of vasomotor reactions the patient may also experience transient polyarthralgia, myalgia, joint swelling, fatigue and malaise. It can be difficult at times to distinguish these effects from the symptoms of active inflammatory disease.

After effects

A meta-analysis of clinical trials found that 11% of all patients had to discontinue gold due to side effects experienced. The majority of side effects occur within the first 12 months of treatment and are not related to cumulative dose. Lockie and Smith (1988) published their experience with gold therapy in 1019 patients; the main reasons for discontinuation of gold was skin reactions (36%) and buccal irritation in 13% of patients. Gold should be stopped if:

1. Total white cell count is $<3.5 \times 10^9$ L or neutrophils $<2 \times 10^9$ L
2. Platelets $<150 \times 10^9$ L
3. Eosinophilia $>0.5 \times 10$/L
4. Proteinuria> 1 g/24 hours
5. There is a severe or progressive rash.

The majority of adverse reactions to gold resolve spontaneously within weeks to months following cessation of treatment.

Auranofin Therapy in RA

Auranofin is used in the long-term treatment of RA. The usual dose is 3 mg twice daily. The absorption of gold from auranofin following a single dose is about 20-25% and there is less retention of gold in tissues than occurs with injectable gold. Nevertheless regular blood monitoring is required.

Alertness

Elderly, renal or hepatic impairment (moderate) history of urticaria, eczema or inflammatory bowel disease.

Symptoms

1. Severe renal or hepatic impairment.
2. History of blood disorders.
3. Significant pulmonary fibrosis and porphyria.
4. Systemic lupus erthematosus.
5. Pregnancy and lactation.

After effects

1. *Integument*

   Rashes. Minor rashes can be treated symptomatically (aqueous cream, 0.5% hydrocortisone cream).

2. *Mouth ulcers*

   Minor ulcers can be treated with Bonjela, Difflam mouth wash or Adcortyl in orabase.

3. *Alimentary canal*

   Diarrhoea is often the commonest side effect and may respond to the introduction of bulking agents such as bran or to a temporary reduction in dosage. Some patients experience nausea and vomiting and it is recommended that this therapy is taken with food.

4. *Kidney*

   Proteinuria

5. *Circulatory system*

   The nurse must observe for any reductions in neutrophils and platelets.

6. *Miscellaneous*

   Rarely colitis, peripheral neuritis, pulmonary fibrosis, hepatotoxicity with cholestatic jaundice and alopecia have all been reported.

   Auranofin should be stopped if:

1. Total white cell count is $<3.5 \times 10^9$ L
2. Platelets $<150 \times 10^9$ L
3. Proteinuria >1 g/24 hours
4. Severe rashes or severe mouth ulcers occur.

D-Penicillamine

D-penicillamine is used in the long-term treatment of rheumatoid arthritis and in scleroderma. Treatment is commenced at 125 mg or at 250 mg daily, increasing at four-week intervals by increments of 125 mg daily to 500 mg or 750 mg. The tablets are best taken as a single daily dose an hour before breakfast and not with iron tablets or milk as they interfere with drug absorption. Toxicity is common with this treatment with about 50% of patients with RA experiencing an adverse effect within the first six months of treatment at 600 mg daily resulting in 25% of patient discontinuing with this treatment. The incidence of toxicity increases as the dose increases.

Alertness

Renal impairment, concomitant nephrotoxic dugs including gold treatment.

Symptoms

1. Systemic lupus erythematosus
2. Renal impairment
3. Pregnancy and lactation.

Clinical effects

1. Digoxin: Levels of digoxin can be reduced by concurrent use of D-penicillamine.
2. Zinc supplements may reduce the absorption of D-penicillamine.

3. Antacids and iron: do not take within two hours as this reduces the absorption of D-penicillamine.
4. Antipsychotic drugs: may increase agranulocytosis.

Integumentary reactions

Urticarial, pruritic, macular and papular eruptions of apparent allergic origin account for the majority of skin reactions. Minor rashes can be treated symptomatically but severe rashes may necessitate stopping D-penicillamine and this will normally resolve the rash. Autoimmune bilious skin eruptions, typical of the pemphigus group, are the most worrying. Patients who have a penicillin allergy are at greater risk of experiencing a skin reaction. Oral ulcers and stomatitis may occur; if severe this will mean a discontinuation of therapy.

Oral mucosal involvement is often present in pemphigus. There is a significant level of mortality amongst patients with D-penicillamine-induced pemphigus. A metallic taste and loss of taste may occur in the first eight weeks of treatment. It may take several months to recover. This can be a particularly distressing symptom for patients and advice may need to be given on food supplementation. Taste disturbance may be a consequence of zinc chelation by the D-penicillamine but zinc therapy does not remove the complaint.

Nausea and anorexia

If these symptoms occur the patient may need to reduce the dose of D-penicillamine and then increase slowly when the symptoms settle. The use of anti-emetics should also be considered.

Circulatory

There can be a gradual or rapid haematological effects at any stage during treatment with D-penicillamine. Thromobocytopenia can be precipitated and serious. Neutropenia due to D-penicillamine as opposed to RA itself is suggested by a rapid or progressive fall in neutrophils and is the commonest cause of death attributed to this treatment. Aplastic anaemia also has a high fatality rate.

Urinary

Proteinuria is common. Increased urinary protein excretion can exceed 0.5 g/day in approximately 9% of patients taking this drug. The presence of HLA B8 and DR 4 antigens coupled with the occurrence of previous proteinuria on gold therapy increases the risk of more severe proteinuria leading to nephrotic syndrome. The majority of patients experiencing severe proteinuria have membranous glomerulonephritis with minimal change neuropathy.

Plasma urea and creatinine may rise intermittently but this does not seem to be related to permanent renal impairment. D-penicillamine should be stopped if over 1 gram of proteinuria occurs in a 24- hour period. Even after the drug has been discontinued the proteinuria may persist for up to 21 months. The appearance of haematuria may indicate rapidly progressive glomerulonephritis – for example, Goodpasture's syndrome or SLE and can be life-threatening requiring steroids and immunosuppressives. However, the presence of haematuria usually does not lead to serious complication and recurrent haematuria is quite common in RA patients without serious renal pathology.

Pulmonary

There have been cases of bronchiolitis obliterans reported. Although this is a rare side effect it may be irreversible and present as dyspnoea late in therapy.

Autoimmune

Myasthenia gravis. This rare complication may evolve after years of uneventful therapy and may take over a year to resolve. Auto-antibodies may appear including anti-striated muscle antibody, anticentromere antibody and anti-glomerular basement membrane antibody. The latter is associated with the life-threatening complication of Goodpasture's syndrome.

Dermatomyositis/polymyositis

Drug-induced SLE usually resolves after stopping D-penicillamine. Pemphigus (oral involvement common) is dangerous and requires cessation of D-penicillamine, glucocorticosteroids and perhaps plasmapheresis. D-penicillamine should be stopped if the following occur:

1. Severe rash.
2. Severe mouth ulcers.
3. Autoimmune side effect.
4. Platelets <150 × 10^9L
5. Total white cell count <3.5 × 10^9 L
6. Neutrophils <2 × 10^9 L
7. Proteinuria> 1 gm/24 hours.

Sulfasalazine EN Therapy

Sulfasalazine is used in its enteric coated form (EN) as second line treatment for RA, spondyloarthropathies and reactive arthritis. To reduce nausea the drug should be commenced at 500 mg daily increasing by 500 mg daily at weekly intervals to the usual maintenance dose of 1 g twice daily. Incomplete responders sometimes require 1 g three times a day. It may take three to six months for a patient to respond to treatment with sulfasalazine. This preparation ranks with the antimalarials and auranofin as the best tolerated of the DMARDs and is a popular first choice of DMARDs amongst rheumatologists.

Alertness

1. Moderate renal impairment.
2. May impair folate absorption.
3. Pregnancy and breastfeeding.

Symptoms

Hypersensitivity; sulfonamide/co-trimoxazole

After effects

These are commonest in the first three months of therapy. Only 20-25% of patients need to stop treatment because of toxicity and only about 5% of patients will experience potentially serious side effects.

Commonest adverse effects are gastrointestinal and on the central nervous system. They include nausea, vomiting, malaise, anorexia, abdominal pain, dyspepsia, indigestion, headache, light-headedness and dizziness. If a patient experiences nausea the dose should be reduced to the previously tolerated dose and further increase should be attempted slowly. Make sure the patient is taking enteric coated tablets and administrating them on a full stomach.

Anti-emetics may be required. With headaches again the dose should be initially reduced and further increases attempted slowly. If headaches are severe the patient may need to discontinue sulfasalazine. Pyrexia is a rare side effect and can be considered only when other causes of raised temperature have been excluded.

Allergy

This is relatively common and may manifest itself in a number of ways ranging from skin rash to fatal multi-organ involvement.

Integument

Rashes. If these are mild, treatment can be continued and the rash can be treated symptomatically. Skin rashes will occur within 5% of cases and take the form of a generalized pruritic maculopapular rash, urticaria or photosensitivity. If the rash occurs within the first two weeks of treatment then desensitization can be tried. Initially given in 1 mg doses sulfasalazine is built up over 25-56 days. Successful desensitization has been achieved in as many as 85% of cases in RA patients. Rarely severe reactions such as Stevens-Johnson syndrome or exfoliative dermatitis can occur. In the case of this happening all drugs should be stopped and urgent dermatological advice should be obtained. Rarely serum sickness occurs.

Breathing

Most toxicity is allergic in origin from sulfasalazine. The commonest presentation is eosinophilic pneumonitis; symptoms include dyspnoea with fever, rash, weight loss, pulmonary infiltration on X-ray and reduced pulmonary function tests. Reduction of symptoms occurs following cessation of therapy.

Circulatory

Leucopenia in 1-5% of cases: It is most likely to occur in the first six months of treatment although it can occur at any time. It is the commonest potential serious side effect associated with sulfasalazine. It indicates the need for continued review particularly as early recognition with dosage reduction or cessation leads to reversal in most cases. Thrombocytopenia is less frequent than leucopenia. Agranulocytosis is rare but can develop rapidly and present with intercurrent infection. It has been recorded in association with a transient but marked plasmocytosis.

Mean cell volume (MCV): Sulfasalazine causes the erythrocyte MCV to rise. It inhibits folate uptake in the small bowel; however it would be rare to see folate deficiency at a dose of 2 g daily.

Abnormalities of red cell morphology: Morphological changes are due to oxidant damage resulting in red cell membrane abnormalities leading to low grade haemolysis. The changes are reversible on discontinuation of therapy but this action would only be necessary if the haemolysis was causing a problem. Significant methaemoglobin can cause headaches and may require a dosage reduction.

Liver

Allergic hepatic responses to sulfonamides are well known; raised hepatic enzyme levels and eosinophilia may cause fever rash and hepatomegaly. Granulomatosus hepatitis may occur necessitating discontinuation of therapy. Minor rises in hepatic enzyme levels during treatment occur in about 3% of cases but do not seem to be productive of severe hepatic reactions and treatment may continue.

Reversible oligospermia. About 70% of men develop oligospermia and abnormal sperm motility. Therefore this drug should be avoided in men wishing to have a family. This temporary reduction in fertility is reversible on stopping sulfasalazine treatment.

Miscellaneous. Sulfasalazine can cause orange staining of soft contact lenses. The urine can become bright orange. Rarely drug-induced lupus.

Sulfasalazine should be stopped if the following occur:

1. Severe rashes.
2. Severe headaches.
3. Deteriorating liver function.
4. Platelets <150 × 10^9 L.
5. Total white cell count <3.5 × 10^9 L.
6. Neutrophils <2 × 10^9 L.

Methotrexate

Methotrexate is an antimetabolite cytotoxic agent used in RA, psoriatic arthritis, myositis and vasculitis. It is given either orally, subcutaneously, intramuscularly or intravenously. Weekly doses are usually between 5 mg and 25 mg.

Rarely the maximum dose can be 30 mg per week. The initial dose will vary depending on the severity of the condition and patient characteristics such as age, renal function and co-morbid conditions. It is preferable to use only one strength (2.5 mg) of tablets and patients should be reminded of the need to check the dose and strength of the tablets with each prescription. It is co-prescribed with folic acid, which may reduce the likelihood of side effects.

Clinical effect is usually evident after two to three months but doses may need to be increased to maintain this effect. A chest X-ray is taken prior to starting treatment as pre-existing lung disease requires special attention. Any patient suspected of alcohol abuse is usually unsuitable for methotrexate therapy. Patients should be advised to limit their alcohol intake to well within national limits.

Alertness

Patients with clinically significant renal impairment.

Symptoms

1. Pregnancy and breastfeeding.
2. Suspected local or systemic infection.

Clinical study

1. Phenytoin, co-trimoxozole, trimethorpim-antifolate effect of methotrexate is increased.
2. Probenecid, penicillin, azapropazone, NSAIDs-methotrexate excretion is reduced. Clinically significant interaction between NSAID and methotrexate is rare. All patients should be advised to avoid over-the-counter medications including aspirin and ibuprofen without the knowledge of the rheumatology team.

After effects

Alimentary canal

These are the most common side effects which include anorexia, nausea, dyspepsia, vomiting, diarrhoea and abdominal pain. It is recommended that the therapy is taken with food. Anti-emetics may be required on the day of methotrexate administration and the following 24-48 hours. If nausea and vomiting are problematic and doubts exist about absorption, methotrexate can be given intramuscularly.

Mouth ulcers

Mouth soreness can present with or without ulceration. The use of folic acid to minimize stomatitis has been recommended. Minor ulcers can be treated symptomatically whilst major mouth ulcers may necessitate dose reduction or discontinuation.

Rashes

Minor rashes can be treated symptomatically whereas severe rashes may necessitate discontinuation of methotrexate.

Circulation

Leucopenia is the most common bone marrow toxicity from methotrexate. This is usually managed by reducing the dosage of methotrexate. Anaemia and thrombocytopenia can also occur. Macrocytosis is a sign of toxicity and may precede bone marrow suppression. Pancytopenia is a rare but potentially fatal complication of methotrexate, which may develop suddenly and without warning signs. It can occur early within one to two months of commencing methotrexate possibly reflecting an idiosyncratic reaction and more commonly later on in the patient's experience reflecting a cumulative effect.

Liver

Hepatoxicity defined as elevation of transaminase occurs within 69-89% of patients. These increases do not usually correlate with the development of cirrhosis. Liver fibrosis and cirrhosis can occur with normal liver enzymes and imaging findings. Current studies in patients with RA suggest that liver biopsies are not cost-effective for at least the first 10 years of methotrexate use in patients with normal liver function.

Allergy

Hypersensitivity reactions include rashes, fever and pneumonitis. Methotrexate pneumonitis occurs in 1-6% of cases. It is a potentially fatal hypersensitivity reaction most frequently seen within the first year of treatment. Many studies suggest that the incidence of pneumonitis is much higher in patients with pre-existing lung disease. Pulmonary function tests may be a useful investigation to detect pre-existing lung disease as methotrexate pneumonitis tends to occur in 48% of patients with pre-existing lung problems. If pre-treatment X-ray suggests abnormal shadowing it may be worth considering a high resolution computerized scan and pulmonary function tests. Airway obstruction may not be a contraindication to the use of methotrexate but the presence of interstitial lung disease is.

Kidney

Acute renal decompensation and renal failure have been reported although rarely, and appear related to high dose therapy with concomitant NSAIDs with underlying renal disease.

Infections

Methotrexate should be stopped immediately if infection is suspected. Significant mortality and morbidity can be associated with viral infections due to herpes zoster/varicella. Patients with a history of contacts with such viral infections need special advice and the consideration of treatment with anti-herpes zoster immunoglobulin.

Folate supplementation

Folic acid has been shown to reduce the likelihood of derangements of liver function tests as well as mucosal and gastrointestinal side effects. No consensus exists amongst specialists as to the optimal dose and timing of folate supplementation and no studies have proven an unequivalent protective effect of folate supplementation on methotrexate-related haematological toxicity.

Folinic acid rescue

In suspected cases of methotrexate overdose or severe haematological toxicity, folinic acid can be given. The initial dose should be at least 20 mg given intravenously. Subsequent doses of 15 mg orally should be given at six-hourly intervals until the haematological abnormalities are improved.

Surgery

Research suggests that continuation of methotrexate treatment does not increase the risk of infection or surgical complications in patients with RA.

Methotrexate should be stopped if the following occur:

1. Major mouth ulcers.
2. Severe skin rashes.
3. Infection is suspected or present.
4. Severe sore throat.
5. Abnormal bruising.
6. Acute pneumonitis.

7. Deteriorating liver function – that is, AST, ALT greater than twofold rise.
8. MCV .105fl -check serum B12, Folate and TFT.
9. Platelets <150 × 10^9 L.
10. Total white blood cell count <3.5 × 10^9 L.
11. Neutrophils <2 × 10^9 L.

Azathioprine Therapy

Pharmaceutical drugs

Azathioprine is an immunosuppressant antimetabolite used in RA, psoriatic arthritis and in vasculitis. It is widely used to suppress transplant rejection. It is given orally in daily doses of up to 3 mg/kg. It can be used in combination with corticosteroids as a 'steroid sparing agent'. It may take three to six months for patients to respond to treatment.

Alertness

1. Thiopurine methyl transferase deficiency.
2. Older people with renal impairment.
3. Localized or systemic infections with hepatitis B and C or a history of TB.

Symptoms

1. TPMT deficiency – can be fatal.
2. Aminosalicylates – for example, sulfasalazine contributes to bone marrow toxicity.
3. Co-trimoxazole and trimethoprim can cause life-threatening haematoxicity.
4. Pregnancy and breastfeeding.

After effects

Alimentary canal

This includes nausea and vomiting; patients may have to reduce the dose and then increase again when symptoms settle. Anti-emetics may be required. It is advisable to take azathioprine on a full stomach.

Mouth ulcers

Minor ulcers can be treated symptomatically whilst severe ulceration may necessitate stopping azathioprine.

Liver toxicity

This is rare but if occurs may be severe and require dose reduction or withdrawal.

Circulation

The most important side effect is reversible marrow suppression. Macrocytosis is common but is not a sign of toxicity.

Lymphoma

Long-term effects may include the induction of lymphoid tumours. In RA the azathioprine-related risk of lymphoma and non-Hodgkin's lymphoma is confounded by an increased relative risk secondary to RA. Overall there appears to be a small added risk of developing malignancy when using azathioprine in RA.

Miscellaneous

Hypersensitive reactions occur early in treatment and may include headaches, confusion and aseptic meningitis.

Clinical study

Allopurinol can increase the myelosuppression of azathioprine. Allopurinol can inhibit the anticoagulant effects of warfarin. Azathioprine can reduce the absorption of some antiepileptic drugs. Azathioprine should be stopped if the following occur:

1. Platelets $<150 \times 10^9$ L.
2. White cell count $<3.5 \times 10^9$ L.
3. Neutrophils $<2 \times 10^9$ L.
4. Deteriorating liver function > twofold rise in AST, ALT.
5. MCV >105 – check serum folate, vitamin B12 and TSH.
6. Abnormal bruising or severe sore throat.

Cyclophosphamide Therapy

Cyclophosphamide is a cytotoxic alkylating agent used for immunosuppression in SLE, vasculitis, resistant rheumatic arthritis, polyarteritis nodosa, Wegener' s granulomatosis, myositis and as a steroid sparing agent. A daily oral dose of 1-2 mg/kg body weight is used, it may be given as pulsed intravenous or oral therapy. Cyclophosphamide reaches its peak of effectiveness after approximately 16 weeks and although remission may be maintained for several years on withdrawal of the drug the majority of patients will experience some recurrence of symptoms. Non-articular complications of RA such as interstitial lung disease, cutaneous ulcers and peripheral neuropathy may respond well to oral cyclophosphamide. Cyclophosphamide should be used with care in patients with diabetes as it may induce hyper- or hypoglycaemia.

Pre-treatment check:

1. Pre-treatment investigations include chest X-ray and a full blood count, including differential white count and platelets, electrolytes and liver function tests.
2. Avoid if possible in men and women of childbearing age. Otherwise offer gamete storage and ensure adequate contraception.
3. Carry out a full infection screen including chest, urine testing and culture, joint examination, sinuses and perineum.
4. Some rheumatology units advise patients on this therapy to minimize or stop alcohol consumption.

After effects

Mouth ulcers

Minor ulcers can be treated symptomatically. Severe ulcers may necessitate stopping cyclophosphamide.

Nausea and vomiting

If gastrointestinal intolerance occurs the patient may require anti-emetic therapy. It is advisable to take this treatment with food.

Alopecia

This is usually mild in dosages less than 100 mg daily.

Infections

Suppression of the immune system exposes patients to the risk of infections. This can be minor - for example, herpes zoster - or major - for example, septicaemia. There is also the increased risk of patients experiencing opportunistic infections.

Haemorrhagic cystitis

Bladder toxicity is due mainly to the effects of acrolein on the urinary metabolite of cyclophosphamide. As well as causing haemorrhagic cystitis and bladder fibrosis it has been associated with bladder carcinoma. Therefore the occurrence of haemorrhagic cystitis is an indication for cessation of therapy. Haemorrhagic cystitis has been reported in about a third of subjects receiving oral cyclophosphamide daily. It is rare in patients receiving intermittent high dose intravenous cyclophosphamide therapy, where it is often co-prescribed with Mesna which reacts specifically with the metabolite acrolein preventing urothelial toxicity. Patients should be advised to take at least 3 litres of fluid per day whilst taking cyclophosphamide therapy.

Malignancy

Kinlen *et al.* (1979) reports a relative risk of 12.8 for all cancers combined and a tenfold increase in bladder cancer for patients receiving at least three months of treatment with cyclophosphamide. There appears to be a relationship of total dose and duration of therapy with the incidence of malignancy. There is also an increased risk of leukaemia.

Fertility

The risk of infertility, azoosperma and amenorrhoea with cyclophosphamide increases with higher dosages, longer duration of therapy and increased age in women. Infertility is not always reversible.

Pulmonary Fibrosis

This can occur with therapy.

Haematological

Abnormalities are detected by blood monitoring. Maximum effect on the marrow dose not occur until 5-10 days after dosage. Recovery is seen within 10-14 days. Macrocytosis is not inevitable but is a sign of toxicity and dosage reduction may be necessary.

Clinical study

Allopurinol can increase the myelosuppression of cyclophosphamide. Cyclo-phosphamide should be stopped if the following occur:

1. Severe mouth ulcers.
2. Platelets <150 × 10^9 L.
3. Total white cell count <3.5 × 10^9 L.
4. Neutrophil <2 × 10^9 L.
5. Macroscopic haematuria or persistent confirmed macroscopic haematuria (infection excluded).

Ciclosporin

Ciclosporin A is a fungal metabolite of *Trichoderma polysporum* and *Ciclocarpon leucidium.* Studies have confirmed potent effects of this drug in several arthritic and other cell-mediated chronic inflammatory conditions. It may be instituted at doses of 2.5 mg/kg/day given in 2 divided doses every 12 hours. The dose may be cautiously increased 25-50% every 2-4 weeks until a maximum of 4 mg/kg/day is reached assuming there are no adverse reactions. Ciclosporin is used in the treatment of RA and Behcet's disease. Clinical improvements in patients receiving ciclosporin are associated with a reduction in C-reactive protein, d-1-acidglycoprotein and platelet counts but not rheumatoid factor or the ESR. When ciclosporin is used in the treatment of rheumatic disease side effects are generally mild and reversible at low doses.

Alertness

1. Uncontrolled hypertension.
2. Use of potassium sparing diuretics.
3. Pregnancy and lactation.
4. Grapefruit juice to be avoided for one hour after administration.
5. Malignancy such as lymphomas.

Symptoms

1. Uncontrolled hypertension.
2. Renal and liver failure.
3. Severe electrolyte imbalance.
4. Suspected systemic infection.

After effects

Minor reactions - can include nausea and vomiting, tinnitus, tremor, paraesthesia and gum hyperplasia.

Major reactions - Nephroxicity. Ciclosporin should be avoided in patients with pre-existing renal disease. Increases in serum creatinine are common during ciclosporin immunosuppression. Ciclosporin therapy in doses of 10 mg/kg/day for even two months may lead to an irreversible loss of more than 10% of renal function in RA patients. Nephrotoxicity is dose-related. Long-term influences on renal function are significant even after cessation of the drug especially in view of the potential for apparent persistent impairment of creatinine clearance.

A study by Tugwell *et al.* (1990) showed that although serum creatinine rose and creatinine clearance decreased over the period of ciclosporin administration the effects stabilized after four months with no further changes. After discontinuation of therapy serum creatinine fell to within 15% of baseline in all except two patients. Dosage reductions of 25-50% are required if serum creatinine increases above baseline by 30-50% or if hypertension cannot be controlled.

Hypertension

Hypertension has always been a factor predisposing to nephrotoxicity in RA. New onset hypertension has been reported in about one third of RA and psoriatic patients. Hypertension should be managed with beta blockers and ACE inhibitors; potassium sparing diuretics should be avoided because ciclosporin may cause hyperkalemia. If hypertension is present prior to commencing ciclosporin the hypertension should be treated first before ciclosporin is introduced.

Hepatoxicity

This can occur with higher dosages of ciclosporin and present with elevated hepatic enzymes and bilirubin.

Clinical study

Diltiazem, ketoconazole, rifampicin and phenytoin may all require adjustment of ciclosporin dose. St John's Wort decreases ciclosporin activity. Colchicine should be avoided and the dose of diclofenac reduced by 50%.

Ciclosporin should be discontinued if the following occur:

1. Elevation of serum creatinine >30% from baseline on two consecutive occasions.
2. Potassium rises to above the normal range.
3. Raised blood pressure - diastolic 95 mm Hg on two consecutive occasions.
4. Ciclosporin levels are >300 mg/ml.
5. Elevated hepatic enzymes - > two-fold increase in AST or ALT or alkaline phosphatase from the upper limit of the normal range.
6. Platelets <150.
7. Abnormal bruising.

Chlorambucil

Chlorambucil like cyclophosphamide is a bifunctional alkylating agent. It is not a cytotoxic but is metabolized to phenylacetic acid - its principal and most active metabolite. It is given in an oral preparation. The intravenous preparation is unstable because of rapid hydrolysis. Chlorambucil can be commenced in doses of 0.1-0.2 mg/kg/day.

Once response or toxicity has occurred it is suggested that a dose of 3-4 mg daily is adopted. Chlorambucil has been used in patients with RA, vasculitis and connective tissue disorders, inflammatory eye disease and amyloidosis.

After effects

The advantages of chlorambucil therapy have to be balanced against the risk of inducing adverse effects. The most common toxicities with chlorambucil use are dose-related bone marrow suppression and infertility.

Bone marrow suppression

Cumulative bone marrow toxicity occurs regularly and often requires discontinuation of the drug.

Neoplasms

Kahn *et al.* (1979) reported that leukaemias occurred in patients with RA who had received a total dose of at least 1 g of chlorambucil and who had been treated for at least six months. Luqmani, Palmer and Bacon's (1990) review of the literature on patients receiving chlorambucil indicated that tumours occurred in less than 1% of patients. The majority of the malignancies were leukaemias with a particularly high incidence of acute myeloid leukaemia. It is not possible to exclude the part that the disease process itself plays in the likelihood of malignancies and Palmer and Denman (1984) found that patients with connective tissue disorders were more sensitive to the leukaemiogenic effects of chlorambucil than patients with other non-malignant conditions.

Infertility

The risks of infertility, azoospermia and amenorrhoea increase with duration of therapy and high doses. Infertility affects men at a greater rate than women; irreversible azoospermia was regularly reported in men who received greater than 400 mg of chlorambucil. In post-pubertal females both gametogenesis and hormonal function are altered adversely and menopause occurs.

Phenylbutazone

Phenylbutazone is not longer recommended or indeed available in many countries for the treatment of arthritis conditions due to the risk of marrow aplasia. Although it is still used in the United Kingdom, it is available only on hospital prescription for patients with ankylosing spondylitis and other spondyloarthropathies. Phenylbutazone is a NSAID but it may also affect the immune response. It is usually prescribed in dosages of 100-200 mg, two to three times a day. It should help with pain and swelling within a week but will take months before an immune response can be assessed.

After effects

Gastrointestinal. Nausea and vomiting.

Mucocutaneous. Skin rashes, mouth ulcers, stomatitis.

Central nervous system. Headaches, dizziness and blurred vision.

Pulmonary toxicity. Care must be taken when prescribing these tablets to patients with asthma or breathing problems as these conditions may be aggravated.

Cardiovascular. Phenylbutazone has fluid retaining properties.

Hepatic. Serious hepatocellular reactions have been reported and hepatic clearance of drugs is reduced as phenylbutazone inhibits oxidase drug metabolism.

Haematological abnormalities

A range of haematological abnormalities have been reported. Long-term recipients may develop iron deficiency anaemia due to chronic blood loss from inflammation of the small intestine. Aplastic anaemia has been associated with the slow metabolism of phenylbutazone suggesting that this adverse effect is related to excessive concentrations of the drug. Thrombocytopenia, effects on platelet function, agranulocytosis, pancytopenia and haemolytic anaemia have also been noted but are still rare although can be fatal.

Clinical study

Phenylbutazone interacts with oral anticoagulants, lithium, oral hypoglycaemic agents, phenytoin, methotrexate (causing pancytopenia probably due to retention of methotrexate due to inhibition of proximal renal tubular secretion), digoxin, aminoglycosides, probenecid and barbiturates.

Dapsone

Dapsone is best known as an anti-leprotic agent but it also has effects on the immune system and has been used for the treatment of RA and psoriatic arthritis. There is a delay of two to three months before onset of action but when successful it will reduce the ESR. The dosage is 50 mg daily for one week increasing then to 100 mg daily depending on the result of the haemoglobin. Occasionally patients may require 150 mg daily.

After effects

Minor side effects such as rashes and gastric upset can occur. Haematological – haemolysis occurs in all patients and gives a drop in haemoglobin of 1-2 g. This therefore requires monitoring before increases in dosage are considered. Patients also develop a pallor which is more marked than the drop in the haemoglobin would produce. It is worth warning patients that they will look pale. Patients of Mediterranean origin will require a G6PD estimation prior to starting treatment since massive haemolysis will occur if there is a deficiency in this enzyme. Agranulocytosis has been reported but is rare.

Minocycline

Minocycline is a broad spectrum antibiotic often prescribed in the treatment of acne. It has also been used in the treatment of RA. Its mode of action is not known but it is thought to be effective at inhibiting the pro-inflammatory enzyme metalloproteinase. It does not become fully effective for two to three months. The dosage is 100-300 mg twice daily.

After effects

An abnormal metallic taste can occur when first starting the treatment. Dizziness, nausea and diarrhoea have also been reported. The treatment is best taken with food. Although severe side effects are rare a drug-induced lupus syndrome and hepatitis can occur that resolves on stopping but necessitates that treatment is monitored with a full blood count and biochemistry monthly and an anti-nuclear antibody six-monthly.

Clinical study

Minocycline should not be taken at the same time as antacids, calcium and iron preparations. A two-hour gap is required between administration of minocycline and these preparations.

Warfarin

If prescribed penicillin, leave off minocycline until the course of penicillin is completed.

Leflunomide

Lefluomide is usually prescribed as a daily dose of 10-20 mg. It is common in the United Kingdom not to use a loading dose of 100 mg daily for three days although this practice is seen in other European countries. Loading doses are associated with an increased incidence of side effects. On doses of 10 mg it will take over six months to reach a therapeutic level. The recommended daily dose is 20 mg. In patients receiving other potentially hepatotoxic drugs, such as methotrexate, or who are renally impaired 10 mg is recommended. Alcohol ideally should be avoided.

Alertness

1. Evidence of Hepatitis B or C infection.
2. History of tuberculosis.

Symptoms

1. Serious infections.
2. Impaired liver function.
3. Severe unexplained hypoproteinaemia.
4. Renal impairment – moderate to severe.

After effects

Many adverse effects can be managed by reducing the dosage or by concomitant administration of symptomatic therapy. If it is necessary to rapidly remove the drug from a patient's system then a washout with cholestyramine will quickly reduce the plasma levels of the drug.

Washout procedure

Cholestyramine 8 mg is administered three times daily. Alternatively 50 mg of activated powered charcoal is administered four times daily. Duration of a complete washout is usually 11 days. The duration may be modified depending on clinical or laboratory variables.

Symptoms of alimentary canal

Data from a US database on the use of leflunomide by 40000 RA patients indicated that gastrointestinal tract-related problems were the commonest reasons for stopping treatment. Problems experienced included diarrhoea, nausea, vomiting, oral mucosal disorders (e.g. aphthous stomatitis, mouth ulceration) and abdominal pain. The prevalence of nausea is less in lefluomide than in patients taking sulfasalazine or methotrexate but the prevalence of diarrhoea is increased and should be treated, in the first instances, rather than stopping the drug. Daily doses of leflunomide 20 mg are associated with more nausea and diarrhoea than a daily dose of 10 mg.

Allergy

In cases of ulcerative stomatitis, leflunomide administration should be discontinued. Minor rashes can be treated symptomatically but severe rashes will necessitate discontinuation of leflunomide. Severe dermatological toxicity has a higher prevalence than is seen in patients taking sulfasalazine or methotrexate. Cases of Stevens-Johnson syndrome or toxic epidermal necrolysis have been reported.

Hypertension

The manufacturers recommend that blood pressure monitoring is carried out on patients taking leflunomide but hypertension does not appear to occur more often with this DMARD in comparison to other DMARDs such as methotrexate, except where loading doses have been used.

Pulmonary reactions

Interstitial lung disease has been reported during treatment with leflunomide. Interstitial lung disease is a potentially fatal disorder, which may occur acutely during therapy. Pulmonary symptoms, such as cough and dyspnoea, should be investigated. Pulmonary infiltration as an acute allergic reaction has been demonstrated in a small number of patients after commencing leflunomide.

Hepatoxicity

Raised liver tests can occur and require monitoring; liver toxicity is more common in methotrexate that with leflunomide. Rare cases of severe liver injury (some with fatal outcome) have been reported after drug treatment with leflunomide. Most cases occurred within six months and were associated with other risk factors for hepatotoxicity. Patients should be asked to limit alcohol intake well within normal limits at four to eight units a week.

Circulatory abnormalities

It is recommended that patients on leflunomide have regular full blood counts although there is no data to suggest an increase in blood dyscrasias due to leflunomide. Leflunomide should be stopped if the following occur:

1. Major mouth ulcers or skin rashes.
2. Respiratory reactions.
3. Liver function deteriorates – > twofold rise in AST, ALT.
4. Platelets <150.
5. Total white blood cell count <3.5.
6. Neutrophils <2.
7. Weight loss of >10% with no other cause identified.
8. Severe persistent headaches.
9. Severe sore throat.
10. Abnormal bruising.

Mycophenolate Mofetil

Mycophenolate mofetil is a potent reversible selective inhibitor of purine synthesis. It has been shown to inhibit T cell proliferation and antibody production by B cells. It is used in the treatment of systemic lupus erythematosus, scleroderma and vasculitis and after organ transplantation.

Mycophenolate is usually taken in capsule form twice a day on an empty stomach. A typical dose may be 500 mg twice a day for four weeks; then, if tolerated, 1 g twice a day thereafter. Alcohol should only be taken in small amounts as it can affect the liver. Mycophenolate should not be taken with cholestyramine or antacids with magnesium or aluminium hydroxides as absorption may be reduced. If being used with azathioprine, lower doses should be used. Pre-treatment assessment includes FBC, urea and electrolytes, creatinine and liver function tests.

After effects

The most common side effects are sickness, diarrhoea, vomiting or abdominal pain. These side effects are usually dose-related and resolve on discontinuation of the drug. Severe neutropenia occurs in 0.5% patients receiving 1 g twice daily compared to 0.8% patients receiving azathioprine; it is most commonly seen within the first six months. There is a slight increase in the risk of skin cancer and patients should avoid exposure to strong sunlight and protect their skin with sunblock or sunscreen. Symptoms of allergy include wheezing, shortness of breath, swelling of the face, lips, tongue or throat. Urogenital reactions include sterile haematuria, urinary tract infection and renal tubular necrosis.

Reasons for stopping treatment

Mycophenolate should be stopped if the following occur:

1. WBC <3.5 × 10/L.
2. Neutrophils < 2.0 × 10/L.
3. Platelets <150 × 10^9/L.
4. Rash or oral ulceration.
5. MCV > 105fl
6. Abnormal bruising or sore throat.
7. Gastrointestinal disturbances – withhold until settle then reintroduce at a lower dose.

IMMUNIZATION

Live vaccines should be avoided in patients taking azathioprine, methotrexate, cyclophosphamide or high doses of corticosteroids. Patients on these medications should be considered for varicella-zoster immunoglobulin after exposure to chicken pox or herpes zoster and should be advised to avoid close contact with patients recently vaccinated with live polio vaccine because of the risk of infections from faecal excretion. Sulfasalazine, D-penicillamine, gold – in general vaccination are safe for patients on these therapies although is probably preferable to avoid live vaccines. Ciclosporin has a poor response to vaccines and patients should ideally have any appropriate vaccinations prior to therapy.

PREGNANCY

Patients planning a family will require advice and support from various members of the rheumatology team. Pregnancy and related issues such as breastfeeding must be addressed when patients are receiving a DMARD. It is necessary to be absolutely certain about the effects of the DMARD on the fetus as well as on lactation. One of the major concerns of any pregnant woman is the risk of congenital malformation due to drugs and this will cause even greater concern to the woman with rheumatoid arthritis.

The most vulnerable period of gestation is during embryonic and fetal development but many congenital abnormalities occur for reasons other than those that are drug-related. Spontaneous and serious malformations occur in 2-3% of all pregnancies and minor malformations occur in a further 6%. Women with rheumatoid arthritis often need to achieve disease suppression to increase their chance of conceiving.

Adequate control of the disease will enable a woman to feel capable of bearing and raising a child. Decisions regarding the withdrawal of potentially toxic drug therapy need to be made in good time since many drugs can affect the vulnerable stages of embryogenesis. Women should discontinue teratogenic agents such as methotrexate, azathioprine and cyclophosphamide at least six months before attempting conception.

Leflunomide, one of the newer DMARDs, has proven teratogenic and fetotoxic effects in animal studies and its active metabolite is detectable in plasma two years after discontinuation of the drug. Consequently the fetus could have *in utero* exposure to leflunomide up to two years after the end of treatment unless an oral cholestyramine regimen, 8 g 3 times daily for 11 days is administered to obtain undetectable plasmatic levels. The manufacturer recommends that for women of childbearing age 'treatment with leflunomide must not be started until pregnancy is excluded and it has been confirmed that reliable contraception is being used'.

Ideally all drug therapy should be avoided during pregnancy and lactation. In ankylosing spondylitis 80% of patients may experience a worsening of symptoms or no alteration of their condition during pregnancy and will require advice on coping with pain. This should include relaxation techniques, the application of hot and cold therapy and the use of diversional techniques. In RA 75% of women experience some remission of their condition during pregnancy but often return to disease status comparable with their pre-pregnant state within eight months of giving birth. Women who develop an exacerbation of their condition post-partum will need to recommence their suppressive drug therapy and individual advice from the rheumatology nurse will be required if the mother is breastfeeding.

Reliable Contraception

Reliable contraception should be advised for all women receiving disease modifying anti-rheumatic drug therapy and all proposed pregnancies will need discussing on an individual basis with the rheumatology nurse.

For those medications known to be teratogenic - for example, methotrexate, azathioprine, leflunomide, chlorambucil and cyclophosphamide - patients must have contraceptive cover and discontinue the drug for six months before trying for a family. In the case of leflunomide the manufacturers recommend that the drug should be discontinued for two years before becoming pregnant. If this waiting time is considered unpractical, prophylactic institution of a washout procedure may be advisable.

SOCIALISM AND DRUG THERAPY

Musculoskeletal disorders are a common cause of long-term disability in the United Kingdom, making up 15% of the workload of GPs. Shared care is the management of the patient between the hospital team - consisting of the rheumatologists, specialist nurse, occupational therapist and physiotherapist - and the primary healthcare team. Drug therapy should be carefully planned between the rheumatologist and the general practitioner. The rheumatologist's main role is to diagnose and establish a treatment plan to be followed in primary care or shared between primary and secondary care. The latter is the case with rheumatoid arthritis, as it is a progressive chronic disease. The specialist nurse acts as a resource for the patients, other professionals and for the primary healthcare team. The nurse is required to be a counsellor, educator and adviser on many aspects of rheumatoid arthritis, particularly drug therapy. Physiotherapists provide a whole range of treatments and aim to maintain and improve function, alleviate pain and motivate the patient to carry out regular exercises.

The occupational therapist provides help and support with problems of daily living and is often instrumental in enabling the patient to develop coping strategies including pacing of activities, joint protection and self-management interventions including relaxation. The primary healthcare team provide health services outside the hospital. They consist primarily of the general practitioner, the practice nurse, the district nurse and the health visitor, all of whom provide valuable information about the patient's family and home situation. They are often the first point of contact the patient makes. An analysis of the work practice nurses do showed they offered practical treatment, health promotion and disease management. When each team complements the other and, provided there is communication and respect for each other's roles, then good quality treatment should be available for all patients. According to Bryan (1995) practice nurses are isolated from colleagues and need a forum for the exchange of ideas, the updating of skills and the acquisition of new knowledge.

In order to try and fill this need, education and opportunities for updating knowledge are provided by the hospital team in the form of study days, visits to the rheumatology unit and the team visiting the general practice. The practice team will then be able to support and advise patients, recognize when side effects require a change of treatment or when the patient would benefit from an early hospital appointment. The early identification of symptoms of the disease, the side effects of drug therapies, the interpretation of the blood results and their implications, plus the psychological effect of arthritis on the patient and their family are important requisites for managing the condition. An evaluation of two study mornings for practice nurses delivered by a specialist nurse in the rheumatology department of a general hospital enabled clinical scenarios to be discussed, increasing the clinical knowledge and consequently reducing the number of telephone calls from practice nurses.

Ignorance

Nurses in the primary healthcare team can play an important part in alleviating confusion in the administration of drugs by using medication cards stating how and when patients are to take the prescribed dose. Pill dispensers are invaluable for patients with poor memory as they have compartments containing drugs for each day on a weekly basis. To obtain the maximum benefit, medicines should be taken in the correct way at the correct time. Some patients will have failing eyesight and be unable to read the instructions.

Name of tablet	Dosage	How/when to take	Special instructions
Diclofenac (Voltarol)	75 mg twice daily	One after breakfast. One after evening Meal.	Always take with food./ Report any sickness/indigestion.
Azathioprine (Imuran)	150 mg daily	Three 50 mg tablets after breakfast.	Take with food. Report any sickness or mouth ulcers.

Fig. 1.2. Medication card.

Many pharmacists will print large labels to overcome this. Patients who have poor hand grip often cannot open the childproof bottles. It is helpful to liaise with the pharmacist who will provide screw tops or ring caps. Patients with reduced manual dexterity frequently can experience difficulties with blister packs. Eye drops also often present a problem. The pharmacist and the occupational therapist can help with the provision of aids to maintain independence. An increased number of tablets taken, and doses involved, often reduces concordance.

Patients with poor memory may find drugs to be taken daily of more benefit. Patients with Sjogren's syndrome can find swallowing tablets difficult. It may therefore be more appropriate to prescribe drugs in liquid form. Many patients do not comply with the drug therapy as they are afraid of the side effects or of being dependent on a drug, particularly analgesia. Confusion often arises as many drugs have three different names, the brand, the pharmaceutical and the generic name. Low patient adherence to therapeutic regimes is an obstacle to achieving therapeutic goals in arthritis management. Improvement of concordance is a prime goal of much patient education and counselling.

SOCIAL GUIDELINES

In GP surgeries most of the drug monitoring is carried out by the practice nurse, guided by protocols issued and agreed with the rheumatologist and GP. The Royal College of Nursing rheumatology forum has produced guidelines for nurses involved in

1. The administration of intramuscular sodium aurothiomalate.
2. The administration of intramuscular methotrexate.

The objective of these guidelines is to ensure that the patient receives the same standard of care and high quality service, independent of the setting in which they are nursed. The guidelines also provide the nurse with a framework for practice. Helliwell and O'Hara (1995) found that the percentage of cases in which the standard monitoring protocol had been complied with within the GP practice was 26% for sodium aurthoimalate, 67% for sulfasalazine and 93% for methotrexate. (The discrepancy relating to the monitoring of sodium aurothiomalate concerned an annual chest X-ray requirement). Havelock (1998), in an audit undertaken of shared care monitoring, found that the majority of patients were being adequately monitored, less than 2% of patients were not being monitored and a significant number of patients were actually receiving more blood tests than had been stated on the agreed protocols.

Computerized Monitoring

Close monitoring of second line drugs ensures their safe use and is a guide to their efficacy. To provide close communication between the general practitioner and the rheumatologist, shared care booklets are often issued to patients from the rheumatology department. The use of a patient-held booklet has been strongly endorsed by the British Society of Rheumatology, the British Health Professionals in Rheumatology and the National Patient Safety Agency. Such booklets can be used to record the blood results serially and monitor the urinalysis for the presence of blood and protein. This enables the rheumatology specialist to examine the trends of the haematological and biochemical results and the presence of abnormalities in the urine. This therefore helps the rheumatology specialist to decide whether to discontinue the drug or reduce the dose. Each book issued should have the contact number of the rheumatology department with which the patient or a member of the primary healthcare team can make contact for advice and support.

Serious side effects can be avoided by the hospital team and the primary healthcare team explaining to the patients the necessity of having regular blood tests and reporting side effects early. In Helliwell and O'Hara's study (1995) there was unanimous agreement by the GPs that shared monitoring cards were helpful. The GPs involved were prepared to keep them updated and were willing to follow suggested protocols. Havelock (1998) also found that this method of documentation worked well within both the primary and secondary care settings. Computerized patient monitoring systems have been gaining popularity in rheumatology as they offer automized validation, improved data capture and immediate result availability.

Lee, Lenert and Kavanaugh (2004) suggest that patient outcome measures could also be completed by patients via a computer program, which would enable the clinician to receive the information required to access the efficacy of drug therapy at the optimal time and prevent the delay that often occurs as a patient waits for an outpatient appointment. Clinicians could then review and compare results with earlier visits and adjust medications accordingly ensuring patients achieve a therapeutic dose at the optimal time.

Practice-based Monitoring

Anxieties over various aspects of GP-based drug monitoring included:

1. Poor facilities for the collection and distribution of blood samples to the hospital laboratory.
2. A perceived lack of time to perform domiciliary visits on those patients unable to attend the practice. One could argue that it may be more appropriate for a community practitioner – for example, a district nurse with expanded knowledge in the field of rheumatology – to carry out such visits.
3. A lack of clarity regarding the responsibility for prescribing and monitoring drug therapy.

One practice cited by Helliwell and O'Hara (1995) was unwilling to monitor second line therapy due to a perceived lack of time to perform the necessary blood tests and a reluctance to accept the responsibility inherent within this practice.

Patients's Experiences

Arthur and Clifford (2004a) compared the satisfaction of patients attending follow- up monitoring care within primary and secondary care locations. They found that whilst patients from both locations were satisfied with the care they received, those receiving specialist nursing care in the secondary care location were more satisfied. Empathy, specialism, information provision, technical aspects, time and continuity of care were identified as being important for patients attending monitoring clinics, with the specialist nurse able to positively influence patients' perceived ability to cope with their arthritis.

COMMUNITY CARES

At the heart of community care is the principle that services should support people in their own home and locality. In certain chronic conditions shared management between the hospital and general practitioner is well established, especially in areas such as diabetics and asthma. The essential features of community clinics are that they are:

1. In a location close to the patient's home.
2. Performed by a consultant specialist.
3. Providing educational opportunities for GPs.
4. Improving communication between the hospital and community staff.

Shared care should include:

1. Effective communication between primary and secondary care teams.
2. .Shared documentation in the form of agreed guidelines.
3. Ongoing support, education and training.
4. An appreciation of both service commitments and difficulties.

COMMON PRACTICE

Most GPs will have only a very small case load (often single figures of patients) of those requiring DMARDs. Therefore the practitioner is unlikely to be familiar with all aspects of this therapy including adverse effects. Although the NHS management executive has decreed that the doctor who has clinical responsibility for the patient should undertake the prescribing, there is a clear need for shared management in both the clinician's and the patient's interest. The British Society of Rheumatology (1992) has recommended the referral of all patients with RA to the hospital setting with the subsequent sharing of treatment between the hospital and general practitioner. GPs feel confident managing the majority of musculoskeletal conditions within their surgery provided they have adequate support with: joint injection techniques, complex cases with poor outcome (including pain syndromes), access to physiotherapy and a multidisciplinary approach to pain control.

GPs' knowledge of their patients with RA in the areas of functional disability, social situation and involvement of health and social care professionals was very variable, with GPs significantly underestimating the functional disability of their patients. The education and training of GPs may not equate with the needs of musculoskeletal patients. A lack of confidence in managing rheumatological conditions is related to lack of experience and poor teaching. This situation is being addressed through the training of GPs with a specialist interest in rheumatology.

SOCIAL CLINICS

Community clinics run by a rheumatologist in a GP' s surgery demonstrated increased patient satisfaction and greater patient access to a specialist, in comparison with traditional outpatient clinics, but only a slight improvement in health benefits. Whilst patients' own costs were lower, NHS staffing and treatment costs were more expensive per patient in outreach than outpatient clinics. The advantages of community clinics cited by both the GP and the specialist included:

1. Shorter waiting time for first appointment.
2. Improved communication between GP and the specialist.
3. Improved patient access to the specialist.
4. Patient convenience.
5. Fewer non-attenders.

Patients in Helliwell's (1996) study reported experiencing a greater satisfaction with their consultants than their counterparts attending a hospital-based clinic. Patients attending the community clinic stated that their questions were always answered.

Potential Problems

1. The opportunity for education as a result of personal contact between the consultant and GP is often infrequent due to GPs engaging in busy surgeries of their own. In a survey by Bailey, Black and Wilkin (1994) GPs were in attendance at only 5% of outreach clinics.
2. Infrequent follow-up intervals (if the specialist attends every four to six weeks then follow-up consultations are delayed).
3. Case mix disruptions may occur and the consultant is not available for other clinical practices whilst in the community. Although fewer patients were seen in the community clinic with a higher old/new ratio than the hospital- based clinic, Helliwell (1996) found no difference in the case mix data between the two healthcare settings.
4. Increase in GP administrative costs and time.
5. Many consultants working within the community have a minimal number of support staff, and the time spent travelling will reduce the number of patients seen per session.

RHEUMATOLOGY NURSE PRACTITIONER

In Norfolk a weekly rheumatology nurse practitioner clinic was established in the general practice to provide the following range of services:

1. Physical assessment of joints.
2. The monitoring of the safety and efficacy of drug treatment.
3. Initiation and interpretation of clinical laboratory data.

4. Referrals to the multidisciplinary team.
5. Liaison between the patient and other healthcare professionals.

The patients were highly satisfied with this service provision. This finding was in accordance with work carried out by Hill (1997). Patients who were allocated to a nurse-led clinic within the outpatient setting were more satisfied with their overall care when compared to a similar group of patients attending the consultant clinic.

Mooney (1996) concludes that the reported patient satisfaction could be attributed to the improved continuity of care the patients were receiving, coupled with the length of consultation time the patient spent with the nurse. The nurse practitioner clinic in the doctor's surgery also had other benefits. These included:

1. If necessary quicker access to a rheumatologist was available and appropriate investigations could be carried out before referral.
2. The number of outpatient visits to the hospital rheumatology clinics was reduced.
3. The provision of specialist advice could be given in a familiar environment.
4. General practitioner visits for rheumatological complaints were reduced.
5. The provision of disease education for patients and the opportunity to discuss issues with the nurse.
6. Less travelling for patients as the clinic was located in a more convenient setting with no parking fees.
7. The patients experienced a continuity of care in their management.
8. The rheumatology nurse practitioner acted as an advocate between the practice, rheumatologist and patient thereby enhancing and improving communication.
9. Early referral could be made to other healthcare professionals.

Dargie and Procter (1993) established a nurse-led arthritis clinic in the health centre in which they worked. One of the primary objectives of this development was to provide an easily accessible service for the assessment, support, education and monitoring of treatment for patients with arthritis and their families. Additional advantages that Mooney (1996) cites included a focus on prevention rather than crisis management. The two individuals involved (a district nurse and a practice nurse) also felt they were able to make full use of their nursing skills which enhanced their own job satisfaction.

Potential disadvantages in the provision of this much-needed service were the lack of community services in some areas - for example, physiotherapist and occupational therapist. The extra time commitment required to provide such a service could also be viewed as a disadvantage in economic terms but the pay-off was that patients received a service that was tailored to their individual needs. The advantage of community clinics to the patient include convenience whilst maintaining a high standard of care from a specialist practitioner. For the hospital one of the benefits is the devolution of some of the care to general practice, creating the scope for more new patients to be seen in the hospital setting.

CLINICAL EFFICACY

Hewlett *et al.* (2000) evaluated the clinical efficacy, cost and acceptability of a shared care system of patient or GP-initiated review in patients with RA. Patients were randomized to either shared care with their GP (no routine hospital review but rapid access on request) or traditional hospital care (regular planned outpatient review usually three- to four-monthly). Patients (or their GPs) in the shared care group could request review by any member of the rheumatology team via the nurse-run telephone helpline. Patients in the study appeared representative of a hospital-based RA population with moderate inflammatory activity, pain and disability. Patients who declined to take part were older and more affected by their arthritis. Results showed no clinical deterioration and some clinical benefit in the shared care group, which was achieved at a 33.5% cost saving, with both patients and GPs experiencing greater satisfaction and confidence in the system.

MANAGEMENT

As is often the case in the management of many chronic conditions, it is the primary healthcare team who are responsible for the early diagnosis and continuing care of patients with established osteoporosis. Osteoporosis is a silent disease that can affect all age groups. Symptoms are not usually present until a fracture occurs, often as a result of minimum trauma. Bone is a living tissue and is being continually regenerated. Cells, called osteoclasts, destroy bone and eat away areas of bone, whilst other cells, called osteoblasts, are bone builders and follow the osteoclasts filling in and repairing the bone.

When this process becomes imbalanced, more bone will be destroyed than is being replaced; then osteoporosis results with bones becoming thinner and more brittle. Peak bone mass is generally achieved between the ages of 25 and 36. Following this peak, bone mass then decreases by 0.3% a year. When women reach the menopause, however, bone is lost more rapidly and may be as much as 5% per year over a 10-year period. Thus 1 in 3 women and 1 in 12 men are likely to sustain a fracture related to osteoporosis by the age of 90 years.

Categories of Osteoporosis

Osteoporosis falls into two categories, primary and secondary. Primary osteoporosis is caused by ageing, the menopause and impaired adult peak bone density. Secondary osteoporosis accounts for 40% of cases in women and 60% of cases in men. It is caused by:

1. Excessive exercise – for example, ballerinas and athletes.
2. Lifestyle factors – smoking and excessive alcohol.
3. Dietary factors – for example, lack of calcium and vitamin D.
4. Anorexia and bulimia.
5. Amenorrhoea for more than six months.
6. Premature menopause in women (whether surgical, natural or radiation-induced).
7. Gut malabsorption and chronic liver disease.
8. Immobilization.

9. High dose corticosteroid therapy (over 7.5 mg daily).
10. Anticonvulsants.
11. Heparin therapy.
12. Genetic links.
13. Cushing's syndrome.
14. Throtoxicosis.
15. Primary hyperparathyroidism.
16. Hypogonadism.

It is important that the primary healthcare team recognize those individuals who are potentially at risk and discuss with them preventive strategies or management of the condition.

Risk Factors for Osteoporosis Fracture

Risk factors do not have adequate sensitivity or specificity to identify people at risk and measurement of bone mineral density is needed to quantify an individual's risk of fracture. Risk factors for osteoporosis:

1. Early menopause (before 45 years).
2. Hypogonadism.
3. Smoking.
4. High alcohol intake.
5. Female sex.
6. Increasing age.
7. Physical inactivity.
8. Low body mass index.
9. Heredity.

Risk factors that act independently of bone mineral density:

1. Prevalent low trauma fracture, particularly vertebral fractures.
2. Family history, particularly of maternal hip fracture.
3. Current smoking habit.
4. Low body weight.
5. Increasing age.

New Findings for Osteoporosis

1. *X-rays*: radiography has an essential role in identifying fractures. However, standard radiographs cannot be used to diagnose osteoporosis on the basis of apparent osteopenia. The appearance of osteopenia on plain X-rays depends on the dose of X-ray used, volume of soft tissues and other variables, in addition to the bone density itself.

2. Dual energy X-ray absorptiometry (DEXA) scanning: this is considered to be the gold standard technique for measuring bone density. Typically scans are taken at the proximal femur and spine. The radiation dose to the patient from this procedure is small, approximately one tenth of a chest X-ray. A T score and a Z score is reported. The T score is the number of standard deviations above/below the peak adult (premenopausal) mean.
3. The Z score is the number of standard deviations above/below the age-matched mean. For every one standard deviation fall in bone density, fracture risk approximately doubles. DEXA scanning is useful for measuring bone density in patients with risk factors in whom results will affect management. This form of scanning may be unreliable in the presence of skeletal deformities for example, scoliosis or degenerative changes.

HORMONE REPLACEMENT THERAPY (HRT)

Once considered to be the gold standard in the treatment of osteoporosis, HRT now only has a role in the treatment of menopausal symptoms such as hot flushes or in women who have undergone an early menopause. Due to concerns in the increased risk of breast cancer and cardiovascular events it is no longer thought to be appropriate to use HRT in the management of osteoporosis.

ANALGESIC MANAGEMENT

Simple analgesics such as paracetamol and non-steroidal anti-inflammatory drugs are sometimes beneficial for chronic pain in osteoporosis. Opiates (e.g. morphine) and calcitonin are sometimes given for the intense pain of a vertebral fracture. If chronic pain is not managed then the patient becomes less active, more helpless and isolated, and self-esteem is greatly reduced.

PREVENTIVE MEASURES

Bisphosphonates

Bisphosphonates act to block the action of osteoclasts. They are poorly absorbed from the gut and readily bind to other agents in the gut and must be taken on an empty stomach.

Etidronate

This is the first of the bisphosphonates to be used commonly in osteoporosis. Due to the risk of osteomalacia, 400 mg is taken daily for two weeks every three months, with the option of taking calcium supplements for the remaining weeks. It must be taken with water in the middle of a four-hour fast to ensure adequate absorption. Contraindications for this drug are severe renal impairment, hypocalcaemia, hypercalcinuria, pregnancy and lactation. Cyclical etidronate reduces bone loss from the spine and vertebral fracture risk in post-menopausal women with osteoporosis.

Alendronate Sodium (Fosamax)

It is taken as a single 5-10 mg daily dose. There is now an option to take one 70 mg tablet per week. Patients should swallow a 10 mg tablet when getting out of bed in the morning half an hour before any food and with a 7o z drink of plain water. It is important that patients should stand or sit for at least 30 minutes after taking the tablet, after which the first beverage, meal or medication of the day may be taken.

Antacids can affect the absorption of this drug. Contraindications are patients with significant renal impairment, pregnancy, breastfeeding women, abnormalities of the oesophagus or upper gastrointestinal problems. The side effects include nausea, dyspepsia and diarrhoea. Treatment is continued long term. Alendronate prevents bone loss from vertebral and non-vertebral sites and decreases the risk of spine and hip fractures.

Risedronate. Risedronate is taken daily as a 5 mg tablet. It should be taken with water 30 minutes before the first drink, food or medication of the day in the middle of a four-hour fast and not before going to bed. Risedronate prevents bone loss from the vertebral and non-vertebral sites and decreases the risk of spine and hip fractures. In 2005, NICE issued guidelines on the secondary prevention of osteoporosis and recommended the use of bisphosphonates for post-menopausal women who had already experienced a fracture.

OTHER SELECTIVE TREATMENTS

Selective estrogen receptor modulators (SERMs). These are synthetic compounds that stimulate oestrogen receptors of bone tissue and work in a similar way to HRT. Oestrogen receptors of breast and endometrial tissue are not engaged and so it is considered that this treatment will not increase the risk of breast or endometrial cancer. Raloxifene is a commonly used SERM. It has been shown to reduce the risk of vertebral fractures but has not been shown to reduce the risk of non-vertebral fractures. Like HRT raloxifene is associated with small increases in the number of thrombo-embolic events but is associated with a reduction in the number of new cases of breast cancer.

Calcitonin

Calcitonin inhibits osteoclast activity slowing the breakdown of bone. It can be given as subcutaneous injections or as a nasal preparation that is associated with fewer side effects. Nausea, vomiting, diarrhoea and dizziness have been reported. Calcitonin has been shown to reduce the risk of vertebral fracture. It has analgesic properties that may be useful in the acute management of vertebral fracture.

Calcium (1000-1200 mg Daily)

Calcium supplementation may be considered where dietary calcium intake is deficient or in cases of malabsorption. It has a less marked effect on fracture reduction than the other antiresorptive agents but it is usually well tolerated.

Vitamin D

Vitamin D is required to facilitate calcium absorption. Most people achieve their vitamin D requirement by the action of sunlight on their skin, in addition to that obtained from food. A French study in nursing homes showed that 800 iu of vitamin D3 and 1.2 g of elemental calcium daily reduces the risk of hip fracture by 43%. There is no evidence that vitamin D and calcium supplementation decreases spine bone loss or the incidence of vertebral fracture. A combination of calcium and vitamin D should be considered in all elderly patients who are housebound or in residential care.

Calcitriol

Calcitriol is an active form of vitamin D which increases calcium absorption from the gut and decreases calcium excretion. It is used when there is active vitamin D deficiency which may be the cause of diminished bone mass. Patients should be monitored for hypercalcaemia whilst taking calcitriol. Calcitriol reduces postmenopausal bone loss and reduces spinal and extra-spinal fractures (excluding the hip).

Formation Stimulating Agents

Fluoride

Fluroide dramatically increases bone mass but evidence of fracture prevention is inconclusive and the therapeutic window is narrow. Consequently it is rarely used in the United Kingdom.

Parathyroid hormone (PTH)

Intermittent injections of recombinant parathyroid hormone have recently been licensed in the United Kingdom. Teriparatide (Forsteo) is almost identical to human parathyroid hormone. It stimulates osteoblast production and is used to reduce vertebral fractures. It is administered as a daily subcutaneous injection.

Strontium Ranelate

Is licensed for post-menopausal women with osteoporosis to reduce the risk of vertebral and hip fracture. It works by suppressing the action of osteoclasts and stimulating osteoblasts. It is administered in powder form mixed with water, taken daily in the evening, in the middle of a four-hour fast.

Unlike bisphosphonates, strontium ranelate has very few gastrointestinal side effects; nausea and diarrhoea can occur on commencement of treatment but are usually mild and short-lived. There is a slight increased risk of blood clots and it would not be recommended to patients already at risk from thrombolytic events.

PRECAUTIONS

Feeding Habit

For optimum bone health a well-balanced diet is recommended with adequate calcium and vitamin D. Calcium requirements change throughout life and an adequate intake is particularly important during childhood, adolescence, pregnancy and lactation. Excessive caffeine, salt and fibre can hamper absorption or increase calcium excretion.

Smoking

Smoking should be discouraged. Smoking alters oestrogen metabolism resulting in a reduction in the biologically active oestrogens which are beneficial to bone health. Women smokers tend to have an earlier menopause. Smokers also tend to have a lower body weight.

Alcohol

Alcohol intake should remain moderate. Excessive alcohol intake has a detrimental effect on bone turnover and formation; it can also effect calcium metabolism.

Exercise

Weight bearing exercise can increase bone formation, improve muscle strength and coordination. Exercise must be sustained as bone loss will resume once exercise stops. Exercise can also improve coordination and reduce the likelihood of falls. Three sessions of 20-minute exercise weekly are required. Excessive exercise will have a negative effect and increase the risk of osteoporosis due to low body weight and the effects on the hypothalamic-pituitary-gonadelaxis.

Falls

Nurses have an input in fall prevention by improving awareness, identifying these who might be at risk and addressing risk factors such as poor eyesight, non-supportive footwear, reduced mobility and unsafe situations in the home environment.

CONCLUDING REMARKS

The primary healthcare team play a valuable preventative role, especially with regard to osteoporosis, emphasizing the importance of a healthy lifestyle from the cradle to the grave. They often have a good relationship with the patients because they see them for a multitude of conditions. The more the practice healthcare team are involved with the hospital care team the better for patients with chronic disease. The efficacy of this arrangement requires continuous audit, with each practice reviewing how it can improve. There are variations in the quality of care of chronic conditions.

It is the responsibility of the primary healthcare team, with the support of the rheumatologist and their team, to ensure the continuity of complex care which is offered to the rheumatology patient. What is regarded as routine in the rheumatology clinic may be rare to the GP. It is therefore important that the appropriate information be given to the patient on ward discharge and after clinic appointments, and to the GP with a diagnosis and details of medication and treatment.

A network of support and communication between rheumatology services, the rheumatology team, the district nurses, the practice nurses and other community-based services will ensure consistent, clinically effective management of the patient and their rheumatoid disease. It is important that the practice nurse team uses the rheumatology nurse practitioner in a coordinating role as a resource and advisor. Management of arthritis is a long-term partnership between the patient and the primary and secondary healthcare team.

2

Pharmacological Control of Diseases

The treatment of rheumatological disorders will involve many components including education, exercise, adaptation and drug therapy. The main objectives of drug intervention will be to: (a) Reduce or alleviate symptoms such as pain and stiffness. (b). Suppress disease activity in chronic disorders such as RA.

ALGESIRECEPTION (PAIN)

The experience of pain can only be defined in terms of human consciousness. As with all sensory experience there is no way of being certain that one person's experience of pain is the same as another's. Pain is a paradox while being an accepted phenomenon experienced by everyone at some time during their life. It is also a unique, subjective and unverifiable personal experience. Pain is one of the cardial symptoms of rheumatoid arthritis and will affect both physical and psychological aspects of functioning. It is the major contributor to the morbidity, disability and socioeconomic cost of musculoskeletal disorders.

There are a number of evidence-based guidelines which can support the clinician in the managment of pain. Acute pain is a transient experience where the source of pain is usually identifiable. Here pharmacological interventions are able to suppress the pain stimuli. In comparison chronic pain is an ongoing experience; often patients may demonstrate an array of associated disorders such as anxiety, depression and insomnia. The successful pharmacological control of pain dictates that nurses possess a comprehensive pharmacological knowledge and effective interpersonal skills to enable the assessment, planning, implementation and evaluation of pharmacological interventions in the management of pain.

Biochemistry of Algesireceptors

The gate control theory originates from Melzack and Wall (1982). It has been most influential in developing our present understanding of the mechanism of pain. Occasionally viewed as oversimplified or too rigid it has provided a framework which demonstrates the complexity of the pain response. The theory is founded on two main concepts.

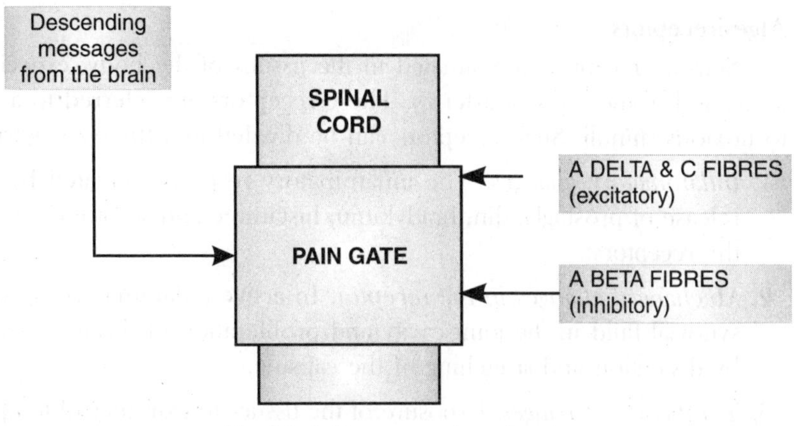

Fig. 2.1. Influences on the pain gate.

1. The transmission of pain messages can be modulated within the spinal cord. This occurs via descending messages from the brain, which inhibit sensations from peripheral and internal sources being experienced as pain. Two neurotransmitters appear to be involved in this process. The release of enkephalin at the spinal cord inhibits the uptake of calcium ions by the spinal cord neurones. This prevents further transmission of the pain message. Serotonin (the second neurotransmitter) is involved in the transmission of pain messages between the raphe magnus and the cord.
2. The transmission of pain messages to the brain can be altered by activating another source of sensory receptor.

When pain impulses enter the dorsal horn of the spinal cord the gate control mechanism is either opened or closed. A number of laminae within the dorsal horn form the gate control mechanism and this specialized area enabled the pain impulses to be modulated. The primary sensory neurones involved in pain transmission are:

1. *Type A delta fibres.* These are small myelinated fibres capable of transporting messages at speeds of 6-30 metres per second.
2. *Type C fibres.* These are very small and have no myelin insulation. Therefore messages are transmitted a lot slower at speeds of 0.5-2 metres per second. Type C fibres transmit slower chronic pain.

Impulses from A delta and C fibres are excitary in nature and promote the release of an exitory neurotransmitter Substance P, which opens the gate facilitating the perception of pain. The opening of the gate allows the transmission of pain impulses along the spinoreticular and reticular thalamic tracts through various junctions into the sensory cortex. Studies of the thalamus and somatosensory cortex show that in arthritis some cells have an abnormally large response to joint stimulation causing changes in the way impulses are transmitted.

1. *Type A beta fibres.* These are larger fibres than other sensory neurones and electrical impulses travel along them at a greater speed. Therefore, the pain stimulus conducted along type A beta fibres will be quicker and faster than transmission along the other fibres. If activated, type A beta fibres will occupy the secondary neurones first thereby blocking other pain messages. These fibres can be activated by rubbing or vibration of the skin and their activation will close the *'pain gate'*. This theory has been instrumental in the development of treatment interventations such as transcutaneous electrical nerve stimulation machines (TENS).

Algesireceptors

Sensory receptors are situated in the tissues of the body, especially in the skin, synovium of joints and in the walls of arteries. These receptors are referred to as nociceptors as they respond to noxious stimuli. Such receptors can be divided into three categories.

1. *Inflammatory changes.* The inflammatory response initiated by tissue damage will cause the release of prostaglandin, bradykinin, histamine and serotonin. This will stimulate a reaction in the receptors.
2. *Mechanical changes in the receptor.* In active inflammatory conditions such as RA, increased synovial fluid in the joint cavity and proliferation of the inflamed synovial tissues causes pain by distention and stretching of the capsule.
3. *Temperature changes.* Exposure of the tissues to extremes of temperature can cause stimulation of the receptors.

Some nociceptor receptors will respond to all the above stimulants. They terminate in the dorsal horn of the spinal cord and transmit their pain messages to secondary neurones. The destination for secondary neurones is the thalamus.

Nervous Regulation

Pain signals terminate in the brain. Fast pain signals are relayed via the brain stem and the thalamus to areas of the cortex, especially the somatosensory area. This region of the cortex can differentiate the area of the origin of the pain message. Slow pain messages are relayed over a wide area of the brain stem and thalamus. The reticuloactivitating system is situated in the brain stem; when it is stimulated it increases the excitability of the brain. This is why people with chronic pain conditions such as fibromyalgia report difficulty in sleeping, resting and relaxing. Also it provides us with a reason why chronic pain is difficult to locate because the pain messages are not relayed to the somatosensory areas. Once stimulation of the receptor ceases so should the pain but pain pathways cannot explain all pain experiences and patients can continue to experience pain after the original cause of the pain has been removed.

Woolf (1994) offers a theory for the development of chronic pain states. He proposes that neurones in the dorsal horn of the spinal cord become hypersensitive and develop alternative synapses with neighbouring neurones. This causes pain messages to go astray. Also, other substances released within the spinal cord can mediate the transmission of pain signals. N-Methyl-D aspartate (NMDA) and neurokinin (NK) cause the spinal cord neurones to become hypersensitive. This can cause the reproduction of *hyperalgesia* (severe pain to a stimulus that would normally only produce mild pain) and *allodynia* (pain in response to a stimulus that would not normally be painful–for example, stroking the skin). Additionally, many aspects of higher levels of processing will have a considerable effect on how unpleasant a pain stimulus is.

The experience of pain is influenced by anxiety, cultural factors, the environment and past experiences. Reassurances that a stimulus will be brief considerably reduces the unpleasantness rating. This assumes that there is an inevitable behavioural sequence that is initiated by a noxious stimulant referred to as the '*bottom up approach*'. However the brain is quite capable of simply ignoring noxious stimuli altogether under conditions of severe stress and it is capable of selecting what sensory information is acted on; this is known as the '*top down approach*'. It is the balance between these ascending and decending processes that determines our perception of pain.

CLINICAL ASPECTS OF PAIN

Once the acute pain impulse has passed the gate control mechanism it enters the reticular activating system, and sympathetic nervous system activity increases aiding the body's flight of fight mechanism. This activity includes:

1. Adrenaline causes increased consciousness and emotion.
2. An increase in hormonal activity. The anti-diuretic hormone increases the reabsorption from the renal tubules retaining water within the body (this can elevate blood pressure).
3. Aldosterone increases the reasorption of sodium from the renal tubules.

Cortisol increases blood glucose levels and the secretion of hydrochloric acid and pepsinogen.

1. An increase in cardiovascular activity. Tachycardia and increased cardiac output.
2. An increase in gastric activity. This leads to a reduction in gastric emptying.

CLINICAL ACTIVITY

The major clinical activity for the management of pain include the following categories of drug therapy: (i) Non-opioid analgesia, (ii) Compound analgesia, (iii) Opioid analgesia, (iv) Antidepressant drugs, and (v) Non-steroidal anti-inflammatory drugs (NSAIDs).

Non-opioid Analgesia

These medications are administered to treat mild–moderate pain.

Acetaminophen–paracetamol

Paracetamol is an undervalued effective analgesic agent. Its mechanism of action remains poorly understood although there is evidence of direct effect on the central nervous system (CNS) rather than on peripheral tissue. It blocks the synthesis and secretion of prostaglandin preventing nociceptor sensitization. It has both analgesic and anti-pyretic properties.

Pharmacokinetics. Taken orally paracetamol is well absorbed from the gastrointestinal tract and inactivated in the liver. A major advantage is the lack of upper gastrointestinal toxicity, especially ulceration and bleeding.

Adverse effects. The well-known hazard of hepatoxicity is virtually only seen in conjunction with a drug overdose and is increased with liver disease and alcoholism. The daily dose should be closely monitored in these situations. The risk of nephrotoxicity with chronic dosing remains uncertain but is probably very small.

Analgesic Compound

Compound analgesia are fixed ratio combinations of non-opioid (aspirin or paracetamol) and opioid analgesia (dextropropoxyphene or codeine) preparations. Compound analgesia contain either a low dose or full dose of opioid analgesic. Thus this range of analgesia is able to bridge the therapeutic gap between nonopioid and opioid analgesia. However it is worth noting that the compound or combined analgesic effect may also result in a combination of the side effects of both analgesia. Examples of these drugs include Co-Codamol and Codydramol.

Table 2.1 Adverse effects of compound analgesia.

Adverse effects
Dizziness
Sedation
Nausea/vomiting
Constipation
Abdominal pain
Rashes
Headaches
Weakness
Euphoria
Dysphoria
Hallucinations
Minor visual disturbances
Abnormal liver function tests

Opioids

This group of drugs is classified in terms of its efficacy into two categories - low and high efficacy opioids. Their role in the management of moderate to severe musculoskeletal pain is controversial. Although certain patients may benefit without experiencing adverse effects and addiction, the question of true efficacy has not yet been answered.

Table 2.2 Opioids.

Efficacy	Drug
Low efficacy	
	Codeine
	Dihydrocoedeine
	Dextropropoxyphene
	Nalbuphine
	Pentazocine
High efficacy	
	Buprenorphine
	Dextromaramide
	Diamorphine
	Dipipanone
	Meptazinal
	Morphine
	Papaveretum
	Pethidine
	Tramadol

Opioid analgesies work by fitting into the opioid receptors of the brain and spinal cord. These receptors are also used by endorphins, the body's own opioid. Opioid analgesia inhibits the transmission of nociceptive messages to the higher centres or through activation of the decending anti-nociceptive pathways. Evidence from Stein (1991) highlights the peripheral analgesic action of these medications in conditions characterized by inflammatory hyperalgesia.

Low efficacy opioids

Effect of codeine phosphate

Primarily prescribed as an analgesic but also used as an antidiarrhoeal drug and a cough suppressant, codeine has been in use for nearly 100 years. Although codeine's analgesic properties are similar to morphine its analgesic action is equal to approximately only 10% of morphine.

Pharmacokinetics. Taken orally codeine is quickly absorbed within the gastrointestinal tract and metabolized within the liver. It has a half life of approximately two to three hours. Excretion of codeine is via the kidneys into the urine. The analgesic effect of 30mg of codeine is suggested to equal 300-600 mg of aspirin. Dihydrocodeine tartrate - this medication possesses a similar analgesic effect to codeine.

Effects of dextropropoxyphene hydrochloride

The analgesic effect of dextropropoxyphene is considerably less than that of codeine. Chemically it is structurally related to methadone but it possesses less analgesic, antitussive and dependency properties. The main use of dextropropoxyphene is as a compound analgesia with aspirin (doloxene).

Pharmacokinetics. Dextropropoxyphene is rapidly absorbed in the gastrointestinal tract and metabolized within the liver. In the event of an overdose the rapid absorption of dextropropoxyphene is able to produce respiratory arrest, hypotension and cardiac dysrhythmia within one hour.

High efficacy opioids

Effects of buprenorphine (temgesic)

This opioid can be useful because of the length (six hours) and strength of its analgesic properties. It also possesses both opioid agonist and antagonist qualities. It is therefore less likely to produce addiction, respiratory depression or affect cardiovascular function than other opioid analgesics.

Pharmacokinetics. Administered sublingually; if swallowed buprenorphine is metabolized within the liver and possesses a half life of five hours.

Effects of dextromoramide (palfium)

A derivative of the opium poppy, dextromoramide is a powerful quick acting analgesia which has been prescribed in Britain since the 1950s for severe and intractable pain. In relation to morphine it has a shorter duration of action and is less sedating. Therefore it is able to produce pain relief without affecting consciousness, mental activity or inducing constipation unlike other opioids. Concomitant administration of tranquillizers such as clopromazine produces a synergistic analgesic effect.

Pharmacokinetics. Taken orally dextromoromide is quickly absorbed and metabolized within the liver. It is short acting with a duration of approximately three hours.

Effects of diamorphine hydrochloride

Introduced 100 years ago diamorphine is indicated for both severe and chronic pain. It is a powerful opioid analgesia and is a derivative of the unripe seed pod of the opium poppy. Although diamorphine mirrors morphine in both its actions and uses it is able to produce enhanced pain relief with less adverse effects such as nausea or hypotension.

Pharmacokinetics. Following administration diamorphine is rapidly converted into monoacetyl morphine and more slowly metabolized into its main active metabolite morphine. Within 24 hours approximately 80% of the dose is excreted via the urine.

Effects of fentanyl

Fentanyl is more efficacious than morphine. It is normally administered as an intra-operative analgesic. Fentanyl can also be used for chronic pain in the form of a patch for transdermal drug delivery (Durogesic). This mode of administration possesses the ability to maintain its efficacy for 72 hours.

Effects of morphine

Derived from opium and in use since the last century morphine is prescribed for moderate to severe pain in both acute and chronic conditions. Morphine acts on the CNS to eliminate pain and it also possesses the ability to transform the unpleasant sensation of pain into a sense of euphoria. Morphine therefore has two major effects, one depressing, the other stimulating. The '*depressing*' effects of morphine can include:

1. Reduction of anxiety and inducing the feeling of euphoria
2. Reduction of the peristaltic activities of the bowel
3. Reduction of the appreciation of pain
4. Some levels of sedation (due to its mildly hypnotic properties)
5. Urinary retention
6. Suppression of respirations and the cough reflex.

The stimulant effects of morphine can include:

1. Increased arousal of the chemoreceptor trigger zone within the brain stem including nausea and vomiting.
2. Increased arousal of the vagus nerve which may produce such effects as bradycardia and hypotension, owing to a parasympathic action in the cardiovascular system

Morphine works on specific opioid receptors situated throughout the body. Such receptors are divided into categories and include delta, kappa and mu receptors. It is the mu receptor on which morphine has its main effects. Mu receptors are associated with analgesia, respiratory depression, euphoria and dependance. The pharmacological effects of morphine on each individual will differ considerably.

Table 2.3 Drug interactions of opioids.

Alcohol: enhances the sedative and hypotensive efects
Antidepressants: may result in CNS excitement or depression
Antipsychotics: enhance the sedative and hypotensive effects
Anxiolytics and hypnotics: enhance the sedative effect
Antihistamines: enhance the sedative effects
Cisparide: may antagonize the effect of gastrointestinal motility
Ulcer healing drugs: may increase plasma concentrations of opiods
Antiemetics: antagonize the effects of gastrointestinal activity

Pharmacokinetics. Following oral administration morphine is absorbed from the gastrointestional tract where is undergoes conjugation in the gut wall and liver with approximately 20% of the dose actually reaching the systemic circulation. Elimination of the majority of morphine is via the kidneys in the urine. Subcutaneous and intramuscular administration is quicker than oral administration.

Table 2.4 Adverse effects of opioids.

Nausea and vomiting
Constipation
Respiratory depression
Hypotension
Dry mouth
Micturition difficulties
Bradycardia/tachycardia
Hallucinations
Mood changes
Addition
Reduced libido

Effects of pethidine hydrochloride

Akin to morphine, pethidine provides rapid pain relief for short durations of time. Prescribed inappropriately it can be addictive.

Effects of tramadol hydrochloride (zydol)

Tramadol can enhance both serotonenergic and adrenergic pathways. It is considered equal to morphine regarding its efficacy when prescribed for moderate chronic pain but is less likely to cause respiratory depression, addiction or constipation than morphine. Tramadol has been shown to reduce NSAID consumption and relieve pain in osteoarthritis of the knee. It is also available in combination with paracetamol.

Pharmacokinetics. Administered orally tramadol is quickly absorbed from the gastrointestinal tract. It possesses a half life of six hours. Tramadol is excreted in the urine and approximately a third of the dose is excreted unchanged.

Antidepressant Therapy

Antidepressant therapy may enhance the analgesic effect induced by other drugs. The presumed site of action is at the spinal cord level. Although there is a beneficial effect on sleep and mood, suggesting action elsewhere in the neuroaxis, debate continues over possible modes of action and whether or not these drugs induce analgesia in the absence of depression in chronic pain generally.

Effects of serotonin uptake inhibitors

These drugs block the reuptake of serotonin. The commonest ones used in clinical practice are fluoxetine, paroxetine and sertraline. They should be avoided in patients with epilepsy, cardiac disease, diabetes mellitus and glaucoma. They may cause drowsiness and abrupt withdrawal can cause a variety of side effects.

Effects of tricyclic drugs

The increased incidence of anxiety and depression amongst chronic pain sufferers is not the only justification put forward for the use of tricyclic drugs. They also appear to have a synergistic effect with centrally acting analgesia and the stimulation of endorphine production. Amitriptyline is of proven efficacy. It is taken in small incremental doses of 10-50 mg, two hours prior to settling at night to avoid the hypnotic effect. Sleep usually improves within two weeks whilst pain relief may take many months. Common side effects include dry mouth and palpitations. Dothiepin has been shown to be of comparable effect in controlled studies. Venlafaxine, an inhibitor of both serotonin and noradrenaline re-uptake, has been shown to be useful in the treatment of fibromyalgia. Pregabalin is of proven benefit in the treatment of neuropathic pain and has also been shown to improve sleep quality and diminish fatigue.

Management of Inflammatory Condition

NSAIDs have become an integral part of the management of inflammatory conditions. The ultimate goal of NSAIDs therapy is to reduce the inflammation that is occurring. The decision to commence therapy will include consideration of the risks weighted against the potential for therapeutic benefit for the patient and will take into account the nature of the underlying condition and the severity of the patient's symptoms. The objective of commencing NSAIDs is to decrease the cardinal symptoms of inflammation which include pain, stiffness, swelling and warmth. These symptoms are not only unpleasant for the patient but also affect physical and psychological functioning. In addition to reducing inflammation this group of drugs also possesses analgesic and anti-pyretic properties, which are necessary to treat the associated features of inflammation. The effectiveness of NSAIDs should be evident within a few days and make an almost immediate impression on the patient's symptoms. The therapy will continue to be effective as long as blood levels of the drug are maintained. If a patient stops taking this therapy for any reason, this action could result in a reoccurrence of the symptoms of inflammation. Although the patient will start to feel better once the features of inflammation are reduced this therapy will not influence the progression of conditions such as RA.

Types of NSAIDs

The NSAIDs can be categorized on the basis of their chemical structure. The older NSAIDs – for example, Indomethian – have excellent anti-inflammatory properties but commonly produce adverse side effects especially on the gastrointestinal system, whereas agents such as the propionic acid derivatives have less likelihood of causing side effects and are generally better tolerated.

Table 2.5 Chemical classification of NSAIDs.

Types of NSAID	*Examples*
Carboxylic acids	aspirin, choline, salicylate, diflunisal
Acetic acids	indometacin, diclofenac, etodaloc
Propionoc acids	brufen, fluribiprofen, fenbrufen, fenoprofen, ketoprofen, tiaprofenic acid
Fenamic acids	piroxicam, phenylbutazone, azapropazone, tenoxicam
Non-acidic compounds	nabumetone
COX 2 inhibitors	rofecoxib, celecoxib

Application of NSAID

The choice between the various NSAIDs is largely empirical. There is a wide individual variability of response. The cause for this variability is not known. If a patient experiences a poor response to one NSAID another agent should be tried. There can be a variation in response even when the NSAID is from the same chemical family. Once a NSAID is commenced the patient should be maintained on it for two weeks before an assessment of its efficacy is made, unless the patient experiences an adverse reaction. The ultimate goal of NSAIDs intervention is to choose a preparation that combines the greatest effectiveness with the least toxicity for each individual patient.

There is significant co-morbidity associated with NSAID use and the clinician needs to balance the benefits against the risks proir to prescribing NSAID therapy. NSAIDs can be divided into two groups according to their half life. Those drugs with a half life of greater than 12 hours - for example, piroxicam, tenoxicam and azapropazone - need only to be administered once or twice daily, which should aid concordance; although it should be noted that these medications with a long half life may build up excessively in the plasma of older patients increasing the potential for the occurrence of side effects. The majority of other NSAIDs are referred to as short half life drugs as their half life is usually less than six hours.

Utilization of NSAID

NSAIDs can be administered in many different ways, including by slow release compounds, topically, intramuscularly and in suppository form.

Topical preparations - These are only available for a few NSAIDs. They are more efficacious than placebo. Local skin sensitivity may occur and there is some absorption into the circulation leading to the rare occurrence of systemic side effects.

Suppositories are rarely used but do reduce the risk of gastric irritation by the direct effect of the drug on the mucosa. However, there is still a risk (although reduced) of gastric ulceration mediated via the circulation. NSAIDs given in suppository form can also cause local irritation in the rectum. Patients can also find them difficult to administer especially if they have reduced manual dexterity.

Effects on alimentary canal

In patients with peptic ulcers disease NSAIDs are contraindicated. If a patient is at risk from GI symptoms the clinician will often prescribing gastric cytoprotection in the form of proton pump inhibitors for example, omeprazole and lansoprazole. These drugs inhibit the gastric acid by blocking the hydrogen potassium adenosine triphosphatase enzyme system (the proton pump) of the gastric parietal cell. They are the treatment of choice for stricturing and erosive oesophagitis and are effective in the short-term treatments of gastric and duodenal ulcers.

Symptoms

In chronic inflammatory rheumatology conditions such as RA, ankylosing spondylitis, psoriatic arthritis and connective tissue disorders (with polyarthritis), patients may well have to take NSAIDs on a regular basis to provide symptomatic relief from the effects of ongoing inflammation. If the patient experiences a reduction in disease activity they may no longer need regular administration of the NSAID therapy and a trial will establish whether this is the case. Patients with OA are better managed on simple analgesia. In the older patient the risk of side effects can outweigh the potential therapeutic effects of a NSAID, although their use may be considered if they experience a flare of inflammatory symptoms. Here the use of NSAID would be considered short term for symptom relief. Other conditions where it may be appropriate to use NSAIDs on a short-term basis include: gout, pseudogout and certain sport injuries.

Biochemical role of NSAIDs

NSAIDs appear to be involved in many of the pathways that are influential in the production of inflammatory mediators as well as actively involved in the interactions between inflammatory cells. NSAIDs are known to play a role in the suppression of prostaglandin synthesis partly by inhibiting the enzyme prostaglandin synthetase H (also known as cyclo-oxygenase). Prostaglandins

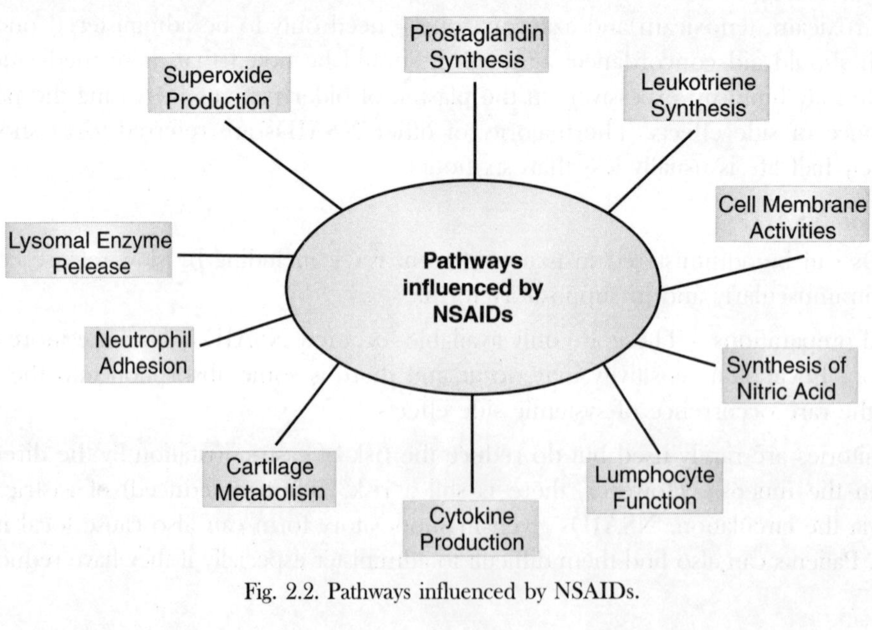

Fig. 2.2. Pathways influenced by NSAIDs.

are synthesized from membrane phospholipids; when induced they act to produce many of the features of inflammation including warmth, erythema and oedema. NSAIDs have effects on various other aspects of the inflammatory response including leukotriene synthesis, superoxide production and cytokine production. The suppression of prostaglandins by NSAIDs may help relieve the symptoms of inflammation but this suppression is also responsible for the major adverse effects of NSAIDs including gastrointestinal disturbances and effects on kidney functioning.

Pharmacokinetics. Most NSAIDs are completely absorbed from the gastrointestinal tract. They bind strongly to plasma proteins and are predominately cleared by the liver. The metabolites are excreted in the urine. Some NSAIDs are absorbed in the active form whilst others, referred to as pro-drugs, are converted by hepatic metabolism to active drugs.

Inhibitory effect

Cyclo-oxygenase-2 (COX-2) inhibitors are a selective type of non-steroidal anti- inflammatory drug (NSAID). They were developed to reduce the gastrointestinal side effects of traditional NSAIDs. Although COX-2 inhibitors are as effective as traditional NSAIDs in relieving pain serious concerns about their cardiovascular safety have arisen. COX-2 plays a beneficial role in vascular health and its inhibition may create an imbalance between thromboxane and prostacyclin. Thromboxane is critical for platelet aggregation and vasoconstriction, and prostacyclin dampens the effects of thromboxane through fibrinolysis and vasodilation. Other mechanisms linking coxibs (or NSAIDs) to cardiovascular events include hypertension from inhibition of prostagalandin-dependent counterregulatory mechanisms and COX-independent oxidative stress. Concerns with trials regarding the safety of COX-2 inhibitors:

1. Trials have been of short duration and not included enough people to observe the incident of cardiovascular events and consequently studies have been unable to demonstrate whether the risk of cardiovascular events associated with coxibs is concentrated in certain subgroups of patients.
2. Most trials have excluded patients with coronary heart disease, and patients taking aspirin, and there has been no consensus regarding cardiovascular outcomes/endpoints.

Objectives

1. Trials of patients receiving valdecoxib undergoing coronary artery bypass grafting have shown an increased risk of cardiovascular events. The risk ratio for confirmed cardiovascular events was 3.7 when the valdecoxib group was compared with the double placebo group. In April 2005 the FDA requested that valdecoxib's manufacturer withdraw the drug from the US market based on insufficient long-term cardiovascular safety data, evidence of an increased cardiovascular risk in short-term studies amongst patients undergoing heart surgery, the risk of life-threatening skin reactions and no proven advantages over other NSAIDs.
2. The Adenomatous Polyp Prevention on Vioxx trial examined rofecoxib 25 mg daily against placebo among 2586 patients with a prior adenomatous polyp. Seventeen per cent were taking daily aspirin at the commencement of the study. Patients taking rofecoxib experienced a 4.61 increase in the risk of congestive cardiac failure. Severe thrombotic cardiovascular events including acute myocardial infarction, peripheral arterial thrombosis, peripheral venous

thromboisis and pulmonary embolus occurred twice as often in patients taking rofecoxib compared to patients on placebo. In 2005 rofecoxib was withdrawn because of its adverse cardiovascualar profile.

3. The Adenoma Prevention with Celecoxib trial demonstrated that high dose long-term celecoxib use was associated with an increase in the risk of thrombotic events. The aspirin user subgroup did not experience a lower risk of cardiovascular events than those not using aspirin. Also patients taking aspirin do not experience a reduced risk of GI toxicity while taking coxibs compared with nonselective NSAIDs. The FDA has requested that the manufacturers of all NSAIDs including celecoxib include a warning highlighting the potential cardiovascular risks of these agents.

COX-2 inhibitors

1. Skin reactions can occur with all COX-2 inhibitors. In the majority of cases these occur in the first month of use and patients with a history of drug allergies may be at greatest risk.
2. COX-2 inhibitors must not be used in patients with established ischaemic heart disease and / or cerebrovascular disease (stroke) or in patients with peripheral arterial disease.
3. Given the association between cardiovascular risk and exposure to COX-2 inhibitors, prescribers are advised to use the lowest effective dose for the shortest possible duration of treatment.
4. Healthcare professionals should exercise caution when presecribing COX-2 inhibitors to patients with risk factors for heart disease such as hypertension, hyperlipidaemia, diabetes and smoking.

Physiological effects of NSAIDs

In general the NSAIDs share a common spectrum of side effects although the frequency of particular side effects varies between different compounds. Mild adverse effects are relatively common, occurring in approximately 10-15% of users. Due to the fact that these drugs are prescribed in large numbers worldwide, a significant number of people will experience more serious side effects. The major side effects occur in several organ systems. These include:

1. The gastrointestinal tract
2. The central nervous system
3. Cardiovascular
4. The haematopoetic system
5. The kidney
6. The skin
7. The liver.

Alimentary canal

The gastrointestinal tract is the system most commonly affected by NSAIDs and often necessitates discontinuation of therapy. Symptoms patients may experience include dyspepsia, epigastric pain, indigestion, nausea and vomiting. Gastrointestinal lesions may range from hypermia to diffuse gastritis, erosions or ulcers. Factors associated with increased risk of ulceration include:

1. Age (over 65 years)
2. Previous peptic ulcer disease
3. Concomitant steroid therapy
4. Heart failure
5. High dose NSAIDs.

Low dose ibuprofen appears to be associated with a low risk of upper gastrointestinal symptoms. Diclofenac, naproxen and indomethacin are associated with intermediate risk of adverse events and azapropazone is associated with the highest risk. If patients are experiencing adverse reactions it will be necessary to carry out an endoscopy to assess any irritation or ulceration in the tract. All patients are advised to take NSAIDs with food.

Excretory system

The synthesis of renal prostaglandins is simulated by vasoconstrictor substances involved in circulating haemostasis such as angiotensin II, norepinephrine and vasopressin. By modulating the effects of these vasoconstrictor substances on the kidney, vasodilatory prostoglandin - especially prostaglandin E2 and prostacyclin - helps to maintain adequate renal blood flow and glomeruler filtration rate. This modulatory effect plays a minor role in controlling renal function in healthy individuals but under circumstances of circulating stress the prostaglandins become essential to the maintenance of adaquate renal function. In this situation treatment with cyclo-oxygenase inhibitors such as NSAIDs may precipitate acute renal failure. Patients at risk from NSAID-induced renal insufficiency include those with:

1. Congestive heart failure
2. Cirrhosis with ascites
3. Neprotic syndrome
4. Age (over 60 yrs)
5. Concurrent diuretic therapy (that causes potassium retension).

Renal side effects from NSAIDs appear to be dose-related and occur more often in older people. NSAIDs can also affect the salt and water balance causing sodium retension which may result in hypertension, oedema or heart failure in predisposed individuals. Drugs such as tiaprofenic acid have also been associated with cystitis. Acute interstial nephritis and nephrotic syndrome have been linked with naproxen, indomethacin, phenylbutazone, fenoprofen and diflunisal. The pathogenesis is unknown. Hypersensitivity phenomena may also be present with nephrotic syndrome including fever, skin rash and eosinophilia.

Integrated system

Cognitive dysfunction, memory loss, inability to concentrate, confusion, personality change, forgetfulness, depression, sleeplessness and paranoid thoughts have all been reported in older patients treated with naproxen and ibuprofen. Headaches and dizziness may occur with indomethacin. Aspirin can affect hearing and cause tinnitus. Several of the NSAIDs have been associated with aseptic meningitis - for example, ibuprofen and sulindec. Patients with collagen diseases seem to be at the highest risk from the complication.

Circulatory system

An increased risk of myocardial infarction was found in patients taking diclofenac and ibuprofen in an observational study. For ibuprofen, one additional myocardial infarction would happen for every 1005 patients aged 65 and over, and for diclofenac one additional myocardial infarction for every 521 treated patients. Ray *et al.* (2002) reported an increased risk of acute myocardial infarction associated with the use of ibuprofen in a high risk population over the age of 50. Patients taking diclofenac had a 55% increased risk of myocardial infarction, which is similar to that reported in a much smaller study of non-selective NSAIDs in patients with RA from the general practice database. There have been suggestions that naproxen has a protective cardiovascular effect but this has not been shown in a meta analysis.

The potential for NSAIDs to cause other cardiovascular events including hypertension is well recognized. At present the Committee on the Safety of Medicines has concluded that there is insufficient evidence to change practice on non-selective non-steroid anti-inflammatory drugs. It advises that prescribing should be based on the overall safety profiles of NSAIDs and individual risk factors. Patients should take the lowest effective dose of NSAIDs or COX-2 inhibitors for the shortest time necessary.

Haematological reactions

The most common reaction is iron deficiency anaemia as a result of gastrointestinal blood loss that occurs secondary to erosion or ulceration. Blood dyscracies associated with NSAIDs therapy are rare but among the major causes of death with this therapy. Agranulocytosis, thrombocytopenia, neutropenia and aplastic anaemia have all been reported. The latter is very rare except with phenylbutazone, the administration of which is limited to hospital prescriptions in the United Kingdom. Inhibitors of platelet aggregation by NSAIDs may cause a mild prolongation of the bleeding time in patients. This becomes a particular concern for those patients receiving anticoagulation therapy or those who have hereditory clotting factor deficiencies.

Digestive gland

NSAIDs can cause a transient rise in liver enzymes and more rarely a hepatic illness such as hepatitis. Older patients with a reduced renal function and receiving high dose NSAID therapy are at greatest risk of adverse liver reactions.

Respiratory system

NSAIDs may precipitate asthmatic attacks in predisposed individuals. In some patients with bronchial asthma the inhibitor of cyclo-oxygenase may reduce bronchodilatory prostaglandins. This diverts arachidonic acid metabolism towards lipoxygenase products (leukotrienes). This activity may precipitate bronchospasm. Pulmonary effects including effusions have been reported with ibuprofen, naproxen and phenylbutazone therapy.

Integument

All NSAIDs can provoke skin reactions including photosensitivity, vesiculobullous eruptions, serum sickness and exfoliative erythroderma. Urticaria has been reported with aspirin, ibuprofen and indomethacin. Phenylbutazone and oxyphenbutazone are the NSAIDs that have the highest incidence of serious or fatal skin reactions including erythema multiforme, exofoliative dermatitis, Stevens-Johnson syndrome and toxic epidermal reaction. Cultaneous vasculitis is a rare occurrence but has been related to the use of indomethacin, fenbrufen and naproxen.

Pregnancy

The inhibitor of prostaglandins synthesis by NSAIDs during pregnancy may cause prolongation of gestation and increase post-partum and neonatal bleeding. NSAIDs should be avoided during pregnancy especially as they may promote premature closure of the ductus arteriosus and impair fetal circulation.

Effects of drug

The major interactions that occur with the administration of NSAIDs are shown below. For safe practice, consult the manufacturer's guidelines before advocating usage.

1. Cytotoxic agents. There is a reduction in the excretion of these agents. The potential interaction between NSAIDs and low dose methotrexate is not a problem at the dose used in patients with RA.
2. Anticoagulant therapy may be enhanced.
3. Uricosurics - for example, probenecid. These drugs increase the plasma concentration of many NSAIDs and delay their excretion.
4. Antidepressant therapy - for example, moclobemide - can acclerate the absorption of NSAIDs.
5. Cardioglycosides. The plasma concentration of cardiac glycoside can be increased, exacerbating heart failure.
6. Anti-epileptic drugs may be enhanced.
7. Diuretics. There is an increased risk of nephrotoxicity and a possible risk of hyperkalaemia associated with potassium sparing diuretics.
8. Lithium. There is an increased risk of toxicity.
9. Muscle relaxants - for example, baclofen. There is an increased risk of toxicity.
10. ACE inhibitors. There is an increased risk of renal damage and an antagonistic hypotensive effect.
11. Beta blockers. NSAIDS are antagonistic of hypotensive effects.

Interactions with specific NSAIDs

1. Azapropazone - this may interfere with the action of oral hypoglycaemic drugs. Also the plasma concentration of azapropazone may increase with the administration of cimetidine.
2. Indomethacin - increases the bioavailability of biophosphates.
3. Brufen - can increase digoxin levels.
4. Fenoprofen - its half life may be reduced following the co-administration of barbiturates - for example, phenobarbitone.
5. Ketoprofen - this may increase the level of sulfonamides in the blood.

5. Diclofenac – the plasma concentration can be increased if taken with ciclosporin.

The administration of two or more NSAIDs will increase the risk of adverse effects occurring.

DISEASE MODIFYING ANTI-RHEUMATIC DRUGS

A number of agents can be used to control rheumatological disorders, particularly RA. Their terminology can be confusing as they are known as either slow-acting anti-rheumatic drugs (SAARDs), second line therapies or disease-modifying anti-rheumatic drugs (DMARDs).

Early Treatment of RA

Historically the pharmacological interventions for patients with early RA were guided by the pyramidal approach. This approach dictated that initial drug interventions were limited to NSAIDs and analgesics which aimed to modify the symptoms of the disease process; hence these drugs were referred to as either symptom modifying drugs or first line therapies, and only when radiological evidence of erosions had been confirmed were DMARDs introduced in the therapeutic regime. However, within the past decade, a consensus of opinion has existed which challenges the traditional pyramidal approach and advocates that DMARDs should be prescribed earlier in the disease process and prior to, rather than following, any structural damage; thus minimizing the disease activity quickly and effectively. The following factors suggest therefore that the pyramidal approach should be inverted:

1. There is evidence that joint destruction and concurrent functional disability occurs within the first 12 months of the disease process. Therefore it is suggested that irreversible destruction of both cartilage and bone, resulting in functional disability, rapidly occurs within the early stages of the disease process, despite the fact that RA is a chronic progressive disease. Consequently it is suggested that the optimum time to introduce DMARDs to gain control or modify the disease process is when the patient initially presents. Also, functional deterioration will occur if patients remain untreated.
2. In comparison with the general population, patients with RA experience increased morbidity and reduced life-expectancy, which has implications not only for the individual patient (pain, disability and loss of self-esteem), but also on society in general relating to loss of earnings, state payments and healthcare costs.

The efficacy of DMARDs in RA proves difficult to validate due to:

(a) The inconsistency of the disease process;

(b) The individual response to DMARDs;

(c) The potential for concurrent toxic side effects.

The limited effectiveness of DMARDs may be attributed to the discontinuation of DMARD therapies because adverse effects are experienced by the patient, rather than inefficacy of the DMARD themselves.

Combination Therapy

Combination therapy (i.e. the use of more than one DMARD) has evolved due to several factors:

1. The change in philosophy entailing a more aggressive treatment regimen in an attempt to gain disease remission as early as possible.
2. The long-term efficacy of a single DMARD regimen has revealed disappointing results.

The aim of combination therapy is to achieve a synergistic effect using two or more DMARDs in order to arrest the disease process and to achieve disease remission with minimal adverse reactions. Combination therapy can be implemented either by:

1. A step down approach may be prescribed; this entails the initial treatment regimen of multiple DMARDs until disease suppression is achieved, then the number of DMARDs can be reduced.
2. A step up regimen where one DMARD is initially prescribed and a second DMARD added if disease remission is not achieved;

Initially combination therapy research studies have targeted patients with established aggressive uncontrolled RA. However, in order to achieve maximum benefits for the maximum of patients, research studies have been extended to include early RA patients.

Mode of Action and Pharmacokinetics of DMARDs

Although DMARDs have different chemical structures, they all possess common properties:

1. All DMARDs (excluding the antimalarials) have the potential to cause adverse effects, which include serious haematological toxicity. Consequently the incidence of adverse effects, which may be major or minor, presents a major constraint on the continual use of these agents, so much so that approximately 20-50% of patients will have to stop their DMARD within one year of commencement.
2. To control the signs and symptoms of RA, which include certain blood parameters denoting inflammation, painful swollen inflamed joints and sometimes slowing down the progression of radiological damage.
3. Their effectiveness is delayed, thus a therapeutic response tends to occur for approximately two to four months.
4. Their mechanism of action is poorly understood.

Prophylaxis of Malaria

Hydroxychloroquine and chloroquine are two drugs that were principally prescribed for both the treatment and prophylaxis of malaria. However, since the 1950s, research studies have revealed the efficacy of both hydroxychloroquine and chloroquine as disease-modifying anti-rheumatic drugs (DMARDs) in the treatment of rheumatoid arthritis and systemic lupus erythematosus. Because of their low toxicity profiles, hydroxychloroquine sulfate and chloroquine sulfate are rated as milder DMARDs, which may be prescribed independently or in combination with other DMARDs when only partial efficacy is achieved. An interesting property of hydroxychloroquine, when used in combination with methotrexate, is its protective mechanism against methotrexate-induced hepatotoxicity.

Symptoms

1. Active RA;

2. Juvenile arthritis;
3. Systemic and discoid lupus erythematosus.

Chemotherapy

Hydroxychloroquine orally 200-400 mg daily with food. The dosage may be reduced to 200 mg daily depending on the clinical response. Maximum dose should not exceed 6.5 mg/kg body weight per day. Chloroquine orally 250 mg daily with food.

Cautions

1. Patients with renal and liver impairment;
2. It may reduce the threshold for convulsions in patients with epilepsy;
3. It may exacerbate psoriasis;
4. Avoid antacids within four hours of administration.

Contraindications

1. Breastfeeding;
2. Pre-existing maculopathy.

Effect on eye

It is generally agreed that retinal toxicity is due not to the cumulative dose but to the daily dose. Therefore, the maximum prescription limit for adults should be 4 mg/kg/day for chloroquine and 6 mg/kg/day for hydroxychloroquine. To prevent potential toxicity, individual physicians' regimen may differ; therefore patients may be prescribed hydroxychloroquine or chloroquine for 5 days of the week or for 7 days a week for 11 months of the year with a 1-month 'drug holiday'. The 'drug holiday' will also allow the clinician to test the efficacy of the drug. Before commencing treatment the patient should be asked about visual impairment that is not corrected by glasses.

The visual acuity of each eye should be recorded using a reading chart. Treatment should only be commenced if there is no abnormality. The Royal College of Ophthamologists recommends annual review either by an optometrist or enquiring about visual symptoms, rechecking visual acuity and assessing for blurred vision using the reading chart. Patients should be advised to report any visual disturbance. The treatment should be stopped if there is any development of blurred visison or changes in visual acuity. If patients are requiring long-term treatment (more than five years) it should be discussed with an ophthalmologist.

Physiological role

The mode of action of both hydroxychloroquine and chloroquine remains unknown, although research studies have suggested that antimalarials affect subcellular organelles such as polymorphs, lymphocytes and macrophages, which produce an anti-rheumatic and immunosuppressive reaction.

Clinical aspects

To avoid gastrointestinal disturbances, hydroxychloroquine and chloroquine should be administered with food. Food is not thought to affect bioavailability. Hydroxychloroquine and chloroquine accumulate extensively in tissues, and both white and red blood cells. These medications possess above average half lives (approximately six weeks); steady concentrations are not achieved for approximately three to four months. Metabolized primarily by dealkylation and thus high renal clearance, approximately 40% of chloroquine and 25% of hydroxychloroquine are excreted unchanged in the urine; therefore therapeutic regimens should be adjusted to accommodate those patients with renal impairment.

Abnormalities

1. Rarely blood disorders (thrombocytopenia, agranulocytosis, aplastic anaemia).
2. Irreversible retinal damage;
3. Skin reactions (rashes, pruritis);
4. Depigmentation or loss of hair;
5. Nausea;
6. Diarrhoea;
7. ECG changes;
8. Abdominal pain;
9. Headaches;
10. Dizziness;
11. Convulsions;
12. Blurred vision;

Symptoms

Psoriatic arthritis (may exacerbate psoriasis).

Sulfasalazine

Prescribed for patients with early RA, and in combination with other DMARDs, sulfasalazine has also been found to have some efficacy when treating sero-negative spondyloarthropathies, juvenile arthritis and psoriatic arthritis. Developed in the 1930s when RA was thought to possess an infective aetiology, sulfasalazine contains the combination of an anti-inflammatory agent (5-aminosalicylic acid) and an antibiotic (sulfapyridine). However, as a result of early conflicting reports regarding the efficacy of sulfasalazine, the drug was originally dismissed by rheumatologists until the 1970s when further studies revealed its efficacy.

Physiological role

The precise mode of action of sulfasalazine is unknown but it is thought that sulfasalazine in some way suppresses relevant immunological processes in the large bowel and, in addition, sulfasalazine possesses the ability to scavenge pro-inflammatory oxygen species released from activated phagocytes.

Clinical aspects

Following oral administration the majority of the dose is absorbed in the large bowel where it reacts with colonic bacteria and separates into 5-aminosalicylic acid and sulfapyridine. Elimination is via the kidneys in the urine.

D-Penicillamine

Used in the treatment of RA for 25 years, D-penicillamine has also been useful in treating progressive systemic sclerosis as well as unrelated rheumatological conditions such as Wilson's disease, heavy metal poisoning and cystinuria.

Studies have revealed D-penicillamine to be effective in the reduction of joint inflammation, joint pain, early morning stiffness, laboratory inflammation indices and the improvements of rheumatoid nodules. However, due to a lack of controlled studies, it has not been possible to determine if D-penicillamine is able to slow progressive radiological damage.

Physiological role

For the past 25 years the precise mode of action and toxicilogy of D-penicillamine has remained elusive, although it is thought to suppress the immune system.

Clinical aspects

D-penicillamine is taken orally one to two hours before food; following absorption, peak plasma levels are present within one to four hours.

Myocrisin

Initially introduced and prescribed earlier in the twentieth century for the treatment of tuberculosis, gold salts were found to be of little value but were found to be efficacious for the treatment of RA. In the treatment of RA, myocrisin (injectable gold) has been shown to impede progressive radiological damage and concomitant fuctional impairment in both short- and medium-term treatment.

Physiological role

To date there is no consensus regarding the precise mode of action of intramuscular gold although it is known to regulate gene transcription and affect polymorphonuclear and synovial cells, monocytes, lymphocytes and immunoglobins.

Clinical aspects

Intramuscular gold is polymeric and water soluble and therefore not absorbed orally. Following intramuscular injection, absorption takes place quickly and gold travels to synovial tissue where it binds to inflamed synovial tissues; consequently the majority of gold is located in the synovial lining cells. Elimination of gold is initially rapid and largely in the urine. However, gold has been detected in tissues some 20 years after therapy.

Auranofin

Introduced in the last decade auranofin is a gold-based drug that can be taken orally. Although considered less efficacious than sodium aurothiomalate, auranofin possesses fewer serious side effects. The same is true when auranofin has been compared with D-penicillamine and methotrexate.

Physiological role

Akin to sodium aurothiomalate, precise mode of action remains elusive.

Clinical aspects

Following oral administration of auranofin absorption is rapid but incomplete as only 20-25% of the gold is absorbed, approximately 15% is then excreted via the kidneys but the majority of gold is excreted in the faeces and consists of both unabsorbed gold and gold that has been absorbed in the gastrointestinal tract and bound to gastric epithelium.

Methotrexate

Initially introduced 50 years ago for the treatment of acute leukaemia, methotrexate is classed as one of the first antimetabolites. The introduction of methotrexate as a treatment for RA took place in the 1950s. However, it was to be some 30 years later, in the 1980s, before methotrexate was used widely by rheumatologists for the treatment of RA and psoriatic arthritis. It is now the commonest DMARD used in the treatment of RA. It is prescribed as monotherapy and in combination with other DMARDs and biological agents.

Physiological role

Methotrexate is a cytotoxic drug which acts as a folate antagonist to cause cell death by affecting the synthesis of DNA. However, its precise mode of action in the treatment RA is not entirely understood.

Clinical aspects

Following administration (oral or intramuscular), methotrexate is absorbed rapidly and almost entirely, taking one to two hours before a peak concentration is achieved. The majority of methotrexate is excreted via the kidneys, and biliary elimination is suggested to account for approximately 10-30% of methotrexate.

Leflunomide

Leflunomide is one of the newer DMARDs that has proven efficacy and safety in randomized controlled trials. It appears as effective as sulfasalazine and methotrexate as a monotherapy and can also be used as combination therapy usually with methotrexate or infliximab. It may have a faster onset of action in reducing synovial vascularity.

Physiological role

It is an isoxazole derivate and its active metabolite inhibits *de novo* pyrimidine synthesis, resulting in inhibition of T cell proliferation.

Clinical aspects

Leflunomide has a long elimination half life of 15-18 days, which results from low hepatic clearance and enterohepatic cycling. Its active metabolite is detectable in plasma up to two years after discontinuation of the drug. If a severe adverse event occurs a washout of the drug may need to be considered using oral cholestyramine, 8 mg three times daily for 11 days.

Azathioprine

Azathioprine is a cytotoxic immunosuppressant drug, originally prescribed to prevent the rejection of transplanted organs. It has also proved to be efficacious in the treatment of RA, vasculitis, systemic lupus erythematosus, polymyalgia rheumatica/giant cell arteritis, Behçet' s disease, polymyositis, dermatomyositis, myasthenia gravis and chronic inflammatory bowel disease.

Physiological role

The precise mode of action of azathioprine is unclear but, as an immunosuppressant, azathioprine interferes with DNA synthesis by either inhibiting cell division or causing cell death in relation to RA. This results in fewer circulating B and T lymphocytes. Azathioprine is also administered as a steroid sparing agent.

Clinical aspects

Following oral administration, azathioprine remains inactive until it is metabolized in the liver into 6-thioinosinic acid and 6-thioguanylic acid. Elimination is via the kidneys.

Cyclophosphamide

Cyclophosphamide, a derivative of nitrogen mustard, is an alkylating agent which was developed 50 years ago as an anticancer drug. Cyclophosphamide has been shown to possess both immunosuppressive and immunostimulatory effects. Its use by rheumatologists for the treatment of RA began in the late 1960s even though research studies revealed conflicting results regarding its efficacy in controlling or slowing down radiographic changes. Cyclophosphamide has proved to be an effective in the treatment of the systemic complications of RA, such as vasculitis, and it has also proved to be effective treatment of systemic lupus erythematosus.

Physiological role

The mode of action of cyclophosphamide is not entirely understood. It is known to cross link DNA and possess properties that act to stop DNA replication and halt cell division. Additionally, it has a toxic effect on resting cells. This is said to account for its quick action, which is accompanied with increased toxicity in comparison with azathioprine.

Clinical aspects

Cyclophosphamide may be administered orally or intravenously. Although metabolized to some extent in both the kidneys and lungs, cyclophosphamide is predominately metabolized in the liver. Cyclophosphamide is excreted via the faeces, expiration, spinal fluid, perspiration, breast milk, saliva and synovial fluid. The majority however is excreted by the kidneys unchanged (less than 20%) and as metabolites in the urine (65%), the dominant metabolite of cyclophosphamide (acrolein) having been highlighted as the major source of bladder toxicity. Hence an increased fluid intake is essential when administering this drug.

Ciclosporin

Ciclosporin is an immunosuppressant initially developed to suppress the rejection of organ transplantation. Its anti-arthritis properties were initially identified in the mid-1970s. Further studies have demonstrated that it is able to reduce bone and cartilage destruction and also to restore T-helper and T-suppressor subsets. Ciclosporin has been used in the treatment of RA and retinal vasculitis including that witnessed in Behçet' s syndrome.

Physiological role

Ciclosporin is able to inhibit T cell response and interaction by blocking IL-2 and other pro-inflammatory cytokines.

Clinical aspects

Taken orally, absorption is noted to be both incomplete and erratic with distribution taking place outside rather than inside the blood volume; hence ciclosporin has been detected in the body fat, liver, lungs, kidneys, adrenal glands, spleen and lymph nodes. Consequently elimination is mainly via the biliary system and, to a much lesser extent, via the urine.

Chlorambucil

Although generally out of vogue for the past 20 years, chlorambucil has maintained its popularity in both France and North America for the treatment of most connective tissue disease and associated inflammatory eye diseases.

Physiological role

Chlorambucil is a cytotoxic drug (a nitrogen mustard derivative) and belongs to the group of drugs known as alkylating agents. Alkylating agents are chemically very active substances which bind with DNA within the cell nucleus resulting in cell death at the point of cell division.

Clinical aspects

Administered orally, chlorambucil is rapidly absorbed and then both chlorambucil and its metabolites are excreted in the urine.

Phenylbutazone

Introduced in 1949 for the treatment of arthritis and gout, phenylbutazone is said to be one of the oldest and strongest non-steroidal anti-inflammatory drugs and is often prescribed for its suppressive properties. However, because of reported toxic side effects, phenylbutazone is currently only prescribed for ankylosing spondylitis under hospital specialist supervision.

Physiological role

Phenylbutazone acts to inhibit the biosynthesis of prostaglandins.

Clinical aspects

Administered orally with food (because of gastric irritation), like other NSAIDs, phenylbutazone is almost entirely absorbed from the gastrointestinal tract (as it is not bound irretrievably by food). Elimination is via the kidneys following conversion by the liver into glucuronides and/or additional metabolites.

Dapsone

Known principally as an antileprotic drug, it has also proved effective as an antimalarial drug, and in the treatment of both RA and psoriatic arthritis.

Physiological role

Related to sulfonamide antibacterials, dapsone produces a depressant effect on the immune system.

Clinical aspects

Taken orally, dapsone is absorbed rapidly, with excretion via the kidneys in the urine.

Minocycline

Minocycline is a broad spectrum antibiotic, prescribed for the treatment of tetracycline-sensitive organisms, certain strains of meningitis, acne and for the treatment of RA.

Physiological role

The mode of action of minocycline is unclear but it is thought to inhibit pro- inflammatory enzymes.

Clinical aspects

Following absorption, minocycline spreads widely throughout the body, including a variable penetration across the meningeal barrier into the cerebrospinal fluid. Excretion of the greater part of the drug is slowly via the kidneys.

Mycophenolate Mofetil

Mycophenolate mofetil (MMF) is a pro drug of the active metabolite of mycophenolic acid. It is a suppressor of T and B cell proliferation and adhesion and inhibits monophosphate dehydrogenase that eventually blocks the progression to DNA synthesis and proliferation.

TREATMENT

This section will focus on the new therapies (biologic therapies) that have been introduced over the last five years and are usually prescribed when traditional DMARDs have failed to adequately control the disease process in inflammatory joint disease. DMARDs have been recognized as playing an essential part in controlling inflammatory joint diseases and are the first line treatment for RA and other inflammatory joint diseases (for example, PsA and AS). DMARDs control the disease by suppressing the autoimmune response and reducing the potential for joint erosions and long-term damage. However it is clear that maintaining control of the disease and reducing joint damage remain suboptimal for many patients despite treatment with DMARDs.

There are now a number of new medications that have a specific and more targeted action on the cell-to-cell interactions of the immune response. These include therapies such as anti-interleukin 1 receptor antagonist (IL-1ra), anti-tumour necrosis factor alpha (anti-TNFα) and more recently anti-cluster differentiation 20 (CD 20) B cell depletion (B cell depletion). A general term frequently used to refer to all of these new biologically engineered therapies is 'biologic therapies'. Biologic therapies follow DMARDs in the treatment pathway and are advocated as appropriate treatment when traditional DMARDs (usually at least two DMARDs) have failed to control the disease.

The initial rationale for placing such therapies at this point in the treatment pathway reflects the need for clinicians to gain a greater insight into using these therapies in routine clinical practice, enabling them to develop a sound knowledge of risks and benefits of such therapies, but also related to the costs of treatment, which can range from approximately £8,000-10,000 per annum per patient. However, more recently there has been an increasing recognition that these therapies are effective in methotrexate naive patients and early aggressive treatment may improve long-term outcomes. Biologic therapies have improved the quality of life for many individuals with moderate to severe RA.

Studies also reveal that not only is progressive joint damage halted, but there is early evidence that repair of joint erosions can be seen; something not experienced with traditional therapies and a factor that adds a very important dimension in treatment for those with RA. Optimal treatment benefits are achieved when biologic therapies are co-prescribed with methotrexate. However, for the practitioners caring for individuals treated with these therapies there are important issues that need to be considered in pre-screening and monitoring of patients receiving biologic therapies. This section will outline the key issues that need to be considered.

Genetically Modified Therapies

The use of biotechnology has enabled scientific knowledge to manipulate and redesign key aspects of the body's own cell-to-cell communications and interactions. A range of new therapies have been, and continue to be, developed for inflammatory joint diseases as a result of this work. A key component of these specifically designed therapies is that of disarming pro-inflammatory cytokines responsible for driving autoimmune responses. Cytokines are proteins or glycoproteins that act as important intercellular messengers travelling through blood and the lymphatic systems to communicate important pro- or anti-inflammatory responses. Therapeutic targets have been identified for the cytokines interleukin 1 (IL-1) and tumour necrosis factor alpha (TNFα) and more recently B cell depletion of CD20 B cells. Further specific cell-to-cell interactions have identified new therapeutic pathways that are currently being researched and it is likely that some of these will be licensed within the next few years. The reader should refer to other key documents in the management of patients receiving biologic therapies. These include:

1. Guidance on eligibility criteria, screening and management of those being treated with anti-TNFα prepared by the British Society for Rheumatology.
2. Guidelines for the patient taking immunosuppressants, steroids and the new biologic therapies. Vaccinations in the immunocompromised person. Immunizations against infectious disease.
3. The British Thoracic Society (BTS) guidelines for the assessment and treatment of tuberculosis in patients due to start anti-TNF therapy.
4. The Royal College of Nursing Guidance on assessing, managing and monitoring biologic therapies for inflammatory arthritis.
5. National Institute of Clinical Excellence (NICE) guidance for etanercept and infliximab for the treatment of RA.
6. Background information on the use of rixutimab in the treatment of non-Hodgkin's lymphoma. NICE proposes to review rituximab for RA in their appraisal programme this year.

This section will give an overview of how these biologic therapies provide a new and more effective approach to treating conditions that are driven by a faulty immune response (autoimmunity). Autoimmune diseases can result in a range of conditions depending upon the tissues affected. Examples of inflammatory joint conditions include RA, PsA and AS.

Types of Biologic Therapies

All biologic therapies have been developed using technology manipulating immunoglobulins. This technology involves designing immunoglobulins so that they can specifically disarm cytokines or their receptors and, as result, stop their normal mode of action. Biologically engineered immunoglobulins are made up of foreign proteins. These proteins may be made up using technology that incorporates either all human or part human/part animal immunoglobulins demonstrating the different components of three biologic therapies and their constituent parts.

Immunoglobulins (antibodies)

Immunoglobulins belong to a family of large protein molecules also known as antibodies. Immunoglobulins are produced by B cells in response to a challenge to the immune system. B cells that have had no antigen/antibody reaction are called 'naive' B cells and therefore produce a good but not highly specific immunoglobulin response (or targeted bullet) on first contact or prior to clonal expansion. Immunoglobulins can become more specific with the ability to recognize an antigen following initial interaction with the antigen resulting in a more rapid and effective response in subsequent challenges to the immune system. When an immunoglobulin develops a specific targeted response it is said to have a 'shared epitope' – that is, a perfect match between the immunoglobulin and the antigen, matching in a similar way to a lock and key, with the lock being the antigen and the key being the immunoglobulin.

Physiological Role

The mechanism of action of traditional DMARDs is not fully understood but it is likely that their effect on immune suppression is suboptimal based upon the limited understanding of their non-specific actions on cytokine production and cell replication. Compared to DMARDs and corticosteroids, biologic therapies have a more specific and targeted effect at cell level. Biologic therapies act by mimicking the normal immune processes, effectively preventing or displacing a cytokine from 'locking' into its defined receptor. This has the consequence of preventing the activation of pro-inflammatory cytokines responsible for ensuring an '*inflammatory cascade*'.

When activated an inflammatory cascade results in the classic signs of inflammation (heat, redness, swelling, loss of function and pain). Biologics have a direct cell-to-cell effect on specific targeted cytokines but also have a less clearly mapped action on other closely related or possibly directly linked communicating cytokines that also have an effect on the inflammatory cascade and ultimately the effects that these responses have on other tissues as a consequence of the cytokine activation. Knowledge of cytokines and their effects continues to be researched with well over 150 cytokines clearly identified. These cytokines are classified into families (for example anti-TNFα or interleukin 1) and are described as either pro- or anti-inflammatory cytokines.

Pro-inflammatory cytokines have been identified as playing an important part in driving an inflammatory response. Research continues on key cytokines and chemokines (smaller molecules than a cytokines but playing a similar role in the immune response) as well as looking at different stages of the cell-to-cell interactions in the immune pathway as the focus for future therapeutic targets. Biologic therapies and other key cells-to-cell interactions that might be amenable to therapeutic manipulation are the focus of research for a number of chronic autoimmune conditions.

As research continues it is possible that current therapies will be identified as effective treatments for a number of other autoimmune conditions.

Table 2.6 Key issues in caring for individuals receiving biologic therapies.

Nurse/practitioner screening prior to starting treatment

Patient has had an opportunity to discuss their treatment options and has made an informed decision about starting therapy. Consent documented for treatment and data collection for BSR Biologics Register data.

Patients need to fulfil the eligibility criteria for treatment and the British Society for Rheumatology according to disease.

Screening for risks of tuberculosis should be undertaken for all patients receiving anti-TNFα therapies according to the BTS guidelines. Those who have risk factors should be referred to the prescribing physician for consideration of referral to respiratory or immunology physician.

If co-prescribed with a DMARD that they are continuing on this treatment and monitoring is satisfactory.

Confirmation that women of childbearing age are not pregnant and (both men and women) are aware they must use an effective contraceptive.

Patients should be free of infections or any potential foci of infection (such as chest infections, indwelling catheter or sepsis of prosthetic joint within the last 12 months). Also ensure there is no recent contact with chicken pox or herpes zoster.

Immunizations should be reviewed and if time allows and patient is well enough ensure immunizations are undertaken 4 weeks prior to starting treatment.

Ensure no malignancy or pre-malignancy state (excluding basal cell carcinoma or malignancies diagnosed and treated more than 10 years previously).

Exclude any history or new symptoms of heart failure or demyelinating disease.

Treatment issues general

Provide written information and contact details for support about their treatment and ensure they know when they should seek urgent advice.

Question to ensure there are no new symptoms or pending investigations awaited (e.g. for possible malignancy).

Training and competencies have been assessed in storage, administration and monitoring issues related to self-administration of subcutaneous therapy if treated with anakinra, etanercept or adalimumab.

Infusion issues (infliximab or rituximab): Ensure that

Full screening has been undertaken and patient has an opportunity to ask questions.

Assessment and disease activity criteria have been fulfilled for data collection (such as the BSR Biologics Register) and review of treatment benefits.

Pre-infusion monitoring has been carried out and these are with in normal range (vital signs – temperature, pulse, respirations, blood pressure) and urinalysis. Patient has been weighted if appropriate.

Have received appropriate prophylactic drug therapy prior to starting an infusion.

Clinical management plans or protocols for infusion regimes are adhered to and recorded.

All infusion-related reactions are monitored carefully and managed promptly according to local and national guidelines. If severe reactions occur (e.g. dyspnoea, hypotension) treatment should stop and appropriate treatment administered promptly.

Recommence infusions following a reaction are started slowly and appropriate medical support and full resuscitation equipment readily available.

Post-infusion: Ensure that

Post-infusion observations are satisfactory, the patient is asymptomatic and all details are recorded.

Information and contact numbers have been provided to the patient about follow-up care and what to do if an adverse event occurs.

Adverse Effects of Therapies

As with all injected foreign proteins (e.g. blood transfusion) there is the potential for the body to develop antibodies to the protein and/or hypersensitivity reactions. The likelihood of an individual experiencing an anaphylactic reaction to an agent is influenced by age, gender, atopy, route of exposure and prior exposure as well as history of prior anaphylactic reactions. A detailed discussion on the incidence of infusion reactions and an algorithm to guide practitioners has been described by Cheifetz *et al.* (2003) in the management of Crohn's disease. Adverse responses for immediate and delayed reactions can be categorized as mild, moderate or severe (according to severity and signs of symptoms).

Instant effects

Cheifetz (2003) suggests reactions should be categorized in simple terms such as immediate and delayed, but also considered based upon immune (antibody/antigen reactions) or non-immune mediated reactions (such as dose-dependent drug toxicity, secondary effects of drug or drug-drug interactions). Severe anaphylaxis type reactions (antibody/antigen reaction) are considered to be driven by IgE-mediated acute hypersensitivity events although Cheifetz *et al.*'s study in the use of infliximab did not support this finding. However anaphylaxis type reactions do appear to occur more frequently in individuals with a history of atopy or previous history of infusion-related reactions. In these cases, retreatment should be commenced at a slower rate initially and then gradually increasing in line with normal treatment regimes if all observations are satisfactory.

After effects

Delayed reactions can occur after 24 hours and up to 14 days after treatment and often present in the form of arthralgia, myalgia, urticarial rash, fever or malaise.

Treatment

Interleukin-1 receptor antagonist (IL-1ra)

Anakinra (kineret)

Taxonomy

Anakinra – a biologically engineered monoclonal antibody working as an interleukin- 1 receptor antagonist. Anakinra is indicated for the treatment of RA in combination with methotrexate when there is an inadequate response to methotrexate alone.

Table 2.7 Screening and management of infusion-related reactions.

General screening

Ensure the patient is free of infection and review any changes in health status (e.g. new investigations for suspected cardiac, neurological or malignancy issues). If of childbearing potential confirm effective contraceptive used.

Check any issues re venous access. The appropriate site for a cannular can sometimes be difficult with consideration required for poor skin integrity, bruising or painful joints.

Practitioners should review blood results before proceeding and check for any documentation of human anti-chimeric antibodies (HACA). The presence of HACA is an additional indicator of potential infusion-related reactions. The co-prescription of a DMARD reduces the risks of HACAs developing.

Check, if prescribed, that the patient is still taking their disease-modifying drugs (usually methotrexate). The risk of infusion-related reactions may be higher if not co-prescribed a DMARD. If stopped DMARD, check the time frame from stopping treatment and refer to prescribing physician.

Enquire on history of atopy or previous infusion-related reactions.

Review baseline blood results (check for any abnormalities and report to prescribing physician or review protocols/guidelines on treatment pathway).

Undertake any data collection and consent that is required (e.g. consent for treatment/or BSR Biologics Register or Disease Activity Assessments).

For individuals on anti-hypertensive agents who are treated with rituximab the prescribing physician may elect to withhold anti-hypertensive medications for 12 hours prior to the infusion.

Pre-administration

If prescribed administer pre-medication prior to starting infusion (e.g. paracetamol and/or antihistamine). If prescribed rituximab review concordance with between infusion oral steroid treatment and ensure 100 mg intravenous methylprednisolone pre-treatment is administered.

Patients should have their general well-being (e.g. flushed or chest pain) and observations recorded prior to the infusion and every 30 minutes during the infusion (blood pressure and pulse).

Advise patient to report if they have unexplained/new symptoms such as breathlessness, pruritis, fever, chills or chest pain.

Infusion-related reactions (acute)

Mild to moderate

If patient experiences symptoms stop treatment and treat according to severity of reaction and according to local protocols. Mild or moderate infusion reactions usually occur within two hours of the initial infusion (although they may commence within a few seconds of commencing treatment). They may necessitate temporarily withholding the treatment but usually respond to reduction in the rate of infusion and possibly treatment with paracetamol and diphenhydramine. If observations are satisfactory and symptoms are mild (e.g. headache) and treatment is not required the physician may decide to re-start the infusion at a slower rate and with frequent monitoring. If the infusion then continues without event subsequent infusions can be undertaken with caution.

Moderate to severe

If the patient experiences significant changes in vital signs (e.g. diastolic blood pressure decreases between 15-20 mm Hg) or the patient experiences symptoms that indicate hypersensitivity or in the treatment of rituximab, severe cytokine release syndrome (e.g. severe dyspnoea, bronchospasm, urticaria, fever, hypotension), *stop* infusion. Treat aggressively according to local anaphylaxis protocol (usually would include pain relief, antihistamines, intravenous saline or brochodilators, corticosteroids and/or epinephrine, oxygen). For those treated with rituximab the infusion should not be recommenced until complete resolution of all symptoms and normalization of laboratory values and chest X-ray findings.

The infusion may be resumed (at the discretion of the physician and consent from the patient) but must recommence at no more than one half of the previous rate. If a severe adverse reaction occurs for a second time the prescribing physician should review the decision to continue subsequent treatments on a case-by-case basis.

Recommencing treatment following an infusion-related reaction

For those patients attending for a subsequent infusion having had a previous serious infusion reaction a physician should be at hand for the next treatment, if the decision to retreat has been taken.

Commence infusions at slower rate and ensure frequent observations of vital signs until patient stable (e.g. every 15 minutes). Gradually increase rate and maintain observations and observe carefully until patient stable.

If a further severe infusion-related reaction occurs, *stop* infusion and treat aggressively according to local protocols. Review with prescribing physician.

Delayed infusion reactions

Usually present with arthralgia, myalgia, urticarial rash, fever and malaise. These are usually managed by prescribing paracetamol, antihistamines and occasionally steroids if required.

Physiological role

In normal joints interleukin-1 receptor antagonist (IL-1Ra) modifies the activity of interleukin 1 (IL-1; a pro-inflammatory cytokine). IL-1Ra binds to interleukin 1 receptors (IL-1R1) and by doing so competitively prevents binding of IL-1 into their receptors reducing the number of receptors available for IL-1 to lock into and cause an inflammatory response. This is the body's normal response to maintain equilibrium and ensure an appropriate level of inflammatory response. The loss of self-tolerance (recognition and acceptance of tissues that belong to self) results in an autoimmunity condition upsetting the usual immunological balance.

Anakinra is the only IL-1Ra therapy licensed for RA. Maximum plasma concentrations of anakinra occur three to seven hours following subcutaneous injection. The plasma half life ranges from four to six hours.

Anakinra has been shown to have limited value for the majority of RA patients, based upon current criteria for evaluating effectiveness of treatment using the Disease Activity Score (DAS 28) although research continues into the therapeutic benefits in other conditions (e.g. osteoarthritis). It has been suggested that in RA despite suboptimal DAS scores, there may be benefits in the reduction of long-term joint damage over time although further research will be needed.

Administration of anakinra

Anakinra 100 mg daily is administered subcutaneously using a pre-filled syringe preferably at the same time each day, and should be co-prescribed with a once-weekly dose of methotrexate (weekly dose of methotrexate may vary according age, and renal and hepatic function).

Treatment

Anakinra should be prescribed for the treatment of RA in combination with methotrexate, where there is an inadequate response to methotrexate alone. In theoretical terms the combined treatment of anakinra with an anti-TNFα was initially thought to be an ideal therapeutic option. However this has been shown to significantly increase risks of infections and therefore co-prescribing is not recommended. No dose adjustment is required for the elderly (more than 65 years of age), or those with hepatic or mild renal impairment. Caution is advised for those with moderate renal impairment.

Anakinra is contraindicated in severe renal impairment. Patients with persistent neutropenia ($<1.5 \times 10^9/l$) should be excluded from treatment with anakinra. The risks related to co-prescription of methotrexate need to be considered in the management of patients with hepatic or renal impairment. Individuals who have hypersensitivity to *E. coli* should not be treated with anakinra.

After effects

Frequent reactions

The most frequently reported side effect (>10%) in all clinical trials has been that of mild to moderate injection site reactions (ISRs). These were usually mild and self-limiting urticarial or maculopapular rashes; however, if severe reactions occur, treatment should be discontinued and appropriate treatment given. Discontinuation due to ISR was seen in only 7% of patients. ISRs most frequently occurred within the first four weeks of treatment.

Blood disorders

Blood disorders (neutropenia) were seen in 1-10% of patients in clinical trials. It is recommended that pre-screening of white count should be undertaken, if neutropenic (below $<1.5 \times 10^9/L$) treatment should not be initiated. Regular blood monitoring (monthly for the first six months and then quarterly thereafter) are advised. If neutrophils drop below (below $<1.5 \times 10^9/L$) treatment should be stopped.

Infections

There is an increased risk of serious infections (particularly upper respiratory tract infections) (1.8% in treated group compared to placebo 0.7%). In clinical trials and post-marketing experience some opportunistic infections have been seen including fungal, mycobacterial and viral pathogens. Infections are commonly bacterial infections; once treated and resolved anakinra can usually be restarted.

Anakinra has not been associated with reactivation of latent tuberculosis. The risks of reactivation of TB appear to be greater for anti-TNFα treatments. The BTS guidance may be helpful in assessing general risk factors related to TB exposure although it is important to note the management guidelines are related to anti-TNFα therapies and not specifically for anakinra. However, the co-prescribing of methotrexate and the potential for patients who fail on anakinra to progress to anti-TNFα makes this a sensible principle to apply in assessing patients.

Lymphomas and malignancy

As there is limited data on the impact of treatment on malignancies it is recommended that individuals with pre-existing malignancy should not be treated with anakinra. Patients treated with anakinra have an increased risk of lymphoma compared to the general (non-RA) population; however, the rate for malignancy is comparable to the RA population (two- to three-fold increase compared to the general non-RA population).

Drug interactions

Drug interactions have not been extensively investigated although in clinical trials non-steroidal anti-inflammatory drugs, corticosteroids and disease-modifying drugs were co-prescribed and no interactions were observed. Studies observing toxic interactions related to methotrexate when co-administered with anakinra did not demonstrate any reductions in clearance rates.

Pregnancy and breastfeeding

As with all new therapies there is insufficient data on the use of anakinra in pregnant or breastfeeding women and therefore treatment should not be prescribed if the patient is at risk of pregnancy, pregnant or breastfeeding. Women of childbearing potential must be advised to use an effective means of contraception when treated with anakinra.

Protection

There is limited data available on the effects of vaccinations in patients receiving anakinra. Live vaccines are contraindicated for both anakinra and methotrexate. Where possible it is helpful to ensure patients have their immune status reviewed for any vaccinations required at least four weeks before starting any biologic therapy. Readers should refer to additional guidance on management of patients who are immuno-suppressed and/or are also co-prescribed methotrexate. There is no data on the secondary transmission of infection by live vaccines.

Anti-tumour necrosis factor alpha (Anti-TNFα)

Anti-TNFα therapies consist of the three currently licensed therapies. These are:

1. Adalimumab (Humira) licensed for:
   (a) RA (severe, active and progressive)
      (i) Dose range 40 mg every other week as a single dose (in combination with methotrexate)
      (ii) Dose may be increased to 40 mg every week (for monotherapy)
   (b) Psoriatic arthritis 40 mg every other week as a single dose.
2. Etanercept (Enbrel) licensed for:
   (a) RA (severe, active and progressive)
      (i) Dose of 50 mg once weekly either as combination therapy with methotrexate or as a monotherapy

(b) Plaque psoriasis

(i) 25 mg twice weekly or

(ii) 50 mg twice weekly for up to 12 weeks reducing to 25 mg twice weekly.

3. Infliximab (Remicade) licensed for:

(a) RA (severe, active and progressive)

(i) 3 mg/kg of body weight in combination with methotrexate (0-, 2-, 6- and 8-weekly thereafter)

(b) AS (severe axial symptoms and elevated serological markers)

(i) 5 mg/kg body weight 0-, 2- and 6-weekly – if no response after three doses no additional treatment.

(c) PsA (active progressive)

(i) 5 mg/kg body weight 0-, 2-, 6- and 8-weekly thereafter.

These three therapies will be discussed individually although the next section will refer to the key issues common to all three anti-TNFα therapies.

Physiological Role

The release and activation of the cytokine TNFα triggers a cascade of responses from other pro-inflammatory cytokines. Tumour Necrosis Factor alpha (TNFα) is a pivotal cytokine in the inflammatory response. Activating inflammation is achieved when the TNFα cytokine is released and locks into a T Cell Receptor (TCR) for TNFα resulting in a cascade of responses from other pro-inflammatory cytokines. There are two types of TCR for TNFα that bind with comparable affinity – p55 and p75. TCR for TNFα can be soluble or tissue-bound. Some anti-TNFα therapies can block soluble and tissue-bound receptors. There are variations in the specific interactions with TNFα tissue and soluble receptors, as well as the structure and composition of each of the anti-TNFα therapies, plasma half lives and routes of administration.

The variations between therapies may also explain some differing benefits in specific disease areas although the full significance of the differing therapeutic options is not completely understood and remains an area of close scrutiny and expert discussions. For example it has been suggested that the ability of etanercept to block lymphotoxin α may explain the benefits achieved in JIA as Lymphotoxin α is identified in inflamed joints of JIA patients.

After Effects

Table 2.6 provides an overview of the issues to consider for each of the biologic therapies. The common points that should be considered regularly are:

1. The need to consider all possible opportunistic infections.
2. Awareness of allergic reactions related to the injected proteins.
3. Clinical indications that might be related to tuberculosis or re-emergence of latent tuberculosis.
4. Indications of chronic heart failure or exacerbations of previously mild disease.
5. Exacerbations or new symptoms suggestive of demyelinating disease.

6. Potential withdrawal of treatment if new malignancies or investigations suggestive of malignancies present.
7. Blood monitoring for abnormalities.
8. The avoidance of pregnancy or breastfeeding whilst being treated with biologic therapies.
9. Care with immunization - live vaccines should not be administered and treatment with other vaccines may be suboptima.
10. Planned management of patients receiving surgery to ensure adequate withdrawal prior to surgery and recommencement of therapy when wounds are free of infection.

Some side effects seen in patients treated with anti-TNFα are only partially explained in research trials and subsequent clinical practice. These include reports of exacerbations of chronic heart failure and demyelinating diseases.

Frequent reactions

As all the anti-TNFα therapies consist of foreign proteins (immunoglobulins) there is a small risk of allergic reactions. Serious allergic reactions have been reported in post-marketing surveillance for all biologic therapies - adalimumab, anakinra, etanercept and infliximab. Treatment should be withheld and appropriate treatment administered.

Infusion or injection site reactions

There is the potential for any foreign injected protein to cause an immune-mediated response. The most common reactions for the therapies administered subcutaneously are injection site reactions (approximately 20%-36% in treated groups compared to 9-14% in control groups). These are usually mild and self-limiting and resolve without treatment. Some patients may benefit from topical hydrocortisone cream if the rash is symptomatic or discomfort persists. A guidance document and training package for subcutaneous administration of biologic therapies and patient self-administration of anti-TNFα therapies has been developed by the Royal College of Nursing as well as by the specific patient education material prepared by the pharmaceutical companies. In the management of the patient it should be remembered that some patients may also be co-prescribed methotrexate either orally or as a subcutaneous injection (by once-a-week injection) and care needs to be taken in ensuring that injection sites are rotated.

Autoantibodies

Auto-antibodies can develop at any time during an individual's lifetime without receiving biologic therapies; however they can also develop during treatment with biologic therapies. The co-prescribing of a DMARD (particularly methotrexate) may reduce the potential of auto-antibodies developing in individuals treated with an anti-TNFα therapy. The presence of auto-antibodies may in rare cases result in the individual developing lupus-like symptoms, which usually resolve on cessation of treatment. The presence of antibodies increases the risk of infusion/injection sites reactions.

Blood monitoring

Many patients receiving biologic therapies will also be co-prescribed a DMARD requiring regular blood monitoring. However, it is important to note that research evidence from anti-TNFα therapies has occasionally included abnormal haematological findings. These include rare cases of pancytopenia and very rare cases of aplastic anaemia (in very rare cases resulting in fatal outcomes) and abnormal hepatic dysfunction.

The severity and incidence of haematological problems varies between each of the therapies although this can be a complex picture with the patient group, co-morbidities, level of disease activity and previous immunosuppression (or current treatment) with disease-mofying drugs. However, clinical practice should include screening prior to starting treatment and regular monitoring, assessment and review of individuals treated with an anti-TNFα even if treatment is not co-prescribed with a DMARD such as methotrexate.

Cardiac

Clinical examination and history taking should evaluate cardiac status. Treatment with anti-TNFα therapies should not be considered for individuals with a New York Heart Classification (NYHC) III/IV. Clinical trials and post-marketing surveillance have identified a risk of mortality for individuals with NYHC III/IV. For individuals with NYHC I/II treatment should be considered with caution.

Infections including tuberculosis

Tuberculosis

All anti-TNFα therapies have the potential to increase the risk of infections, including the re-emergence of latent tuberculosis (TB). All patients should be closely questioned and examined for possible TB infections prior to starting treatment. This should include detailed questioning of those who travel to areas of high prevalence of TB or who have had close contact with relatives who travel to areas with a high prevalence of TB. Referral to a respiratory or immunology physician may be required for those who have close contact with TB or relatives with TB or have previously been infected with TB (treated or partially treated), and those who have symptoms suggestive of TB (fever, weight loss).

Further assessment and possible prophylactic regimes to treat previously untreated or latent tuberculosis may be required well before the patient can commence anti-TNFα. The British Thoracic Society (BTS) have developed guidelines specifically to address these issues. Reactivation of TB is at its highest in the first 12 months of treatment although for individuals treated with infliximab the majority of cases occurred within 3 cycles of treatment (median range approximately 12 weeks). Practitioners undertaking assessments and screening of patients prior to starting treatment and during regular monitoring and review should bear in mind the risks of infections including TB, and be vigilant in getting prompt treatment for infections.

Other infections

A rigorous approach for all patients treated with anti-TNFα should be applied when screening and reviewing patients, particularly in relation to the risk of opportunistic infections. As TNFα is implicated in the regulating of temperature the normal immune response may be compromised and the usual signs of infection may not be present; for example, one of the classical signs of infection such as a raised temperature may be absent. The co-prescribing of additional immunosuppressant therapies (such as steroids or DMARDs) compounds this problem. Preliminary data from the BSR Biologics Register show that infection rates are broadly similar across all anti-TNFα therapies and broadly similar to the control groups with active disease (treated with DMARDs) but not receiving biologic therapies. Serious infections (including bacterial, mycobacterium, viral and fungal) have been identified, with some fatalities occurring with all anti-TNFα therapies. A range of infections have been seen including:

1. Upper and lower respiratory infections, including rarely pneumonia.
2. Tuberculosis (chiefly although not completely related to re-emergence of latent tuberculosis).
3. Sepsis/septic arthritis, cellulitis, fungal dermatitis.
4. Varicella.
5. Reactivation of hepatitis B (in chronic carriers).

Treatment should not be started in the presence of a serious infection. If serious infections occur, treatment should be discontinued and only recommenced once the infection has been adequately treated and the infection has resolved. There remains uncertainty about the effects of treating individuals who have HIV with anti-TNFα and therefore therapy should not be recommenced if HIV is identified until additional evidence is available.

Lymphomas and malignancy

There is a theoretical risk of malignancy (as yet unproven) as a result of tumour necrosis factor alpha blockade. Observational studies will provide important evidence of the long-term risks of patients receiving biologic therapies, and preliminary data published in 2006 do not indicate any major influence on mortality in the early years after first anti-TNFα use. The RA population who have severe disease have a two- to three-fold increase in the risk of malignancy as a result of their condition and traditional (non-biological) medications. As there is limited evidence at present, treatment with anti-TNFα should not be commenced where malignancy is suspected until appropriate investigations have been carried out. In addition those with pre-existing malignancies or previous history of malignancy (within the last 10 years) should only be treated with caution and following a review of the potential risks and benefits to the individual.

Neurological

Treatment with anti-TNFα therapies has exacerbated demyelinating diseases (such as multiple sclerosis). Treatment for individuals with a strong family history, pre-existing or recent onset central nervous system or demyelinating disorders should be considered with caution. If neurological symptoms develop, treatment should be stopped and investigations undertaken.

Pregnancy and Breastfeeding

As clinical trials on humans have not been carried out to evaluate the potential risks related the pregnancy or the unborn child none of the biologic therapies should be prescribed if attempting to conceive (or father) a child. The co-prescription of a DMARD such as methotrexate should mean that individuals prior to starting treatment with a biologic should already have been using an effective contraception (adequate contraception should be continued for at least six months after stopping biologic therapy). Preliminary observational data shows that there are no major congenital malformations or evidence of maternal harm in patients treated with all biologic therapies. Immunoglobulins can be excreted in human milk and therefore a biologic should not be administered to mothers who are breastfeeding.

In some circumstances the clinician together with the patient may need to make a careful risk-benefit analysis for both mother and child to decide the best option for management (either stop the biologic or advise that the child is bottle-fed). For individuals who have been prescribed methotrexate it is recommended that at least a six-month period should elapse before attempting conception; the same length of time is recommended for infliximab (which is co-prescribed with methotrexate). Five months free of treatment is recommended on cessation of adalimumab. Withdrawal of methotrexate co-prescription must also be considered and manufacturer advice varies on the length of time from withdrawal of methotrexate before attempting to conceive (between three and six months).

Vaccination

Live vaccines must not be administered to individuals receiving anti-TNFα therapies (with or without methotrexate or another DMARD). It is therefore useful, if patients are well enough, that time should be allowed to review and plan any necessary immunizations at least four weeks before starting treatment.

One study suggests that pneumococcal vaccinations may result in a poor response and others have demonstrated anti-TNFα treated patients to have similar response to controls following vaccination with pneumococcal vaccination although methotrexate does seem to reduce immune response. If immunizations are to be given they should ideally be administered four weeks prior to starting treatment or at least six months after the last infusion of infliximab or two to three weeks after the last dose of etanercept.

There is no current advice for adalimumab on the time to elapse before it is safe to administer a vaccine. Additional guidance can be sought from the centre for disease control and prevention for those who are immunosuppressed. Specific guidance on immunization for TB is included in a guidance document produced by the BTS.

Specific Information on Anti-TNFα

Adalimumab (humira)

Adalimumab is a recombinant human monoclonal antibody and works by binding specifically to TNFα and preventing the normal TNFα function of locking into p55 and p75 cell surface receptors. The bioavailability following a 40 mg subcutaneous dose was 64%. Absorption and distribution is slow with a peak serum concentration reached at approximately 5 days after administration and a half life for adalimumab of approximately 12-14 days.

Indication for use

Adalimumab is indicated for the treatment of adults with moderate to severe RA following inadequate response to DMARDs and for active and progressive psoriatic arthritis (PsA) when previous response to DMARDs has been inadequate.

Administration

Adalimumab is administered by subcutaneously pre-filled syringe in a dose of 40mg every other week. Adalimumab can be prescribed as monotherapy but for optimal benefit should be prescribed with methotrexate (for RA patients). Those individuals already receiving methotrexate should continue with their treatment when prescribed adalimumab. For individuals with RA on monotherapy an increase in dose to 40 mg every week may be indicated.

Specific additional comments

For those with renal or hepatic impairment, there is no specific guidance for individuals receiving adalimumab. If co-prescribed with methotrexate guidance from the British Society for Rheumatology monitoring guidance (2006) and advice from the National Patient Safety Agency should be reviewed.

Etanercept (enbrel)

Etanercept is a human TNFα receptor p75 fusion protein using recombinant deoxyribonucleic acid (DNA) technology. Etanercept is slowly absorbed following subcutaneous injection reaching maximum concentration approximately 48 hours after a single dose. The half life is approximately 70 hours.

Indication for use

Etanercept can be prescribed as monotherapy or in combination with methotrexate for the treatment of adults with active RA who have failed to respond adequately to DMARDs and PsA who have failed or are contraindicated/intolerant of DMARDs. It can also be prescribed for adults with RA who have active and progressive disease not previously treated with methotrexate.

Administration

Etanercept 50 mg is administered by subcutaneous injection once weekly in 1 ml of water. For moderate to severe plaque psoriasis the usual dose is 25 mg twice weekly or if needed etanercept can be administered at a dose of 50 mg twice weekly for up to 12 weeks reducing to 25 mg twice weekly.

Specific additional comments

For those with renal or hepatic impairment, there is no specific guidance for individuals receiving adalimumab. If co-prescribed with methotrexate guidance from the National Patient Safety Agency should be reviewed.

Infliximab (remicade)

Infliximab is a monoclonal antibody that binds with high affinity to both soluble and transmembrane TNFα inhibiting the normal activity of TNFα. Infliximab is administered by intravenous infusion usually over a two-hour period. The median terminal half life at 3, 5 and 10 mg/kg doses ranged from 8-9. 5 days. In most patients infliximab can be detected in the serum for at least 8 weeks after a single dose of 3 mg/kg.

It may be that repeated treatments with infliximab might increase this time frame, although the data was studying Crohn's disease with doses of 5 mg/kg with infliximab detected in serum for 12 weeks after administration.

Indication for use

Infliximab is indicated in the treatment of active RA when the response to DMARDs (including methotrexate) has been inadequate and for severe progressive RA not previously treated with DMARDs. For those with RA, infliximab should be prescribed in combination with methotrexate. It is also indicated for the treatment of PsA when the disease is active and progressive and has failed to respond adequately to DMARDs and in the treatment of AS for those who have axial symptoms and have failed to respond adequately to conventional therapy. It is also licensed for the treatment of Crohn's disease.

Administration

Infliximab is administered by intravenous infusion initially, usually over a two-hour period at a dose of 3 mg/kg of body weight for RA and 5mg/kg body weight for AS and PsA. Infusions are administered following the first infliximab treatment at two and six weeks and then every eight weeks thereafter.

After effects

The long half life of infliximab means that monitoring of side effects or re-emergence of latent TB should be continued for up to six months after cessation of treatment.

Infusion reactions

Infliximab is administered intravenously usually by infusion over a two-hour period. Infusion-related reactions (non-specific symptoms such as fever, chills) are reported as ranging from 5-15%. Serious infusion reactions (chest pain, hypertension, hypotension or dyspnea) occur in less than 1% of patients treated with infliximab. Sany *et al.* (2005) studied infusion-related reactions and found that pre-medication with betamethasone provided no additional benefit in reducing infusion-related reactions.

Infusion-related reactions are more common on the third and fourth infusion and approximately 3% of patients discontinued treatment due to infusion reactions. Individuals receiving a diphenhydramine as a pre-medication had a higher rate of infusion-related reactions. Although the majority of infusion reactions are mild and do not result in discontinuation of treatment, a small percentage (<1%) can be severe and include anaphylactic reactions that will require emergency treatment including adrenaline, antihistamines and corticosteroids. If patients are re-treated after a prolonged period of time they should be monitored carefully.

The episodic treatment (rather than regular maintenance treatment) with infliximab appears to exacerbate the risk of infusion-related reactions. The risk is reduced if patients continue their DMARDs therapy. It is recommended that infliximab is administered particularly in the first year of treatment, usually over a two-hour period and post-infusion observations continue for one to two hours. There are some differing infusion regimes evolving with developing clinical expertise. Shortened infusion and observation times have been used for those who are established on treatment, and for those who do not have a history of atopy or have experienced a previous infusion reaction. Buch *et al.* (2004) suggests that the risks of infusion-related reactions were high up to and including the fifth infusion and as a result subsequent infusions can be administered at a faster rate. Delayed hypersensitivity reactions (myalgia, arthralgia fever, rash or facial oedema, dysphagia) are uncommon and appear to be exacerbated by increasing drug free intervals of less than a year.

Auto-antibodies

When methotrexate is co-prescribed with infliximab the risk of antibodies against the infliximab monoclonal antibody (human anti-chimeric antibodies, HACA) is reduced. Infusion-related reactions are seen more frequently in individuals who have developed antibodies to infliximab.

Rituximab (mabthera) B cell CD20 depletion

Physiological role

Rituximab is a genetically engineered anti-CD20 monoclonal antibody that acts by selectively depleting the number of circulating pre B and mature B cells. B cells expressing CD20 antigen develop during a specific phase of the B cells' maturity. This means that the overall integrity of the immune system is not compromised and is designed to sustain remission. A number of research studies have focused on the role of B cells as a key player in the immunopathogenesis of RA. Apart from producing immunoglobulins, B cells may have additional stimulatory effects in inflammatory arthritis and may include:

1. Acting as antigen-presenting cells and encouraging co-stimulation of T cells.
2. Activating T cells.
3. Secreting pro-inflammatory cytokines and chemokines.
4. Producing rheumatoid factor (RF). Positive RF is linked to aggressive disease and may be implicated in enhancing stimulus to T cells.

Methods for use

Rituximab has been used to treat non-Hodgkin's lymphoma (NHL) for a number of years but it is possible that this licence could be extended to include the treatment of RA in patients who have failed anti-TNFα and are RF-positive. Although as yet unlicensed at time of going to press rituximab is being used for patients with severe RA that has failed to respond to anti-TNFα, and therefore it is important for practitioners to be aware of this treatment option. Treatment with ritixumab should always be given with a pre-medication (pain reliever and anti-histamine). Steroids should also be prescribed if rituximab is not administered as a combination with cyclophosphamide, doxorubicin hydrochloride; oncovin (vincristine) and prednisolone (CHOP) or cyclophosphamide, vincristine and prednisolene (CVP) chemotherapy. Rituximab has a long half life (up to 152.6 hours) after the first infusion and has been detected for three to six months following treatment.

Contraindications

Hypersensitivity to any component of rituximab or murine proteins.

Administration

An intravenous infusion of methylprednisolone (100mg), paracetamol and antihistamine should be administered before each treatment of an intravenous infusion of rituximab of a fixed dose of 1000mg rituximab. Treatment usually consists of 2 infusions administered 2 weeks apart (day 1 and day 15). A brief course of oral glucocorticoids should also be prescribed between first and second infusion.

After effects

The research and clinical experience of rituximab is extensive in the field of NHL although clinical experience in the treatment of RA and other inflammatory arthritidies is in its early years to date. Early studies appear to show that there are a reduced number of adverse events compared to those seen in patients with NHL (36% of RA-treated patient compared to 75% of those with NHL) for their first infusion and 10% during the second infusion. It remains to be seen, with greater clinical experience, if the wider aspects of the side effects profile for RA will vary from NHL.

Cardiovascular

In clinical trials 18.8% of patients' cardiovascular events were reported with the most frequent reports related to hypotension and hypertension. Individuals treated with anti-hypertensives may need to have their treatment withheld for 12 hours prior to infusion. Rixutimab should be used in caution in individuals treated with cardiotoxic chemotherapy or who have cardiovascular disease as exacerbations of angina, arrhythmia and heart failure have been seen.

Haematological/immunoglobulin

Haematological abnormalities (e.g. thrombocytopenia, neutropenia, anaemia) occurred in a minority of patients and were usually mild and reversible. Blood monitoring should be undertaken and if neutrophils are below $1.5 \times 10.^{9}/l$ or platelet counts are below $75 \times 10.^{9}/l$ treatment should be withheld until a medical opinion/review is undertaken.

Autoantibodies

There is a theoretical risk that human anti-chimeric antibodies (HACA) may develop from the biologically engineered rituximab, made up of human/mouse chimeric antibody. To date the incidence is low (<1%) and rarely associated with any clinical symptoms.

Infections

Bacterial, viral and to a lesser extent fungal infections have been identified in patients receiving rituximab, with severe infections occurring in 3.9% of patients in clinical trials. Some of these infections occurred during the treatment period (1.4%) and others presented during the follow-up period (2.5%).

Very rare cases of hepatitis B reactivation have been reported in individuals who were also receiving cytotoxic chemotherapy whilst being treated with rituximab.

Infusion-related reactions

The most common adverse events during or following rituximab treatment are mild to moderate infusion reactions (occurring in approximately 9% of patients) and are related to the rate of the infusion, usually occurring within two hours of the initial infusion. Patients with HACA titres may have allergic or hypersensitivity reactions when treated with other diagnostic or therapeutic monoclonal antibodies. Data in the treatment of NHL state that the incidence of infusion-related reactions are seen in approximately 50%. However, this data may not reflect the RA experience as some reactions seen in NHL are related to tumour lysis syndrome (TLS) and cytokine release syndrome (CRS), which may be specific to NHL and the specific tumour burden. Infusion reactions include fever, chills, rigors and sometimes hypotension and dyspnoea and generally resolve quickly although in post-marking surveillance some fatal outcomes have been reported for those with severe reactions. It is therefore essential that full resuscitation support is readily available.

Drug interactions

No data are currently available on drug interactions and the knowledge of agents capable of causing depletion of normal B Cells is as yet not known.

Pregnancy and breastfeeding

Women (and men) of childbearing potential should use an effective method of contraceptive during treatment and up to 12 months from cessation of treatment as rituximab has a long half life (although proportional to the dose administered). Women should not breastfeed whilst on rituximab.

Immunization

B cell depletion typically lasts for several months following treatment and it is therefore important to consider the immunization status of individuals prior to commencing treatment with rituximab and to complete all immunizations at least four weeks prior to the first administration of rituximab. Live vaccines should not be administered.

Chemotherapies

Patients should have had the opportunity to discuss all aspects of their treatment and be able, where appropriate, to select the treatment that best suits their needs. Risks and benefits should be explained and consent should be documented. All advice and information provided should be supported by written information and clear guidance on when to stop treatment and seek urgent medical advice.

The RCN guidance on training patients in the self-administration of subcutaneous therapies and detailed information on the management of patients prior to infusions will be a useful tool for practitioners. Individuals treated with anti-TNFα therapies should be easily identified and treated promptly for infections. As a result specific patient alert cards have been developed for the patient to use to inform healthcare professionals and other care workers or dentists to inform them of key issues related to anti-TNFα therapies. These alert cards can be accessed easily. Biologic therapies will play an increasingly important role in the treatment of a range of autoimmune conditions and as such it is imperative that practitioners in all care settings have the expertise to support the patient in the assessment, management and ongoing monitoring of biologic therapies.

PHARMACEUTICAL USES

Corticosteroid drugs are synthetic derivatives of the body's naturally occurring corticosteroid hormones, which are produced in the cortex of the adrenal glands. The principal corticosteroid of the human adrenal cortex is cortisol (hydrocortisone) derived from the hydroxylation of cortisone.

The adrenal gland production of corticosteroids is controlled by the hypothalamus and the pituitary gland via a negative feedback system. Corticosteroids consist of androgenics, mineralocorticoids and glucocorticoids.

Glucocorticoids

These are responsible for:

1. Carbohydrate, protein and fat metabolism;
2. The maintenance of blood sugar levels;
3. The body's response to both physical and psychological stress;
4. The suppression of inflammation and the immune response.

Much research has taken place to isolate the glucocorticoid that possesses precise anti-inflammatory and immunosuppressive properties but to date it remains elusive. Therefore, the role of systemic steroids in the management of rheumatological disorders is complex because, whilst they are very effective anti-inflammatory and immunosuppressive agents, they are also associated with potentially serious side effects even with relatively low doses, which may cause significant morbidity. Systemic steroids currently prescribed in the treatment of rheumatological disorders include hydrocortisone (cortisol), prednisolone, cortisone and methyl prednisolone.

Physiological role

Corticosteroids act by binding to specific cytoplasmic receptors which then enter the nucleus where they cause the production of certain mRNAs (ribonucleic acid) coding for various proteins. The beneficial effects of systemic steroids are partly related to increased production of lipocortin and decreased production of the inflammatory response – for example, cytokines, prostaglandins and leukotrienes. The exact effects on the lymphocyte function are still poorly understood. It is not yet possible to separate different corticosteroid effects by using different corticosteroid analogs, indicating that their actions are similar and their relative potencies relate to their structure and plasma half life.

Hormonal Therapy

In a review of clinical trials of corticosteroids in RA it was concluded that in both short- and long-time studies corticosteroids are effective inflammatory agents, significantly better than placebos and NSAIDs in the relief of pain and stiffness. Work by Kirwan (1995) has demonstrated that patients with early RA who received prednisolone 7.5 mg daily for two years experienced less radiological joint damage (in the hands) than did those individuals receiving placebos. A systematic review from the Cochrane database concluded that prednisolone in doses of less than 15mg daily may be used intermittantly in patients with RA, especially if the condition could not be controlled by other drugs. The review highlighted that the lower the dose of steroid the less risk of adverse side effects.

Bone protection therapy should be commenced at the initiation of treatment with steroids. However, the problem with using steroids for control of synovitis is that benefits are not sustained unless increasing doses are employed. This increases the risk of steroid side effects and it often becomes very difficult to decrease or stop the steroids.

Corticosteroid Sparing Agents

This term is used to describe drugs which are prescribed in order to make it easier to reduce the corticosteroid dose, while at the same time controlling the underlying disease. Azathioprine, cyclophosphamide and methotrexate can all be used for this purpose.

After Effects of Corticosteroids

Immune system

Corticosteroids are powerful immunosuppressants, which affect both humoral and cell-mediated immune responses. Although corticosteroids can be valuable in the treatment of autoimmune diseases they can also be potentially life-threatening because high levels of corticosteroids will also reduce the formation of antibodies and the white cell count of patients, therefore permitting the invasion of bacteria or viruses and the spread of infection. Patients on long-term corticosteroids may also be predisposed to staphylococcal, gram-negative, tuberculous and listeria infections.

Glucose metabolism

The problem of glucose intolerance with patients receiving corticosteroids was initally highlighted in the late 1940s. The suggested mechanisms of glucose intolerance are as follows:

1. Decreased insulin secretion;
2. Increased hepatic glucose production;
3. Impaired peripheral glucose metabolism.

Intake methods

1. Orally;
2. Intravenous pulses;
3. Intra-articular injection;
4. Intra-muscular injection;
5. Soft tissue injection.

Clinical aspects

Following oral administration of prednisolone, absorption takes place quickly within the gastrointestinal tract and is rapidly metabolized within the liver, with excretion via the urine. Its action on the tissues is of a much longer duration than its presence in the blood.

Therapy

The factors that dictate the treatment regimen include the actual disease, its severity and the clinical response. Kirwan (1994) divides the treatment regime into three dose-related ranges:

1. Low daily doses (up to 15 mg) to treat polymyalgia and symptomatic RA.
2. High daily doses 20-60 mg for serious disease – for example, temporal arteritis, dermatomyositis, lupus erythematosus.
3. Very high doses (pulsed methylprednisolone) for acute or life-threatening crisis.

Effects of Corticosteroids

The symptoms of PMR and temporal arteritis (TA) often overlap. No controlled trials have been conducted to provide guidance on the best treatment dosage with corticosteroids. Although practice will differ, PMR is often treated with an initial dose of 15 mg prednisolone daily reducing slowly over 18-24 months governed by blood inflammatory parameters and clinical symptoms. One potential problem is to commence too high a dose of prednisolone accompanied by too rapid a reduction in dosage overtime. This will often result in an exacerbation of original symptoms. The treatment of TA requires a much higher dose of steroid, usually 60-100 mg daily.

Collagenosis

The use of steroids will be largely determined by the nature of the collagenosis complication rather than by the collagenosis itself – for example, in the case of SLE many individuals with mild disease manifestations may not require steroids.

Inflammatory muscle diseases

Dermatomysitis and polymyositis patients are treated with medium to high daily dose steroids initally. This will often be accompanied with a cytotoxic agent such as azathioprine or methotrexate.

Vasculitis

Management of vasculitis involving major organs (e.g. Wegener's granulomatosis, rheumatoid vasculitis, polyarteritis nodosa) will necessitate the use of high dose steroids orally or intravenous pulse therapy.

Bone Mineral Metabolism

Studies have revealed that cytoplasmic glucocorticoid receptors are present in bone cells, which appear to influence calcium intake and excretion and the factors that regulate hormones, cytokines and growth factors.

Therefore, prolonged steroid therapy possesses multifaceted adverse effects on bone mineral metabolism which precipitate growth retardation in children, osteoporosis, long bone and vertebral fractures. The adverse effects of steroids that affect bone material metabolism include their effects on:

1. *Osteoblasts*: steriod therapy has been shown to reduce circulating osteocalcin levels which affect osteoblast activity thus inhibiting bone formation. Studies have also revealed that bone matrix synthesis is affected within 24 hours of the administration of steroids.
2. *Osteoclasts*: steriod therapy is thought to have a dual accelerating effect of bone reabsorption (1) by directly stimulating osteoclasts or (2) indirectly under the influence of parathyroid hormone (PTH).
3. *Gastrointestinal absportion of calcium*: the consensus of opinion is that steroids suppress the intestinal absorption of calcium although the reason remains unclear.
4. *Excretion of calcium*: a decrease in the reabsorption of calcium within the kidneys has been shown to occur in patients. Although glucocorticoid receptors are present in the kidney their precise mode of action on calcium reabsorption/excretion is unclear.

5. *Vitamin D*: extensive research regarding the effects of steroids and 25- hydroxyvitamin D levels has revealed conflicting evidence; hence the suggested osteopenic effects of steroids in relation to 25-hydroxyvitamin D levels remains controversial.
6. *Parathyroid hormone (PTH)*: PTH-regulated blood phosphorus and calcium levels via its actions on the intestines, bone tissue and kidneys. A normal concentration of calcium (2.2-2/6 mmol/l) is essential for normal physiology. However, numerous studies have demonstrated that steroids stimulate the parathyroid glands and increase PTH levels, which stimulate osteoclasts to accelerate bone resorption.
7. *Growth retardation*: since the early 1950s growth retardation in children has been a recognized side effect of long-term steroid therapy, although the precise mechanisms remain elusive. It is suggested that steroid therapy has a direct effect and an indirect consequence on both the osteoblast and cartilage cells, which consequently suppresses the growth of linear bone and impedes epiphyseal closure. The prevention of growth retardation is helped by prescribing an alternate-day regimen.

Studies of bone loss in adults treated with cortisteroids do not clearly suggest a threshold dose below which osteoporosis can be avoided, although there does appear to be a consistent relationship between doses above 7.5 mg daily and the rate of bone loss.

Peptic Ulceration

Evidence from controlled studies of corticosteroid therapy suggests that the increased risk of peptic ulceration is considerably lower than is widely believed. However, corticosteroids may exacerbate the ulcergenic properties of NSAIDs.

Atherosclerosis

Prolonged corticosteroid therapy may accelerate the development of atherosclerosis, with lower limb atherosclerosis occuring in as many as 60% of corticosteroid-treated patients with RA. The evidence linking corticosteroid therapy with atherosclerosis is still considered controversial but, in time, even a small effect could have considerable clinical significance.

Reducing the Dose of Corticosteroids

High doses of corticosteroids are often used for only short periods and can be reduced rapidly with little adverse effect. The exact mode of reduction will depend on the control of the clinical situation. Patients treated with moderate to low doses of corticosteroid over a longer duration of time may develop a corticosteroid withdrawal syndrome when their treatment is reduced. This can include the symptoms of myalgia, fatigue and nausea and has been reported in as many as 70% of patients treated with 30 mg prednisolone daily for longer than three months. One approach to try and avoid this is to reduce the corticosteroid treatment very slowly, often by as little as 2.5 mg daily every two months down to 7.5 mg daily, then introduce further graduated reductions of 1.0 mg daily or by 1.0 mg on alternate days.

The administration of exognous steroids may result in the suppression of the body's own (endogenous) steroid production. Therefore, when exogenous steroids are withdrawn, the adrenal gland may fail to produce adequate cortisol, resulting in a steroid crisis with a failure of cortisol production. This in turn results in hypotension, hypoglycaemia and electrolyte imbalance. All patients on steroid treatment should carry a steroid card with them. The withdrawal of steroids can lead to a marked flair in symptoms of the disease process, which may necessitate the patient being recommenced on steroids.

Pulsed Corticosteroids

Pulse therapy involves the intravenous infusion of a large dose of corticosteroid (usually 500 mg – 1 g of methylprednisolone) over 30-60 minutes. There are many different treatment regimens in practice but most include a course of three pulses on alternate days, followed by a resting phase of around six weeks. It may be used to bridge the time interval between initiation and response to DMARDs or, when a patient has an acute flare of their RA, to induce remission. A review of the use of pulsed methylprednisolone shows few or minor side effects. Those who have experienced severe adverse effects of the cardiovascular system, or experienced infection, had existing compromised cardiovascular and immune systems as a result of their disease or due to concomitant drug therapy.

Intramuscular Corticosteroids

Intramuscular injections of corticosteroid may be used to reduce the symptoms of a flare in RA, or be administered prior to the initiation of DMARD where it is known that these medications take many months before benefit can be assessed. It is important to give the injection deep into the muscle to prevent muscular atrophy occurring. Choy *et al.* (1993) found that intramuscular methylprednisolone was superior to equivalent oral doses in their study of corticosteroids and gold therapy.

Intra-articular/Soft Tissue Injections of Corticosteroids

Intra-articular and soft tissue injections are interventions used in the management of rheumatology disorders that offer an effective supplement to systemic therapies in the control of inflammatory joint disease. In comparison to oral corticosteroids, they are well tolerated and safe if administered appropriately. The Royal College of Nursing Rheumatology Forum has produced guidelines on the use and administration of intra-articular injections, including contraindications for the administration.

GOUT

Gout is caused by inflammation induced by microcrystalline uric acid in the form of monosodium urate monohydrate in joints and soft tissues. A usual but not essential prerequisite for the condition is a raised plasma level of urate, which is derived from endogenous and dietary nucleoproteins.

Hyperuricaemia

Uric acid is the product of purine metabolism and hyperuricaemia can occur either from an over-production of uric acid (20-25% patients) or when the kidneys fail to excrete uric acid (75% of patients). Excessive hyperuricaemia may then result in the formation of uric acid crystals, which are deposited in various parts of the body, more commonly in the joints of the foot, hand and knee. However, uric acid crystals may also form torphi in soft tissues or as stones in the kidney.

Prevalence

The prevalence of gout is 5-10 per 1000 adults in the United Kindom. It can occur in young men with a family history of gout, middle-aged men with hypertension, obesity and cardiovascualr disease and in older men and women secondary to diuretic use.

Clinical features

70% of patients present with an acute attack of gout, which is typically characterized by a sudden onset of podagra (an acute, red, painful swelling at the metatarsophalangeal joint of the great toe), although both the knee and hand may also be affected. Patients with acute attacks may also present with a raised erythrocyte sedimentation rate, pyrexia and leucocytosis.

Precipitating factors

1. A sudden raised level of uric acid (greater than 430 umol/1 resulting either from an oversecretion from the liver or impaired kidney excretion).
2. Obesity.
3. Alcohol.
4. Hypertension.
5. Severe dietary restriction.
6. Diuretics.
7. A sudden fall of uric acid levels following the administration of allopurinol or uricosuric drugs.
8. Trauma.
9. Surgery.
10. Severe systemic illness.

If episodes of acute gout reoccur and remain untreated, gout develops into a chronic condition in which erosions of both cartilage and bone may occur along with deposits of trophi. Trophi are composed of sodium biurate and appear as small hard chalky or cheesy deposits which may discharge and ulcerate. Consequently, chronic gout may result in joint deformities and subsequent functional impairment.

Symptoms

The examination of synovial fluid by polarized light microscopy is used to identify monosodium uric acid crystals. The identification of these crystals provides the clinician with a definite diagnosis of gout, thus excluding either sepsis or pseudogout. The crystals present in pseudogout are calcium pyrophosphate. The correct diagnosis is paramount to ensure that patients are not prescribed treatment unnecessarily.

Diagnosis features

1. Hyperuricaemia (although not all patients with hyperuricaemia have gout).
2. X-rays in early acute attacks are unremarkable. However, the appearance of classic 'punched out' erosions (radiolucent urate tophi) appear later with chronic gout.

3. Following susbequent episodes of gout, examination of the patient may reveal tophaceous deposits in the helix of the ear, bursae, tendon sheaths and kidney parenchyma.

Biological therapy

The management of gout is planned in two stages: initially the management of the acute inflammatory joint symptoms, and secondly the long-term management of persistent hyperuricaemia.

First-stage treatment

The aim of initial therapy is to suppress the inflammatory process induced by the deposit of uric acid crystals, which may be present in and around joints, tendons and in the kidney parenchyma. Hence, non-steroidal anti-inflammatory drugs are the first line of treatment. The NSAID prescribed will depend on both the patient's tolerance and the clinician's preference. The initial dose required is usually twice the normal maintenance dose for three days, reverting to the maintainance dose until resolution is achieved, usually by 14 days. In those patients for whom NSAIDs are contraindicated, colchicine is prescribed but is often not tolerared at its full dose. Oral or parental steroids are also effective but septic arthritis must be excluded before their use. If using oral prednisolone it can be prescribed at 30 mg daily tapering to zero over 10 days.

Colchicine

Derived from the autumn crocus *colchicum autumnale*, colchicine is an ancient remedy which has been prescribed since the eighteenth century for the treatment of gout.

Physiological role

Colchicine is an alkaloid that halts both the inflammatory response and the deposit of uric acid crystals, although its precise mode of action is complicated and not entirely understood.

Clinical aspects

Following administration, colchicine is absorbed through the upper small bowel, metabolized by the liver and then excreted both in the bile and intestinal secretions. As 20% is excreted unchanged in the urine, the dose of colchicine should be adjusted accordingly for those patients who have pre-existing renal disease.

Therapy

Initially patients are prescribed 1mg, followed by 0.5mg every 2 hours which usually has an effect within 24 hours of commencement of therapy. However, because the therapeutic regimen may mirror the toxic dose, dosage continues until either:

1. The maximum total dose is attained (10 mg).
2. The gout subsides.
3. Side effects are experienced (diarrhoea and vomiting).

NB: colchicine 0.5 mg 2-3 times daily may also be prescribed concurrently with initial long-term treatment of hypouricaemic drugs to prevent further attack of acute gout.

Drug interactions

Ciclosporin: increased plasma levels - therefore possible risk of both nephroxity and myotoxicity.

Second-stage therapy: long-term treatment

The aim of long-term treatment of gout is to prevent the possible consequences of further episodes of acute gout (joint deformity, loss of joint function, and kidney damage). The hypouricaemic drug most commonly prescribed is allopurinol, which is a xanthine-oxidase inhibitor which acts to inhibit the production of uric acid.

Other hypouricaemic drugs include probenecid and sulfinpyrazone, which are both uricosuric drugs that block the tubular reabsorption of filtered urate in the kidneys and therefore increase the excretion of uric acid via the kidneys. Second-stage therapy, therefore, is prescribed to prevent recurring acute attacks of gout by maintaining uric acid levels that fall within normal blood plasma parameters. Indications for the prescription of second-stage therapy:

1. Hyperuricaemia with recurrent episodes of acute gout.
2. Visible tophi and/or erosions revealed by radiological examination.
3. Associated renal impairment.

Allopurinol

Allopurinol acts by inhibiting xanthine oxidase, the product that enables the conversion of xanthine and hypoxanthine to uric acid. Because of its low incidence of side effects, allopurinol is the most widely used drug for the long-term prophylaxis of gout. Its added advantage is that it can also be prescribed for those patients with renal impairment or those who have kidney stones (which are conditions contraindicated for treatment with uricosuric drugs).

Treatment regimen

A dose of 100 mg once daily initially for one week (initial dose for both the elderly and patients with renal dysfunction should be reduced to 50 mg to prevent toxicity), increasing over two to three weeks to a maintenance dose (dependent on either blood plasma or urinary uric levels) of 200-600 mg daily in divided doses. Tablets should be taken after food.

Uricosuric drugs

Probenecid - treatment regimen

1. Week one: 250 mg twice daily.
2. Week two: 500 mg twice daily.
3. Maintenance dose up to 2 g daily in divided doses according to uric acid levels. Tablets are to be taken after food.

Sulfinpyrazone - treatment regimen

Initially 100 mg - 200 mg daily, increasing over two to three weeks to 600 mg daily. Maintenance dose (when uric acid blood plasma falls within normal levels) may be possible to reduce to 200 mg daily. Tablets to be taken with either food or milk. Compliance of long-term therapy is dependent on patient education regarding the differing actions of both initial and long-term therapies. To prevent further attacks of gout, patients should receive health promotion regarding the risk factors that can precipitate an acute episode of gout.

Benzbromarone is more potent than probenecid or sulfinpyrazone and it can be used in patients with renal impairment and in conjunction with allopurinol. It does increase the risk of stone formation so a high fluid intake is important.

Diet and lifestyle

Dietary advice can prevent futher attacks of gout.

1. Weight reduction.
2. Saturated fats replaced with unsaturated fats.
3. Relative increase in protein intake.
4. Many beers contain guanosine which is converted into uric acid by gut bacteria.
5. Alcohol taken without food is catabolized to lactate and other ketones that compete with urate for excretion via the renal tubule.
6. Alcohol also decreases the conversion of allopurinol to the effective metabolite oxipurinol.
7. Fortified wines may contain oxalates, which contribute to the formation of stones.
8. Coffee and tea are diuretics and can interfere with assays for urate.

METHOD OF PRESCRIBING

The Crown Report on Prescribing, Supply and Administration of Medicines provided the framework for the development of prescribing for nurses. The report describes two types of nurse prescribers.

1. Independent prescribers who diagnose patients and then prescribe an appropriate medication.
2. Supplementary prescribers who are responsible for the care of the patient once the diagnosis and treatment plan has been determined by an independent prescriber. For example, a patient with inflammatory symptoms will be reviewed by the rheumatologist. Once a diagnosis of RA has been established and the need for DMARDs established the rheumatologist could refer the patient to a rheumatology nurse for the instigation of a DMARD and the necessary ongoing surveillance.

The recommendations of the Crown Report were implemented in three stages.

1. Patient group directions (PGDs) for the supply and administrations of medicines.
2. Independent prescribing for nurses.
3. Supplementary prescribing for pharmacists and nurses.

Basic requirement

The National Prescribing Centre (NCP) (2005) have suggested the following criteria that may help to identify nurses who would benefit from becoming prescribers. These include nurses who:

1. Run their own clinics or services, which would include most rheumatology clinical nurse specialists.
2. Work in isolation from other prescribers (although to a lesser extent with supplementary prescribers who need to work in partnership with an independent prescriber).

3. Could complete episodes of care if they were able to prescribe.
4. Are likely to be able to prescribe for more than one medical condition.
5. Hold additional qualifications whereby their professional expertise would facilitate prescribing for specified medical conditions.

Instructions

The first extended nurse prescribers qualified in 2002. The main goal was to have competent nurse prescribers who could treat patients in four specific areas:

1. Minor illness;
2. Minor injuries;
3. Health promotion;
4. Palliative care.

These areas were chosen as they represent the commonest reasons for patients requiring input from their GPs and the introduction of the extended nurse prescribers (ENPs) formulary was to enable nurses to prescribe in these four domains, in a safe, quick and effective manner to benefit the patients and free up GP time. Consequently the initial nurse prescribing courses were aimed at nurses working in walk-in centres or running minor injury clinics. The ENPs formulary is a list that contains over 70 symptoms and specific conditions for which the nurse prescriber can prescribe. Acute musculoskeletal conditions for which an ENP can prescribe include:

1. Acute chronic back pain (uncomplicated);
2. Neck pain (uncomplicated);
3. Soft tissue injury;
4. Sprains;
5. Low back pain;
6. Knee pain;
7. Osteoarthritis;
8. Rheumatoid arthritis.

Independent prescribing is appropriate in the following circumstance:

1. The nurse works remotely from a doctor, seeing patients independently for those conditions listed in the nurse prescriber extended formulary (NPEF).
2. The doctor could see and treat other patients while the nurse sees some patients.
3. The nurse is competent to assess, diagnose and make treatment decisions for the patient.

It is not suitable for prescribing for complex medical conditions or for patients with several co-morbidities. From spring 2006, qualified extended formulary nurse prescribers and pharmacists who are independent prescribers will be able to prescribe any licensed medicine (except controlled drugs) for any medical condition. Extended formulary nurse prescribers will be known as nurse independent prescribers (NIPs) and supplementary prescribers as nurse supplementary prescribers (NSPs).

Specific instructions

The Department of Health define supplementary prescribing as a voluntary prescribing partnership between an independent prescriber and a supplementary prescriber to implement an agreed, patient-specific, clinical management plan with the patient's agreement. Once the patient has been diagnosed by the independent prescriber (who in this situation has to be a doctor) a clinical management plan (CMP) is required that has to involve both the independent and supplementary prescriber (in this case a rheumatology nurse). The Department of Health (2002) state that the CMP should:

1. Be patient-specific.
2. Be agreed by both the independent and supplementary prescribers before supplementary prescribing begins and supported by the intended recipient.
3. Specify the range and circumstances within which the supplementary prescriber can vary the dose, frequency and formulation of the medicines identified.
4. Specify when to refer from supplemntary prescriber to independent prescriber.
5. Contain relevant warnings about known sensitivities of the patient to particular medicines, and include arrangements for notification of adverse drug reactions.
6. Contain the date of commencement of the arrangment and a date for review (not normally longer than one year for patients with chronic conditions).

Supplementary prescribers can prescribe all medications currently prescribed by doctors with the exception of unlicensed medicines.

Supplementary prescribing is appropriate in the following circumstances:

1. Patients with long-term conditions, who can be managed by a nurse or pharmacist and some allied health professionals (AHPs) between reviews by the doctor.
2. Nurses or pharmacists (and some AHPs) who are competent to manage the patient's condition.
3. There is a close working partnership between the independent prescriber (doctor) and the supplementary prescriber, who have access to the same patient record.

Supplementary prescribing is not suited to an urgent prescribing situation because an agreed CMP is required before prescribing can begin.

Method of Teaching

Potential nurse prescribers must undertake a Nurse and Midwifery Council (NMC) approved course at level three (first degree level) consisting of at least 26 taught days plus 12 days learning and assessment in practice working with a designated medical practitioner.

Professional Activities

As with all areas of practice, nurse prescribers must act within their own level and area of competence acknowledging the scope and limitations of their role. In order to promote safe and effective practice there must be clearly defined policies, support mechanisms and audit systems. The National Prescribing Centre (2002) have produced a useful resource reminding nurses of their legal and professional accountability and duty of care.

Management

National research evaluating the first two years of extended formulary prescribing demonstrated that it was well received by nurses, patients and doctors and had a positive impact on patient care because of improved access. Nurse prescribers felt that:

1. They were confident in their prescribing practice;
2. Extended prescribing had a positive impact on patient care and enabled nurses to make better use of their skills;
3. The limited nursing formulary imposed unhelpful limitations on their practice;
4. They had received support from their medical practitioner.

An audit of the prescribing practice of rheumatology nurses in one rheumatology department over a 10-month period demonstrated that the majority of patients requiring a prescription were patients with RA (81%) and the commonest drug prescribed was methotrexate (35%). All patients were satisfied with specialist nurses issuing a prescription. The specialist nurses have a greater sense of job satisfaction and are autonomously able to complete episodes of care. It also provides medical staff with more time to manage patients with complex medical needs.

Clinical management plan

Name of patient: Mrs B. Davenpot

Patient identification e.g. ID number, date of birth: 04/04/1944

Patient medication sensitivities/allergies: penicillamine

Current medication: diclofenac 75mg BD; co-codamol.

Medical history: Hypertension

Independent prescriber: Dr H.

Contact details: tel./e-mail/address

Supplementary prescriber: Nurse X.

Contact details: tel./e-mail/address

Condition(s) to be treated: active rheumatoid arthritis

Aim of treatment: to suppress disease activity

Treatment plan

1. Commence methotrexate 7.5 mg weekly for eight weeks.
2. After eight weeks increase methotrexate by increments of 2.5 mg every fortnight until a dose of 15 mg in reached.
3. Commence folic acid 5 mg daily.
4. Observe full blood count (FBC), AST and ALT weekly for the first month and then monthly.
5. Observe for dry cough or breathlessness.

Indication: methotrexate

Preparation: oral

Dose schedule: as above

When and who to refer to: Dr H. if change, no improvement (after four months) or if adverse event occurs. Guidelines supporting treatment plan:

1. Inhouse protocols.
2. Current BNF.
3. Oxford Handbook of Rheumatology.

Review and monitoring requirements: weekly for the first month and then monthly.

Process for reporting adverse drug reactions (ADRs): yellow card.

Documentation and record keeping: hospital case notes and patient's computerized record.

Name and agreement of independent prescriber/supplementary prescriber, date, date agreed with patient.

INDEPENDENT PRESCRIPTION

If the treatment of RA is to be effective, patients must have the ability to adjust drug administration according to the changing activity of their disease. It is vital in patients with chronic inflammatory conditions who take medications for a long period of time as there is a clear need to assess both the safety and efficacy of the drug regime, especially as it is the patient who will be responsible for administering prescribed medications at home. Healthcare professionals can no longer assume that people are passive recipients of care, and patients will require information to administer medicines safely and effectively. Beardsley, Anderson Johnson and Kabat (1983) define drug self-administration procedures as specific education strategies that contain the necessary knowledge and behavioural components to effect better compliancy.

Case for Self-medication

The conventional system of drug administration from a ward trolley often neglects to address the need for individual education nor does it seek to prepare patients to administer drugs within their own home environment following discharge. All the nurse has to do is to administer and sign that the prescribed drug has been given. This is despite the fact that health education has become a major part of the nurse's role. This traditional system frequently provides dosages at incorrect times and unfortunately is not error-free. It should no longer be acceptable to the profession and we should aim to provide patients with a personal pharmacy.

Webb (1990) states that the traditional approach of batch processing suits staff convenience rather than being governed by patient need; whereas a holistic humanitarian approach to nursing care necessitates considering people as individuals and providing care in partnership with them. An important but often neglected area of discharge planning is whether a patient is being discharged home without adequate knowledge of their tablets or with containers that they cannot open. The NMC has recommended that self-administration projects be established during a hospital stay to provide patients with the necessary knowledge and confidence to continue with a high degree of compliancy on discharge. Many patients, particularly those in the older age group, are taking medications when they are admitted to hospital. Quilligan (1990) offers a framework for ensuring that older patients are given enough support in learning about their tablets. It follows the structure of the nursing process:

Patient assessment

1. Does the patient understand their condition and the purpose of their medications?
2. Does the patient want to learn more about their medications and are they able to do so?
3. Can the family be involved in the process?

Planning and implementation

1. Use realistic joint goals on agreed topics and if possible conduct the sessions when the family can be present.
2. Limit the teaching to 3 topics in a 15-minute-maximum session.
3. Discuss the patient's worries and check previous knowledge.

Observations

1. Ongoing assessment of managing tablets and the teaching programme is required.

Benefits

Patients have the chance to familiarize themselves with their medications and the opening of the containers/packaging. They would be able to discuss any needs or concerns whilst the nurse would have the opportunity to assess patient knowledge, evaluate previous teaching and identify those patients who will need ongoing support from the community services. A programme of self-medication may help to reduce the number of drug errors as it would incorporate extensive patient education and provide the opportunity for supervised practice of drug administration. Nurses reported that those patients who had self-medicated in hospital had a better understanding of all aspects of their drug regime.

Work by Scrivin and Bryan (1987) found that patients were very appreciative of a self-medication programme and medication errors reduced from 17.9% to 6.9% after the programme had been introduced. Also, if patients experience accidental overdoses it is safer for this to occur in the hospital rather than the home environment. Self-medication returns control to the patient, promoting comfort and demonstrating trust. A self-medicating patient will not be woken at 6 a.m. or stay awake until after midnight for sedation, or seek out a nurse for analgesia. Corrigan (1989) found that patients would question if the tablets looked different and were likely to refuse steroids until after breakfast. Webb (1990) found that most patients took in the process of self-medication very quickly, feeling unrushed over administration and enjoyed having something positive to do.

Points to Remember

1. Planning – this is often considered the most important stage and cannot be rushed. It is vital to the overall success of the programme to gain the cooperation and support of all those involved including the ward staff, doctors and (crucially) the pharmacist.

2. Assessment – in some units this will be a dual process involving both the pharmacist and the nursing staff. The patient will be assessed on their understanding as to the purpose of their medications, their safety in administration (i.e. timing and dosage), potential side effects to report and their psychomotor skills in the handling and reading of their medications. It is at this stage that the nurse will enter into a therapeutic partnership with the patient, offering support and education on an individualized basis to both the patient and their family as required.
3. Accountability can be a major concern to nurses when they consider implementing a self-medication programme. It can generate concerns of losing control over drug administration whilst still holding the same degree of responsibility. In fact, the nurse is not losing any element of control but is able to utilize the skills of educator, guide and supporter in a more constructive manner to ensure that patients are safely and knowledgeably administering their tablets. It could be argued that there are more concerns about accountability in the traditional system of drug administration as nurses could be engaging in the provision of medicines without ascertaining the patient's knowledge and understanding of the therapy. Bird (1990) states that the anticipated problems that nurses fear of patients forgetting to take their tablets, taking too many tablets or gaining access to other people's drugs seldom occurs in practice.
4. Implementation – in some units a staging process is used. The assessment process will determine the level the patients enter the programme and what degree ofsupervision is required – for example:
   (a) Level One – the nurse administers the medicines providing full explanation.
   (b) Level Two – the patient administers the medicines with nurse supervision.
   (c) Level Three – the patient administers the medicines without supervision.

   Some areas have introduced memory aids in the form of calendar cards, which have been shown to lead to fewer errors when compared with written or verbal instructions.
5. Evaluation – various degrees of supervision have been used, generally including at least daily discussions and observations of the patient's progress. Evaluation should include the obtainment of both the patient's and the nurse's views on progress.

SUBSTITUTES

Alternative and complementary medicine is the aggregate of diagnostic and therapeutic practices and systems that are separate from, and in contrast to, conventional scientific medicines. A third to almost 100% of patients with RA admit to using some form of complementary therapies. Reasons for using complementary therapies include dissatisfaction with conventional medicines and the need for psychological support. Many nurse practitioners actively recommend complementary treatments to patients.

Nutrition

The extensive research into dietary treatments of RA has been reviewed by Buchanan *et al.* (1991) and the results remain inconclusive. Individual dietary manipulation may be beneficial to selected patients but it is difficult to generalize from these findings and further study with sound methodological design is required. Patients are extremely interested in this area and invest both time and money, striving to obtain relief from their symptoms. Dozens of publications vie with each other to suggest yet another method of ridding sufferers of pain and inflammation - often contradicting each other. Trials have shown that the green-lipped mussel extract (Seatone) does not work but its sales continue. Surveys have found that many patients with RA consume diets that are marginally inadequate in several essential nutrients.

There is no sound evidence that the pharmacologic doses of vitamins used to treat arthritis have definite therapeutic efficacy. However, there is work to suggest that vitamin C, and perhaps vitamin E as an oxidant, could slow the progression of osteoarthritis. RA is associated with moderate hypochromic normocytic anaemia, which is caused by a reduction in endogenous iron metabolism rather than dietary deficiency. Abnormalities in iron metabolism tend to be corrected as disease activity is suppressed.

Diets rich in fish oil containing N-3 fatty acids, along with a reduction of N-6 fatty acid intakes, have been associated with improvements in pain and stiffness. Some authors claim that fish oils will help inflammation by reducing arachidonic acid and leukotrienes production. Dietary fish oils may prove to be more effective when used in combination with specific anti-inflammatory agents.

Dietary advice

For patients with OA:

1. Reduce carbohydrates and fats.
2. Keep weight down.
3. Exercise regularly.

For patients with RA:

1. Eat a well-balanced diet that contains minerals and vitamins.
2. Eat more fruit and vegetables.
3. If weight loss is experienced, increase carbohydrate intake.
4. Seek advice regarding drug therapy.

For patients with gout:

1. Advice on weight management.
2. Reduction of high purine foods.
3. Reduction in alcohol intake.

Massage

Body massage is a method of manually manipulating the tissues of the body by either stroking, percussion or applying pressure. These actions increase the circulation of blood and lymph, and induce relaxation by soothing sensory nerve endings in the skin. There have been verbal reports from patients with RA and fibromyalgia that this form of intervention reduces pain and enhances sleep but evaluation from a critical perspective remains limited.

Aromatherapy

Aromatherapy is the use of essential oils to promote healing in both a physical and a psychological manner. Methods of application include massage, bathing, compresses and steam inhalations. Aromatherapy professes to provide pain relief, reduce inflammation and to maintain joint mobility in patients with RA. Advice must be sought from a qualified aromatherapist before treatment can be instigated.

Reflexology

Reflexology is a form of massage applied to the reflex areas present in both the feet and hands. It is suggested that these reflex areas correspond with a specific area of the body and can stimulate the natural healing powers of the body. Reflex areas suggested to ease the symptoms of RA include:

1. The diaphram to reduce muscle tension.
2. The shoulder and arm to enhance circulation and nerve conduction.
3. The hip and leg to aid healing.
4. The spine to improve flexibility and to balance the nervous system.
5. The solar plexus to promote deep breathing and relaxation.
6. The parathyroid glands to maintain homeostasis with both calcium and potassium levels.
7. The liver to remove toxins from the body.
8. The adrenals to stimulate cortisone production, maintain mineral balance and enhance muscle tone.

Acupuncture

Acupuncture is believed to change the vital energy which flows in the body and connects the internal organs with the superficial parts of the body. It is currently used for musculoskeletal conditions by orthodox, complementary and alternative practitioners.

Herbal Medicine

Herbal derivatives are widely used in orthodox rheumatology - for example:

1. Salicylates (willow bark).
2. Colchicine (Autumn crocus).
3. Opiates (opium poppy).
4. Quinine (cinchana bark).

Controlled trials of other herbal treatments have proved inconclusive. Herbal preparations may be taken as tablets, infusions - such as tea - tinctures or added to baths. Popular herbs for rheumatism include aloe vera, comfrey, devil's claw, feverfew and evening primrose oil.

Naturopathy

This is a mixture of traditional folk wisdom, empiricism and selections of biomedical science. The basic principles are:

1. The healing power of nature is fundamental;
2. Treat the cause rather than the effect;
3. Ill health results from a lowering of resistance due to diet, stress etc.

Holism

Treatment involves a combination of counselling, nutritional advice, herbal medicine and homeopathy remedies.

GLUCOSAMINE

Biological compounds such as hyaluronans, chrondroitin sulfate and glucosamine are now used routinely in the treatment of osteoarthritis. Although not yet proven these compounds may be potentially chondroprotective, in that they may favourably modify the natural progression and course of OA. Glucosamine is a natural substance and a building block of the glycosaminoglycans (CAG) and glycoproteins in the ground substance of the articular cartilage. Glucosamine is usually taken in a daily dose of 1500 mg. Adverse effects are likely to be mild and predominantly affect the GI tract and are reversible on discontinuation of glucosamine.

Concerns about glucosamine affecting glucose homeostasis, increasing insulin resistance and/ or impairing insulin secretion appear not to be founded. Research by Scroggie, Albright and Harris (2003) and Tannis *et al.* (2004) found that normal doses of glucosamine supplemetation did not cause glucose intolerance in healthy adults. Randomized trials comparing glucosamine to NSAIDs suggest that glucosamine produces similar symptomatic benefits as NSAIDs but with fewer adverse effects. A meta-analysis comparing the structural and symptomatic efficacy of glucosamine in relation to placebo found an improvement in symptoms as well as in joint space narrowing. Two randomized controlled studies have demonstrated that patients prescribed glucosamine had no significant progression of cartilage loss, suggesting that glucosamine may indeed modify the natural radiological progression of OA of the knee.

Literature Available

Glucosamine is a dietary supplement that may be effective in the treatment of osteoarthritis.

Symptoms of osteoarthritis

Healthy cartilage covers the ends of your bones within your joints and helps with smooth and pain-free movement. Osteoarthritis occurs when the healthy cartilage starts to wear out causing the bones to rub together. The joint lining may become inflamed, leading to swelling. Cartilage is made up mainly of water, cells called chondrocytes and other substances such as collagen and proteoglycans. These substances give the cartilage its elastic properties and the ability to absorb shock, especially in hips and knees. As we grow older we produce less of these substances and the cartilage wears down.

Symptoms of glucosamine

Glucosamine is a natural substance that helps to provide the building blocks within cartilage. It helps to produce and maintain healthy cartilage. Exactly how it benefits the cartilage in osteoarthrits is unknown. Studies have looked at the effect of glucosamine in the treatment of osteoarthritis. It has been found to reduce pain and improve function in some people. There is also some evidence that it may slow the progression of the condition.

As with other forms of medication and supplements, everyone can respond differently, and there is no guarantee that it will help. There is no proven evidence that glucosamine helps in other types of arthritis. Glucosamine is generally well tolerated. Common side effects include nausea, diarrhoea and stomach upset which resolve on stopping glucosamine. There is very little information available about whether there are any increased side effects taking it long term. Always read the information provided with glucosamine. The current recommended dose is 1500 mg (1.5 g) of glucosamine sulfate per day. Some doctors will prescribe it. It is also available over the counter in pharmacists and health food shops. Benefit can be variable and it is worth taking for at least six months before assessing whether it has been helpful.

CAPSAICIN

Capsaicin (trans-8-methyl-*N*-vanillyl-6-nonenamide) is the alkaloid which makes chillies hot. Interest has centered on the use of capsaicin as a topical analgesia for a variety of conditions characterized by pain that are not responsive to analgesia or NSAIDs. Local application of capsaicin to the peripheral sensory endings in the skin results in depletion of substance P from the neurone, both peripherally and centrally. One randomized placebo-controlled trial in osteoarthritis has reported an improvement in pain levels.

3

Clinical Aspects of Hypertension

MEANING OF HYPERTENSION

1. Stage 1 (mild) hypertension is common worldwide and affects more than 60 million Americans, with approximately 2 million new cases diagnosed annually.
2. The prevalence of high BP in African Americans is among the highest in the world, and organ damage occurs earlier in these patients than in whites.
3. The JNC has reclassified BP into three stages; stage 4 has been deleted. When systolic BP and diastolic pressures fall into different categories, the higher category should be selected to classify the patient's BP status.
4. Proper technique for BP measurement is essential for adequate diagnosis and patient care.
5. Except when the initial BP reading corresponds to stage 3, the diagnosis of hypertension grading is based on an average of two or more readings taken at each of two or more visits during the next 1-8 wk, after the initial reading.
6. In addition to classifying stages of hypertension on the basis of BP readings, it is generally accepted that systolic hypertension is as important as diastolic hypertension.
7. The results of the Hypertension Optimal Treatment (HOT) trial indicate that, in patients with diastolic BP mean 105 mmHg, it is safe to decrease the diastolic BP to <90 mmHg; further lowering to <85 mmHg did not result in a significant decrease in mortality or event rates but was not harmful. In diabetic patients, a decrease in diastolic BP to ≤80 mmHg resulted in a 51% reduction in major cardiovascular events compared with target group ≤90 mmHg.
8. *Isolated systolic hypertension in older patients.* Approximately 64% of Americans over the age of 65 yr are hypertensive. Isolated systolic hypertension is defined as systolic BP > 140 mmHg, diastolic BP < 90 mmHg present in the absence of target organ damage and coexisting disease. Isolated systolic hypertension in individuals aged over 65 yr is a major cause of stroke, left ventricular failure (LVF), and mortality from IHD.

9. In approximately 95% of individuals, hypertension exists without a known cause and is termed primary (essential) hypertension. In the remaining 5%, hypertension is secondary to known causes.

Table 3.1 Causes of secondary hypertension

Cause	%
Renal parenchymal disease	3
Renal vascular disease	1
Cushing' s syndrome	0.1
Pheochromocytoma	0.1
Primary hyperaldosteronism	0.1
Coarctation of the aorta	0.1
Estrogens	0.4
Alcohol	0.2 or more

NATURAL THERAPY

Natural therapy should be tried rigorously before drug therapy in all patients with mild hypertension. Nondrug therapy–low-sodium diet, weight reduction, cessation of smoking, reduction in alcohol intake, removal of stress and/or learning to deal with stress, relaxation, exercises, and a potassium-enriched diet–may result in adequate control of hypertension in up to 40% of patients with stage 1 or isolated systolic hypertension in the elderly. In addition, low saturated fat intake is often necessary because of coexisting hyperlipidemia, which increases risk.

Preparation

1. Serum cholesterol and high-density and low-density lipoprotein (HDL and LDL) cholesterol.
2. Complete blood count and platelets.
3. Urinary and plasma metanephrine and normetanephrine if pheochromocytoma is suspected.
4. Electrolytes, blood urea, creatinine, estimated GFR, and uric acid.
5. Chest radiography and electrocardiography.
6. Urinalysis.

Selection of Drugs

1. It is important for the physician to consider the efficacy and the pharmacologic and adverse effects of the antihypertensive agent to be chosen, as well as the cost to the patient and the ability of the drug to prolong life.
2. The physician should strive for monotherapy in the treatment of mild primary hypertension whenever possible. If the choice of initial agent is based on age and ethnicity, about 45% of patients attain goal BP with monotherapy.
3. The combination, however, of two agents at low dose may achieve the therapeutic goal with less potential for adverse effects.

4. It is advisable to give a trial of individual agents before using combinations of drugs or fixed-dose combinations.
5. The patient should have a thorough understanding of the problems associated with drug therapy to facilitate acceptance and compliance during medication changes.

Recommendations

Mild primary hypertension in younger patients

There is abundant evidence from RCTs that whites and African Americans differ in their response to antihypertensive agents. It may appear old-fashioned to recommend beta-blockers and diuretics as first-line agents in the new millennium, but the scientific evidence gleaned from RCTs reemphasizes their efficacy, safety, and costs.

1. A metaanalysis by Staessen et al. 2001 and ALLHAT indicates that new drugs are not more effective, they do not have more beneficial effects, and, importantly, calcium antagonists significantly increase the risk of HF.
2. Alpha-blockers increase the risk of HF and have deleterious effects on the cardiovascular system. Thus, they have little or no role in the management of hypertension as emphasized in the second edition of this book in 1998. A warning was issued by the American College of Cardiology following the ALLHAT findings in 2000.
3. An RCT by Materson and colleagues and the HANE Study both indicate that beta-blockers are effective in younger whites.

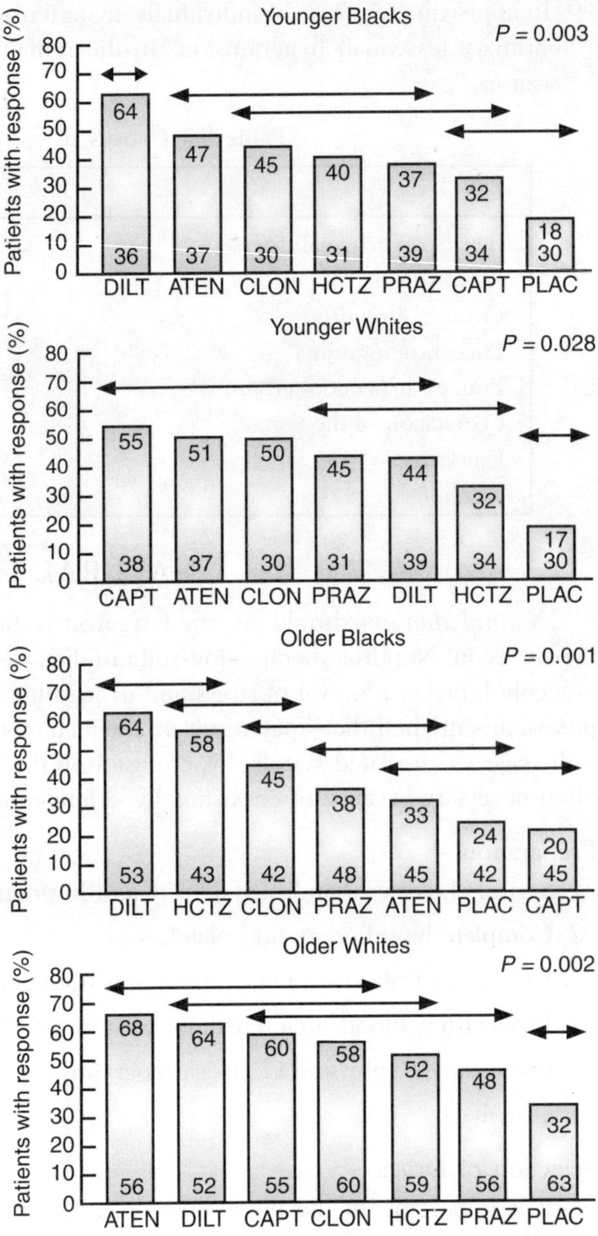

Fig. 3.1. Younger black patients, younger white patients, older black patients, and older white patients with responses in each of the study groups. The arrows group the drugs whose effects do not differ from each other by more than 15%.

4. ACE inhibitors are more effective than diuretics in younger whites.

For younger black patients

Calcium antagonists are the most effective agents, followed by beta-blockers; the latter may be tried first because these agents are safe and inexpensive and are particularly useful in patients with diabetes.

Table 3.2 Choice of drug for the treatment of mild primary hypertension based on age and ethnicity

White patients younger than age 65

1. Beta-blocker: bisoprolol, carvedilol, or metoprolol (Toprol XL) preferred over other beta-blockers: monotherapy success in ~65% in this age and ethnic category: *Warning:* avoid atenolol
2. ACE inhibitor: monotherapy success in ~62%; often requires combination with diuretic to achieve goal BP
3. Diuretic: monotherapy success in ~32%
4. Beta-blocker or ACE I/ARB + diuretic
5. Calcium antagonist: success in ~58%

Black patients younger than age 65

1. Calcium antagonist: monotherapy success in ~70%
2. Beta-blocker (bisoprolol, carvedilol, metoprolol extended release [Toprol XL]): success in ~51%; avoid atenolol
3. Calcium antagonist + beta-blocker success in >90% (avoid verapamil)
4. Diuretic: success in ~48%

White patients older than age 65

1. Diuretic: monotherapy success in ~68% (first choice because of safety)
2. Beta-blocker: carvedilol, bisoprolol, metoprolol extended release; avoid atenolol; success expected in ~72%; second choice because safer than calcium antagonists in the elderly
3. ACE inhibitor: success in ~62%
4. Calcium antagonist (avoid verapamil): success in ~72%

Black patients older than age 65

1. Diuretic: success in ~64%; preferred over calcium antagonist heart failure risk
2. Calcium antagonist (avoid verapamil): success in ~85%
3. Diuretic + calcium antagonist

Mild hypertension in older patients

Recommendations for white patients older than age 65

There is no sound evidence to support the WHO, JNC, and other national committees' choice of calcium antagonists for the management of isolated hypertension in elderly whites. Beta-blockers are as effective and are safer and much less expensive. Beta-blockers are safe in patients aged 65-79 yr; and the safety and ability of these drugs to prevent HF and death have been endorsed in two RCTs: CAPRICORN and COPERNICUS. In the excellent meta-analysis by Staessen and associates, the authors concluded: "All antihypertensive drugs have similar long-term efficacy and safety," but in the same article indicated that new drugs significantly increased the risk of HF. HF is a common cause of morbidity, hospitalization, and death; effective control of hypertension is necessary to stem the epidemic of HF. In the Intervention as a Goal in Hypertension Treatment (INSIGHT) study, among primary end points HF and MI were statistically more frequent in the Adalat XL group.

In the Prospective Randomized Amlodipine Survival Evaluation (PRAISE) study, pulmonary edema was more frequent in the amlodipine-treated group. In a non-Q-wave MI study, diltiazem

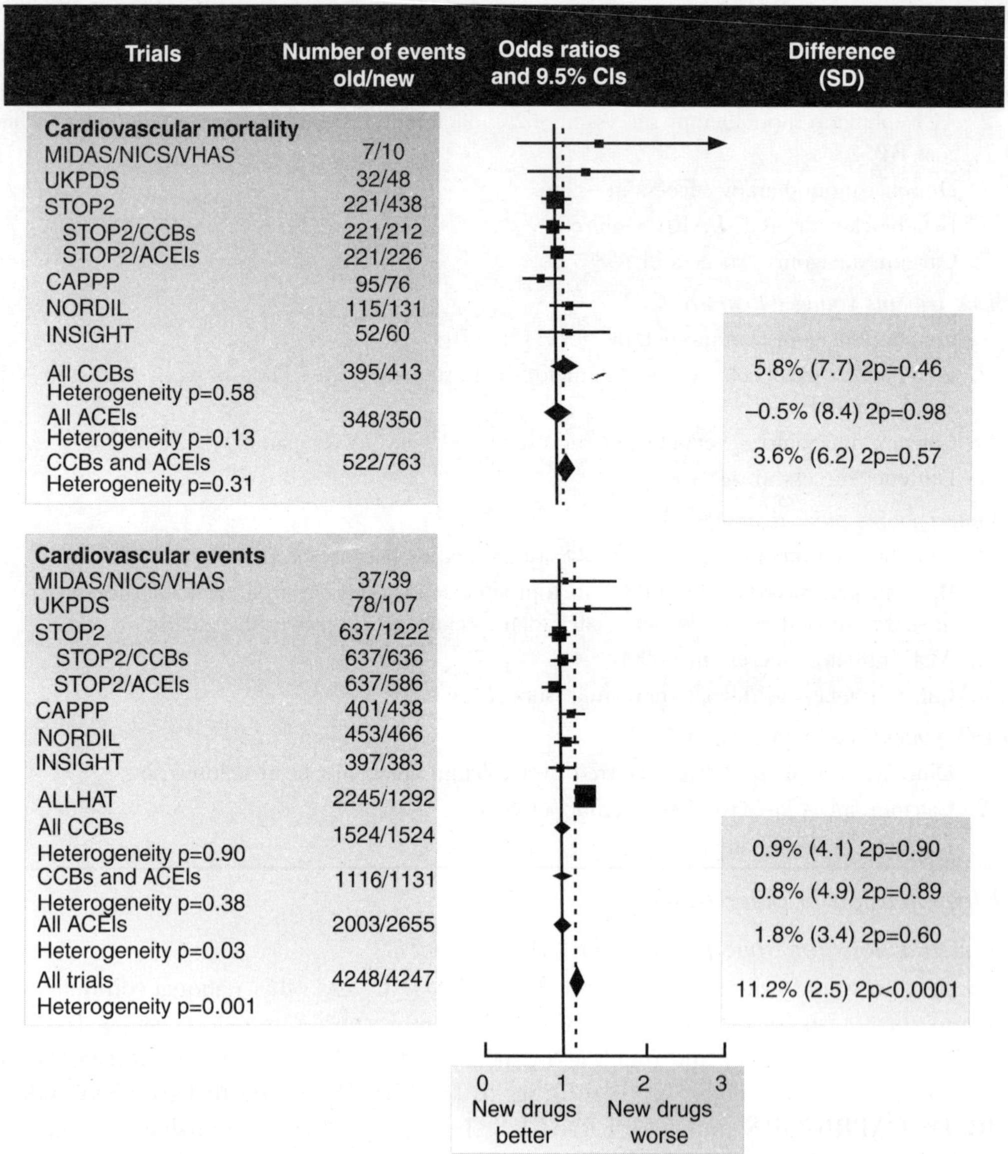

Fig. 3.2. Effects of antihypertensive treatment on cardiovascular mortality and all cardiovascular events in trials comparing old with new drugs.

increased the risk of HF. These findings are not surprising because calcium antagonists possess significant negative inotropic effects. Elderly hypertensive patients are at a higher risk of HF and MI than younger patients; thus, although calcium antagonists are highly effective antihypertensives in the elderly, beta-blockers, diuretics, and ACE inhibitors are preferred because these drugs have proved useful in the prevention and management of HF. The combination of ACE inhibitor and beta-blocker is not considered complementary, because these drugs do not enhance antihypertensive effects. However, they are both cardioprotective and deserve a trial.

1. The JNC advises a diuretic, as preferred, based on the results of the Systolic Hypertension in Elderly Programs (SHEP) study.

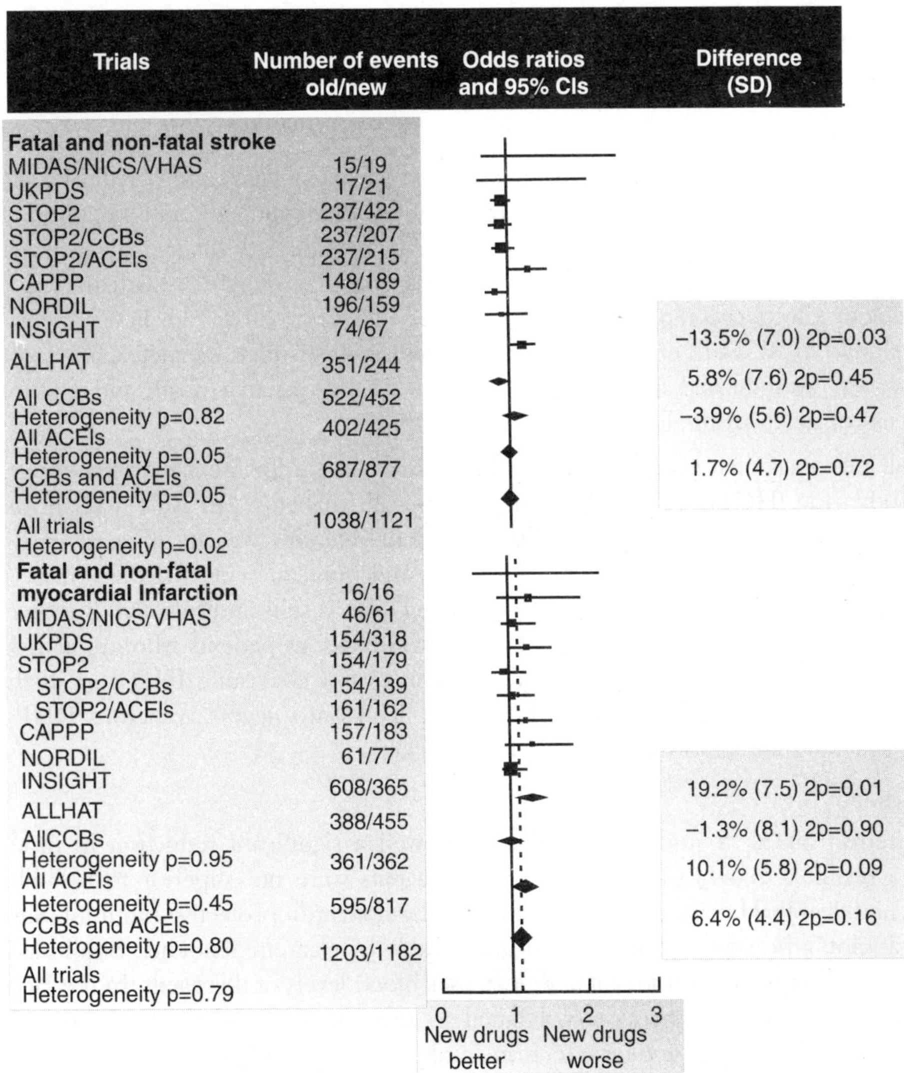

Fig. 3.3. Effects of antihypertensive treatment on fatal and nonfatal stroke and myocardial infarction in trials comparing old with new drugs.

2. In the SHEP study, atenolol was the second agent used, after a diuretic. After 5 yr of follow-up, IHD was reduced by 25%, with a major reduction in stroke. The Swedish Trial in Old Patients with Hypertension (STOP) in the elderly showed a significant reduction in mortality and strokes as a result of beta-blocker therapy.
3. A calcium antagonist, long-acting dihydropyridine (DHP), is recommended by the JNC as an alternative to diuretics based on the results of the Systolic Hypertension in Europe (SystEur) trial: nitrendipine, a long-acting DHP, was combined when required with diuretic or enalapril. The treated group showed a 42% reduced occurrence of all strokes; total mortality and cardiovascular mortality rates were not significantly reduced ($p = 0.07$; $p = 0.22$). The trial was not done solely with nitrendipine; follow-up was not sufficiently long to determine the absolute safety of DHPs in the elderly, who may have undetected IHD and are at high risk of HF and IHD events.

Recommendations for elderly black patients

A diuretic and calcium antagonist have been shown to have better antihypertensive effects than beta-blockers and ACE inhibitors and are recommended. However, caution is required in patients with ejection fraction < 45% because the risk of HF may increase with long-term calcium antagonist therapy. Thus, a diuretic is the first choice in elderly patients of African origin. The following RCTs are relevant to decision-making strategies in the elderly 65-80 yr old. Few patients over age 80 were included in RCTs to enable us to draw conclusions; a subgroup metaanalysis of RCTs in patients over age 80 indicated no treatment benefit for cardiovascular death and a nonsignificant relative excess of death from all causes.

The calcium antagonist isradipine caused an increase in cardiovascular events compared with hydrochlorothiazide (HCTZ), particularly in patients with diabetes. An RCT with nisoldipine in patients with hypertension and non-insulin-dependent diabetes was terminated early because nisoldipine-treated patients had a higher risk of fatal and nonfatal acute MI than observed in the enalapril group: 25/235 versus 5/235 ($p < 0.00$ 1). These results indicate that sustained-release calcium antagonists may increase the risk of cardiac mortality in patients who are at risk of IHD events, particularly in those with diabetes, dyslipidemia, and coexisting IHD and in the elderly with silent ischemia or undetected IHD. Long-term RCTs are necessary before DHPs may be considered safe alternatives to diuretics and beta-blockers.

Beta-blockers

The Swedish STOP 2 study in the elderly showed a significant reduction in mortality and strokes as a result of beta-blocker therapy. Newer agents were not superior to beta-blockers or diuretics. The beta-blockers used were atenolol and the noncardioprotective pindolol. Some RCTs used propranolol, a beta-blocker influenced unfavorably by cigarette smoking. Smoking interferes with hepatic metabolism of propranolol and decreases blood levels of this agent. In the Beta-Blocker Heart Attack Trial (BHAT), propranolol failed to prevent fatal and nonfatal MI in smokers. Propranolol and oxprenolol lose their effects in smokers; metoprolol and timolol have been shown to prevent fatal MI, nonfatal MI, and sudden cardiac death in smokers and nonsmokers. In addition, the physician must be aware of the reasons for poor drug control of BP in the treated patient.

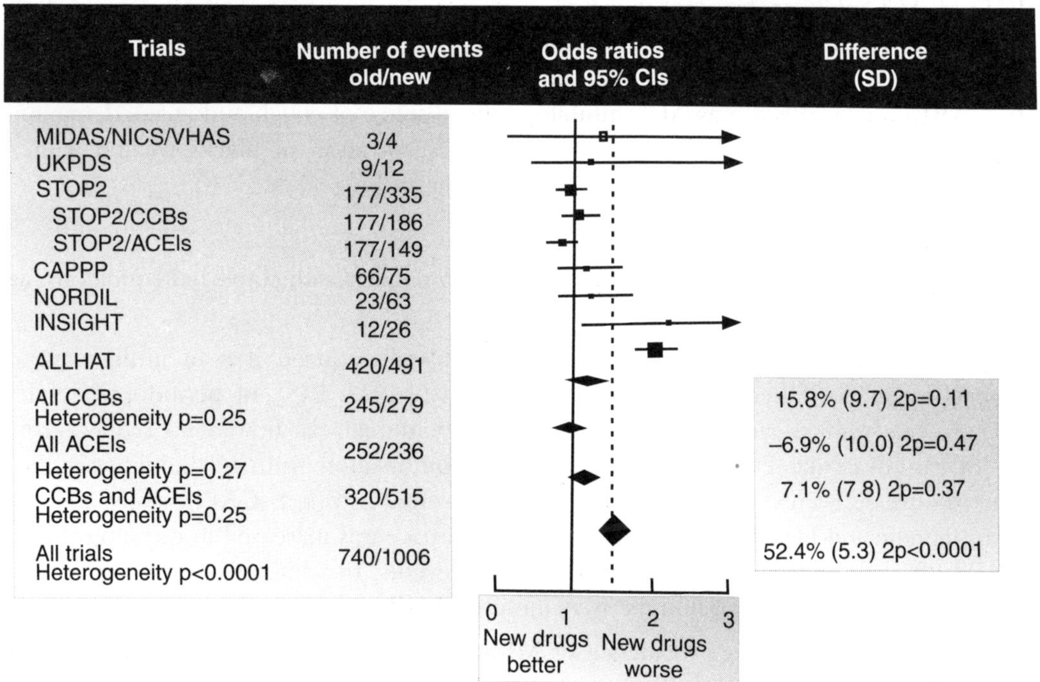

Fig. 3.4. Effects of antihypertensive treatment on fatal and nonfatal congestive heart failure in trials comparing old with new drugs.

Table 3.3 Reasons for poor drug control of blood pressure

1. Poor compliance
   Inconvenient drug dosing, e.g., b.i.d. or t.i.d.
   Inability to purchase drug (financial)
   Adverse effect of drug
2. Related to drug
   Inappropriate drug selection or drug combination
   Interactions with NSAIDs, nasal decongestants, cocaine
3. Volume overload
   Diuretic required
   ↑Na^+ intake
   Renal failure
4. Weight gain
5. Alcohol intake >4 oz/d
6. Renovascular hypertension
7. Other causes of secondary hypertension
8. Cuff size
9. White coat syndrome
10. Pseudohypertension

RCTs of *ACE inhibitors* have not shown these drugs to be superior to beta-blockers or diuretics. Importantly, RCTs have indicated that ACE inhibitors rarely achieve goal BP without the addition of a diuretic. In the large PROGRESS trial, only 42% achieved goal BP; 58% required addition of diuretics. ARBs are as effective as ACE inhibitors; the absence of cough and angioedema and the better patient tolerance are advantages; angioedema is common in blacks treated with ACE inhibitors, and if costs are not a concern ARBs are preferred.

Chemotherapy

In the presence of coexisting disease and target organ damage with comorbid conditions, agents are recommended as follows:

1. *Stroke or transient ischemic attack.* BP must be cautiously reduced, if at all, in the acute stage. A beta-blocker or diuretic or combination is advisable. An RCT of perindopril 18 among 6105 patients with previous stroke or transient ischemic attack showed no reduction in the risk of cardiovascular events or stroke, but the combination with indapamide significantly reduced adverse events. In the large Captopril Prevention Project (CAPP) study of captopril versus diuretic and beta-blocker, fatal and nonfatal stroke was increased in the captopril group ($p = 0.044$), with no significant differences in other events. In addition, some captopril-treated patients required addition of diuretic to achieve goal BP.
2. Diabetic patients have a high incidence of MI and cardiac death; thus, both ACE inhibitors and beta-blockers are strongly recommended. The UKPDS study showed beta-blocker therapy to be as effective as captopril in reducing the risk of macrovascular and microvascular complications in type 2 diabetes. Importantly, glycemic control was not different between groups.
3. *Angina, after MI, or suspected IHD.* Beta-blockers are strongly recommended. The JNC also recommends calcium antagonists, but the author qualifies this with the following statement: use diltiazem sustained-release preparation (e.g., Cardizem CD) or amlodipine if beta-blockers are contraindicated. Other DHPs are advisable only if a beta-blocker is being used in combination. A DHP used without a beta-blocker may increase mortality if unstable angina or acute MI supervenes.
4. *Severe depression.* Diuretics or ACE inhibitors are recommended; a calcium antagonist may be used in the absence of diabetes. Beta-blockers may increase depression, but this effect is rare and does not contraindicate the use of beta-blocker therapy if needed. Reserpine, methyldopa, clonidine, and other central alpha-agonists are contraindicated.
5. *Atrial fibrillation or other arrhythmias.* Beta-blockers and ACE inhibitors are strongly advisable. Hypertension is the leading cause of atrial fibrillation and HF.
6. *Aortic and other aneurysms.* Beta-blockers are strongly indicated because they decrease ejection velocity and shearing stress in arteries; thus, they may provide some protection from rupture or further expansion of the aneurysm.
7. *Diabetes type 2 without proteinuria.* The JNC recommends diuretics. In the Antihypertensive and Lipid-Lowering Treatment to Prevent Heart Attack Trial (ALLHAT), diuretics were as effective as ACE inhibitors and calcium antagonists in decreasing CVD outcome.

8. *Dyslipidemia.* During the years 1988 to 2001 prior to the results of ALLHAT, which indicated that the alpha-blocker doxazosin caused a high incidence of HF, the JNC recommended alpha-blockers with no alternatives. Many internists agreed and promoted this fallacy.
9. Alpha-blockers may increase cardiovascular mortality and have not been shown to prevent or reduce LVH. Alpha-blockers were recommended by the JNC because these agents do not unfavorably alter lipid levels, but alpha-blockers have undesirable cardiovascular effects and have not been shown to prolong life. Patients with dyslipidemia are at high risk of IHD events. Because beta-blockers are strongly recommended by the JNC for patients at high risk of IHD events, these agents are the logical choice. If beta-blockers are contraindicated, an ACE inhibitor and low-dose diuretic are advisable.
10. *Congestive heart failure (CHF) classes I-II and III*: Diuretics plus ACE inhibitor plus a beta-blocker. In this subset, beta-blockers have been shown in RCTs to prolong life and to reduce the risk of hospitalization. Thus, beta-blockers are strongly indicated for patients with LV dysfunction, ejection fraction 25-45%.
11. *Preoperative hypertension.* Beta-blockers are recommended by the JNC. Atenolol and bisoprolol have proved successful in RCTs.
12. Renal insufficiency, except renovascular hypertension, creatinine clearance < 265 μmol/L, 3 mg/dL: ACE inhibitor plus loop diuretic is indicated, but a beta-blocker and/or calcium antagonist may be necessary to achieve goal BP; see later discussion of *metolazone.*
13. *LVH.* Beta-blockers and/or ACE inhibitors are preferred. Avoid alpha-blockers. Calcium antagonists and alpha-blockers have not been shown to reduce LV mass consistently.
14. *Liver disease.* Avoid methyldopa and labetalol; the latter may cause hepatic necrosis.
15. *Stasis edema or varicose veins with edema.* Avoid calcium antagonists, particularly the DHPs.
16. Renal vascular hypertension. Avoid ACE inhibitors and ARBs.
17. *Gastroreflux syndrome*: Avoid calcium antagonists because they may increase reflux.
18. *Gout.* Avoid diuretics.
19. *Sensitivity to sulfonamides.* Avoid all diuretics except ethacrynic acid and amiloride.
20. *Severe peripheral vascular disease.* Avoid beta-blockers; these agents are not contraindicated in patients with mild or stable peripheral vascular disease because these patients often have coexisting IHD, and beta-blockers may ameliorate symptoms and prolong life in this subset.
21. *Renal calculi.* Avoid triamterene because this agent may precipitate calculi.
22. *Prostatism.* An alpha-blocker has a role except when IHD or HF is present.
23. *Migraine.* A nonselective beta-blocker is the most logical agent because vaso-constriction may relieve symptoms.
24. *Osteoporosis.* Thiazides have been shown to increase bone mass.
25. *Mitral valve prolapse.* Beta-blockers are the obvious choice.
26. *Hyperthyroidism.* Beta-blockers are recommended for the management of arrhythmias; propranolol is the most effective.

27. *Cyclosporine-induced hypertension.* Calcium antagonists are of value.
28. *Pregnancy.* Avoid ACE inhibitors and ARBs. Methyldopa, atenolol, or propranolol may be indicated at certain periods. For preeclampsia, a diuretic should be avoided. For eclampsia, hydralazine or labetalol is recommended.
29. *Subarachnoid hemorrhage.* Nimodipine or a beta-blocking agent such as labetalol should suffice.
30. *Silent myocardial ischemia is commoner than nonsilent ischemia* (angina): Beta-blockers are strongly indicated.

BETA-BLOCKERS

A beta-blocker is strongly recommended for the management of

1. All hypertensive patients, white or black, aged < 70 yr with no coexisting disease.

Dosage

In the elderly, it is preferable to start with smaller doses, for example, bisoprolol 5 mg, metoprolol succinate extended release (Toprol XL) 50 mg, carvedilol 12.5-25 mg, or timolol 5 mg daily. It is not advisable to use a dose higher than that given in Table 3.4. If BP goal is not attained, a low-dose diuretic should be added or the beta-blocker should be discontinued and the patient commenced on a diuretic or other agent.

Physiological Role of Beta-Blockers

Beta-blockers lower BP by

1. Releasing norepinephrine from sympathetic neurons.
2. Decreasing central vasomotor activity.
3. Decreasing cardiac output.
4. Inhibiting the release of renin, thus decreasing angiotensin and aldosterone.

Individual Beta-Blockers

Drug name:	*Metoprolol succinate*
Trade name:	Toprol XL
Supplied:	50, 100, 200 mg
Dosage:	50-200 mg once daily

Drug name:	*Metoprolol tartrate*
Trade names:	Betaloc, Lopressor
Supplied:	50, 100, 200 mg
Dosage:	50-100 mg twice daily

Toprol XL has a 24-h duration of action. A combination of Toprol XL and low-dose HCTZ 6.25 mg should be appropriate for the control of more than 80% of patients with stage 1 isolated hypertension. *Metoprolol tartrate* is given 50-100 mg twice daily.

Drug name: *Atenolol*

Trade names: Tenormin, Totamol

Supplied: 25, 50, 100 mg

Dosage: the drug is not recomended by the author.

Drug name: *Bisoprolol*

Trade names: Zebeta, Monocor, Emcor

Supplied: 5, 10 mg

Dosage: 5-10 mg once daily (rarely 15 mg) once daily

Table 3.4 Doses of beta-blockers

Drug	*Before addition of diuretics (mg daily)*	*With diuretics (mg)*
Acebututolol	300–400	400
Atenolol	50–100	25–50
Bisoprolol	5–10	5–10
Carvedilol	12.5–50	12.5–50
Metoprolol	100–3 00	100–200
Nadolol	40–80	20–80
Propranolol LA	120–240	120–160
Timolol	10–20	5–15

Ziac is a combination of bisoprolol 2.5-, 5-, and 10-mg tablets with HCTZ 6.25 mg. This combination showed response rates of up to 84% in two well-controlled RCTs. Ziac has been approved by the U.S. Food and Drug Administration (FDA) as initial therapy; Monozide 10 = bisoprolol 10 mg, HCTZ 6.25 mg, in the United Kingdom.

Drug name: *Carvedilol*

Trade names: Coreg, Eucardic

Dosage: 12.5-25 mg twice daily

FREQUENT URINE

The Systolic Hypertension in the Elderly Program (SHEP), in which a diuretic was used in patients aged 65-85 yr, showed a 36% reduction in the risk of stroke ($p = 0.0003$), a 27% decrease in IHD event rates, and a 54% decrease in the risk of LVF. In the SHEP trial, a low-dose diuretic was used as initial therapy, and a beta-blocker was added if necessary.

There is no question about the efficacy of diuretics in patients with mild to moderate hypertension, and, when they are combined with other anti-hypertensive agents, they can be used in all types and degrees of hypertension.

Recommendations for the use of diuretics are as follows:

1. Low-dose diuretic therapy is strongly recommended as first-line therapy in older patients (age > 70 yr) of any ethnic group.
2. If the diuretic was not the first-choice agent, it is logical to add a diuretic if goal BP has not been achieved.
3. Diuretics are highly recommended in patients with coexisting disease for special situations, in particular in combination with ACE inhibitors for HF. Renal failure with increased volume overload is a common cause of resistant hypertension that requires the addition of a diuretic to enhance the efficacy of other agents.

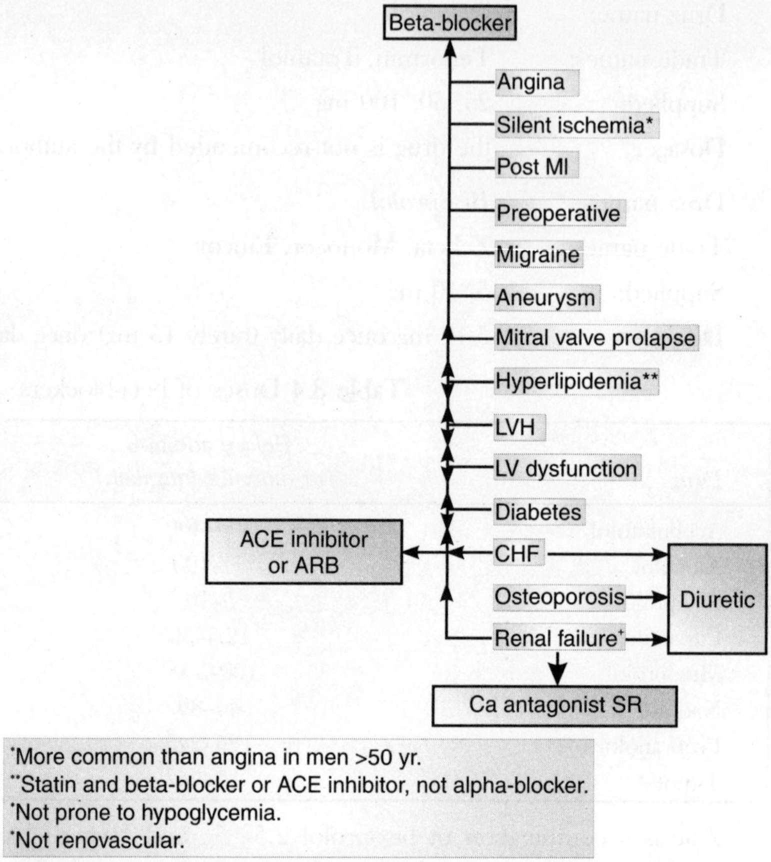

Fig. 3.5. Choice of antihypertensive agent in patients with coexisting diseases.

4. Metolazone is a thiazide diuretic that has a unique property of retaining effectiveness when other thiazides become ineffective in patients with renal failure GFR < 30 mL/min. Dosage 2.5 to 7.5 mg daily.

Drug name: Hydrochlorothiazide

Trade names: HydroDIURIL, Esidrex, Hydro-Saluric

Supplied: 25, 50 mg

Dosage: 12.5-25 mg each morning; max. 50 mg daily

The 50-mg dose was necessary to achieve control of BP in one study, and it was commonly used from 1965 to 1985.

Biochemical Action

The exact mechanism of action and antihypertensive effect are unknown, and the effect is believed to be related to decrease in vascular volume, negative sodium (Na^+) balance, and arteriolar dilation, causing a decrease in total peripheral resistance.

Advice, Adverse Effects, and Interactions

Symptoms

1. Hypersensitivity to thiazides or sulfonamides.
2. Anuria or severe renal failure.
3. Pregnancy and breastfeeding.
4. Concomitant lithium therapy.

Abnormalities

The adverse effects include the following:

1. Dehydration and electrolyte imbalance are the most common adverse effects. The dose is too large if signs of dehydration or orthostatic hypotension develop, or if the patient has increased urea > 10.0 mmol/L or an increase of 7.0 mmol/L from baseline, chlorides < 94 mEq (mmol)/L, total carbon dioxide > 32 mEq (mmol)/L, and uric acid > 7 mg/dL (420 μmol/L).
2. Hypokalemia occurs in a significant number of patients receiving thiazides and contributes to increased risk of cardiac arrest. Chlorthalidone causes a greater loss of potassium (K^+) than for equivalent doses of HCTZ. The incidence of hypokalemia can be decreased by the use of low-dose thiazide regimens with K^+-sparing diuretics. It is advisable to use the following:
   (a) A thiazide-K^+- sparing diuretic, such as amiloride with HCTZ, if renal function is normal: for patients aged < 75 yr, serum creatinine level < 1.3 mg/dL (115 μmol/L); for patients aged > 75 yr, serum creatinine level < 1.0 mg/dL (88 μmol/L).
   (b) If mild renal dysfunction exists, hyperkalemia may occur with the use of K^+-sparing diuretics; therefore, a plain thiazide is recommended without potassium supplements.
3. If the physician does not wish to use the combination of amiloride and HCTZ (Moduretic) or the combination of triamterene and HCTZ (Dyazide) when the serum creatinine concentration is normal, a thiazide can be used and the K^+ rechecked in months to evaluate whether the patient is a K^+ retainer or has a tendency to lose K^+. If the serum K^+ level is less than 3.5 mEq (mmol)/L, the deficiency should be corrected with potassium chloride (KCl) and the thiazide replaced with the combination of amiloride and HCTZ (Moduretic). The possibility of alternate-day therapy may be considered because the antihypertensive effect is significant and the risk of electrolyte imbalance is greatly reduced.
4. Gastrointestinal effects include anorexia, gastric irritation, intrahepatic cholestatic jaundice, and pancreatitis.
5. Central nervous system effects include dizziness, vertigo, paresthesia, and headache.
6. Hematology and blood disturbances include leukopenia, rare agranulocytosis, thrombocytopenia, aplastic anemia and hemolytic anemia; increased serum cholesterol, decreased HDL cholesterol, and increased blood viscosity.
7. Cardiovascular effects include orthostatic hypotension, low cardiac output, and arrhythmias from hypokalemia, especially if the patient is taking digitalis.

8. Hepatic coma may be precipitated, so do not use in patients with impending hepatic coma or moderate to severe hepatic dysfunction.
9. In diabetes, insulin requirement may be increased; latent diabetes may become manifest; rarely, hyperosmolar, nonketotic hyperglycemic diabetic coma may be precipitated. If the patient has ever had this condition, avoid all diuretics.
10. For use in pregnant or nursing mothers, thiazides cross the placental barrier, appear in breast milk, and can cause fetal or neonatal jaundice, thrombocytopenia, decreased vascular volume and placental perfusion, and acute pancreatitis.
11. Precipitation of gout and hyperuricemia is a well-known complication of all diuretics. If gout or hyperuricemia occurs, this is treated in the usual fashion, and the diuretic is discontinued. There is no reason to add allopurinol to the regimen to prevent further episodes of gout. This is commonly done and adds to polypharmacy and expense.
12. Thiazides decrease calcium (Ca^{2+}) excretion, but hypercalcemia rarely occurs.
13. Acute allergic interstitial pneumonitis is an extremely rare, but life-threatening, complication.
14. Drug interactions with lithium, steroids, and oral anticoagulants have been reported.

Drug name: *Bendrofluazide*

Trade names: UK Aprinox, Berkozide, Centyl, Neo-Naclex, Urizide

Supplied: 2.5, 5 mg

Dosage: 2.5-5 mg daily

Advice and adverse effects

Recommendations are the same as for HCTZ.

Drug name: *Triamterene and Hydrochlorothiazide*

Trade name: Dyazide

Supplied: Tablets containing 50 mg triamterene and 25 mg hydro-chlorothiazide

Dosage: 1 tablet each morning

Advice and adverse effects

Contraindications are the same as those outlined for HCTZ; special note should be made of the following:

1. Triamterene is a K^+-sparing diuretic and is contraindicated in patients with past or present renal dysfunction or renal calculi.
2. The use of Dyazide (triamterene and HCTZ) is not recommended if the serum creatinine level is > 1.3 mg/dL (115 μmol/L) in patients aged < 70 yr. For patients > 70 yr of age, the lower creatinine value of not more than 1.0 mg/dL (88 μmol/L) should be used to prevent hyperkalemia. Do not use along with KC1, ACE inhibitor, or any other K^+-sparing regimen. Do not use with indomethacin: renal failure may be precipitated.

Drug name: *Hydrochlorothiazide and Amiloride*

Trade names: Moduretic, Moduret (C)

Supplied: Tablets containing 50 mg hydrochlorothiazide and 5 mg amiloride

Dosage: Half a tablet daily

Dosage

The manufacturer indicates up to four tablets daily. The physician must insist on a lower dose of diuretics. A more appropriate combination, co-amilozide, available in the United Kingdom and Europe, contains amiloride 2.5 mg, HCTZ 25 mg.

Advice and adverse effects

Do not use along with potassium supplements.

Drug name: *Furosemide*; *Frusemide*

Trade name: Lasix

Supplied: Tablets: 20, 40, 80, 500 mg

Injectable: 20 mg in 2-mL ampule; 100 mg in 10-mL ampule

Benefits and harms

Furosemide is less effective than thiazides in mild to moderate hypertension, and therefore it is not advisable to use this drug unless the patient has significant renal dysfunction, such as a creatinine level > 2.3 mg/dL (203 μmol/L). Because furosemide inhibits the reabsorption of a very high percentage of filtered sodium, it is more effective than thiazides in patients with reduced glomerular filtration rate (GFR).

As with thiazides, the combination with a K^+-sparing diuretic has proved to be useful in causing further diuresis and in reducing K^+ loss. Ziac (US) or Monocor (UK) is a combination of bisoprolol 2.5, 5, and 10 mg with 6.25 mg HCTZ. This is an appropriate combination of a diuretic and a beta-blocker. BP is lowered to goal levels in about 80% of patients. This combination-type therapy with Ziac has shown consistently low discontinuation rates because of adverse effects, compared with enalapril.

Drug name: *Eplerenone*

Trade name: Inspra

Dosage: 25 mg once daily, max. 50 mg

PHARMACOLOGICAL STUDIES

ACE inhibitors and ARBs play a major role in the management of hypertension, particularly in patients with target organ damage and coexisting diseases. The major advantage of ARBs over ACE inhibitors is the absence of cough and very rare occurrence of angioedema; these agents are as effective as ACE inhibitors for the control of hypertension and HF. The failure of losartan, however, to control BP in scleroderma renal crisis with responsiveness to lisinopril has been reported.

1. In patients with target organ damage and coexisting conditions and special situations, particularly HF, LVH, diabetes with proteinuria, and selected cases of nephropathy (excluding renovascular hypertension), these agents are indicated. They are, however, not superior to other agents in diabetic patients, as is often claimed.

Physiological Functions

ACE inhibitors prevent the conversion of angiotensin I to angiotensin II; ARBs block the effects of angiotensin II on blood vessel walls and do not interfere with the breakdown of bradykinin. These actions of ACE inhibitors and ARBs result in

1. Arteriolar dilation, which causes a fall in total systemic vascular resistance.
2. Attenuation of angiotensin potentiation of sympathetic activity and release of norepinephrine. The diminished sympathetic activity causes vasodilation, reduction in afterload, and some decrease in preload. Also, heart rate is not increased by these agents, as opposed to other vasodilators.
3. Reduction in aldosterone secretion. The latter action promotes Na^+ excretion and K^+ retention.
4. Blocking of angiotensin-mediated vasopressin release, which appears to be important in HF.
5. Converting enzyme (the same as kinase II), which causes degradation of bradykinin. The accumulation of bradykinin stimulates release of vasodilator prostaglandins that contribute to the decrease in peripheral vascular resistance. Thus, indomethacin and other prostaglandin inhibitors reduce the effectiveness of ACE inhibitors.

Symptoms

ACE inhibitors retain K^+, so these drugs should not be given with potassium supplements or K^+-sparing diuretics, Aldactazide (spironolactone and HCTZ), Dyazide (triamterene and HCTZ), Maxzide (triamterene and HCTZ), or Moduretic (Moduret) (amiloride and HCTZ). Hyperkalemia may also occur with renal failure. Captopril causes proteinuria in about 1% of patients; this finding has occurred mainly in patients with preexisting renal disease who are taking doses of captopril in excess of 150 mg daily. Proteinuria caused by diabetes is not a contraindication because the drug has been shown to diminish proteinuria in some diabetic patients. Adverse reactions reportedly include severe pruritus and rash in 10% and loss of taste in 7% of patients, as well as mouth ulcers, neurologic dysfunction, and gastrointestinal disturbances. Occasionally, tachycardia, increased angina, and precipitation of MI may occur.

Neutropenia and agranulocytosis are rare and occur mainly in patients with serious intercurrent illness, particularly immunologic disturbances, and in those with an altered immune response, in particular collagen vascular disease. Precipitation of renal failure in patients with tight renal artery stenosis has been reported. Uncommon side effects include fatigue, Raynaud' s phenomenon, cough and/or wheeze, myalgia, muscle cramps, hair loss, angioedema of the face, mouth, or larynx, impotence or decreased libido, pemphigus, hepatitis, and the occurrence of antinuclear antibodies.

Alarming effects

The use of ACE inhibitors and ARBs should be avoided in the following clinical situations:

1. Hypotension: systolic pressures < 95 mmHg.
2. Moderate or severe aortic stenosis.
3. Severe renal failure; serum creatinine level > 2.3 mg/dL, 203 μmol/L. If the use of an ACE inhibitor is necessary, alter the dose and dosing interval.
4. Hyperkalemia or concomitant use of K^+-sparing diuretics or potassium supplements.
5. Tight renal artery stenosis or stenosis in a solitary kidney.
6. Immune problems, in particular collagen vascular disease and autoimmune disease; ACE inhibitors are necessary. Discontinue the following drugs, which alter immune response: steroids, procainamide, hydralazine, probenecid, tocainide, allopurinol, acebutolol, pindolol, and others.
7. Patients known to have neutropenia or thrombocytopenia.
8. Severe carotid artery stenosis.
9. Restrictive, obstructive, or hypertrophic cardiomyopathies, constrictive pericarditis, cardiac tamponade, and hypertensive hypertrophic cardiomyopathy of the elderly with impaired ventricular relaxation.
10. Pregnancy and breastfeeding.
11. Women aged < 45 yr who may wish to become pregnant; these drugs should be avoided.
12. Chlorpromazine therapy because severe hypotension may occur.
13. Uric acid renal calculi: Captopril increases uric acid excretion. The drug has been proposed as strongly indicated in hypertensive patients with gout or hyperuricemia. However, excretion of increased uric acid may predispose to accretion of renal calculi. Further, the usual measures of alkalinizing the urine are not acceptable in hypertensive or cardiac patients; probenecid interacts with captopril and so may allopurinol.

Drug name:	*Captopril*
Trade name:	Capoten
Supplied:	12.5, 25, 50 mg
Dosage:	Hypertension: 12.5 mg twice daily for a few days and then three times daily for about 1 wk, and then 25 mg three times daily; max. 75-100 mg daily.

Dosage

The 75-mg dose appears to be as effective as higher doses. Do not exceed 150 mg daily. A twice-daily maintenance dose is effective for hypertension. If the BP is not excessively high (systolic > 180 mmHg, diastolic > 100 mmHg), it is advisable to discontinue diuretics 24-48 h before commencing the ACE inhibitor and to restart them a few days to weeks later, depending on BP response.

In renal failure, increase the dose interval depending on creatinine clearance (GFR):

GFR 75-35 mL/min dose every 12-24 h

GFR 34-20 mL/min dose every 24-48 h

GFR 9-6 mL/min dose every 48-72 h

Drug name: *Enalapril*

Trade names: Vasotec, Innovace (UK)

Supplied: 2.5, 5, 10, 20 mg

Dosage: Hypertension: 2.5 mg once daily for 1-2 d, and then 5 mg increasing over weeks or months to 10-20 mg; max. 30 mg (rarely 40 mg)

Dosage

Occasionally, the daily dose must be given in two divided doses to achieve 24-h control of BP. The usual recommended initial dose in patients not taking a diuretic is 5 mg once daily, except in the elderly or in those with renal impairment, and in suspected high renin states as seen with renal artery stenosis, prior diuretic use, and low-sodium diets. Here 2.5 mg is advisable. Note that the serum creatinine level may be normal in patients aged > 70 yr with renal impairment, and caution with dosage is necessary.

Table 3.5 ACE inhibitors

Drug	*Trade name*	*Dosage (mg)*	*Max/d (mg)*
Benazepril	Lotensin	5-30	30
Captopril	Capoten	25-100 (2)	100
Cilazapril	Inhibace, Vascase	1-5	5
Enalapril	Vasotec, Innovace	5-20 (1-2)	30
Fosinopril	Monopril	10-30 (1-2)	30
Lisinopril	Prinivil, Zestril	5-30 (1)	30
Moexipril	Perdix	7.5-15 (2)	15
Perindopril	Coversyl	2-8	8
Quinapril	Accupri, Accupro	5-40 (1-2)	40
Ramipril	Altace, Tritace	1.25-15 (1)	15
Trandolapril	Mavik, Gopten	1-4 (1)	4

The dosages of ACE inhibitors and ARBs are given in Tables 3.5 and 3.6, respectively. The maximum doses given in these tables are approximately 20% lower than those stated by the manufacturer. The author does not recommend the use of the manufacturer's maximum for BP control. The goal BP should be achieved with approximately 75% of the maximum dose; if this is not achieved, the drug should be combined with a small dose of another agent.

Synergistic effect

Capozide. 15-30 mg contains 15 mg captopril with 30 mg HCTZ for twice-daily dosage. These preparations carry the disadvantages of fixed combinations.

Vaseretic. 10-25 mg contains enalapril 10 mg with HCTZ 25 mg and is effective when given once daily. The combinations are not recommended if there is moderate renal impairment: serum creatinine level > 2.3 mg/dL, 203 μmol/L.

Hyzaar. This is a combination of losartan 50 mg and HCTZ 12.5 mg. This agent should not be prescribed as initial therapy but only after a trial of a diuretic or ACE inhibitor independently. This drug is contraindicated in patients with renal vascular hypertension.

Table 3.6 Angiotensin II receptor blockers

Drug	*Trade name*	*Dosage (mg)*	*Max/d (mg)*
Candesartan	Atacand, Amias	4-10	32
Losartan	Cozaar	25-100 (1, 2)	100
Valsartan	Diovan	80-160 (1)	160
Irbesartan	Avapro, Aprovel	75-225 (1)	300
Telmisartan	Micardis	20-80 (1)	80
Eprosartan	Teveten	300-400 (2)	800

SUBSTITUTE OF CALCIUM

Calcium antagonists are recommended by the JNC for the following situations:

1. In older African Americans or if diuretics are contraindicated or cause adverse effects.
2. In younger African Americans if beta-blockers are contraindicated or not well tolerated.
3. In older white patients with isolated systolic hypertension if beta-blockers or diuretics are contraindicated or cause adverse effects provided that left ventricular function is normal.
4. As second-line therapy in combination with beta-blockers or other agents in patients with angina or renal dysfunction.

The calcium antagonists are used with caution in the following situations:

1. Patients with IHD, with silent ischemia, and after MI: diltiazem is indicated if beta-blockers are contraindicated. DHPs may be used in this subset of patients if a beta-blocker is being used concomitantly.
2. Patients with hypertension and diabetes: RCTs have shown an increase in mortality in patients treated with extended-release calcium antagonists. The isradipine and nisoldipine studies confirm an increased risk of IHD events caused by calcium antagonists in diabetic patients. ACE inhibitors and beta-blockers are preferred.
3. Patients with HF, LV dysfunction, or ejection fraction < 45%.

Drug name: *Amlodipine*
Trade names: Norvasc, Istin (UK)
Supplied: Tablets 5, 10 mg
Dosage: 5 mg, max. 10 mg once daily; initial 2.5 mg in the elderly

Drug name: *Nifedipine*
Trade names: Procardia XL, Adalat CC, Adalat LA (UK), Adalat XL (C)
Dosage: 30 mg daily increasing to 60 mg, max. 60 mg

DHPs cause a significant increase in sodium excretion, and a diuretic may not be necessary or may not lower BP further. The rapid-release capsule of nifedipine is no longer recommended. The extended- release preparation has been used worldwide since the 1980s. The drug has a role in renovascular hypertension and in the perioperative management of pheochromocytoma. Nifedipine and other DHPs may cause rebound hypertension. A case of an increase from 170/100 to 300/200 mmHg has been reported on sudden cessation of nifedipine therapy. The INSIGHT trial showed nifedipine equivalent to diuretic in preventing stroke, but it increased the risk of HF significantly.

Drug name:	*Diltiazem extended release*
Trade names:	Cardizem CD, Tiazac, Adizem-XL (UK), Viazem XL
Supplied:	Cardizem CD, Tiazac: 120, 180, 240, 300 mg
Dosage:	120 mg increasing to 240 mg daily, max. 300 mg

Diltiazem is a weak vasodilator, 50% less potent than DHPs. Caution is needed when this drug is combined with a beta-blocker because bradycardia or LVF may ensue. Interactions occur with digoxin and amiodarone. A diltiazem-lithium interaction causing psychosis has been reported. The NORDIL trial indicated that diltiazem was as safe as a beta-blocker or diuretic, but there was a small increase in the combined end point of MI and HF.

Drug name:	*Verapamil extended release*
Trade names:	Covera-HS, Chronovera, Isoptin SR, Securon SR (UK)
Supplied:	120, 180, 240 mg
Dosage:	120 mg once daily, max. 240 mg. 180-240 mg daily; reduce the dose in liver disease, renal failure, or the elderly

Caution. Short-acting formulations are not recommended, Verapamil should not be used in patients with cardiomegaly or LV dysfunction. Also avoid in HF, conduction defects, and sick sinus syndrome. Constipation and bradycardia are limiting considerations, particularly in patients aged > 70 yr. Combination with a beta-blocker is contraindicated.

Drug name:	*Felodipine*
Trade names:	Plendil, Renedil
Supplied:	Extended-release tablets: 2.5, 5, 10 mg
Dosage:	2.5-10 mg once daily

BASIC DRUGS

Drug name:	*Clonodine*
Trade name:	Catapres
Supplied:	0.1, 0.3 mg
Dosage:	0.1 mg at bedtime; increase to twice daily with a larger dose at night; maintenance 0.2-0.8 mg

Caution. Clonidine is contraindicated in patients with depression. Rebound hypertension can pose a major problem. Newer agents have rendered centrally acting drugs obsolete.

Drug name:	*Methyldopa*
Trade name:	Aldomet
Supplied:	125, 250, 500 mg
Dosage:	250 mg twice daily, increasing over days to weeks to 250 mg three times daily; max. 500 mg two or three times daily

Caution. This drug should be avoided in patients with active liver disease, depression, or pheochromocytoma. Hemolytic anemia is a well-known complication, and myocarditis causing death has been reported. The drug has a role mainly in pregnancy and in special situations as combination therapy when other agents have failed or are contraindicated.

ALPHA$_1$-BLOCKERS

The ACC has issued a caution regarding the use of alpha-blockers.

Drug name:	Prazosin
Trade names:	Minipress, Hypovase (UK)
Supplied:	1, 2, 5 mg
Dosage:	0.5 mg test dose, then 1-2 mg twice daily, max. 10 mg daily

Dosage

Withhold diuretics for 24 h. For mild hypertension, start a 0.5-mg test dose at bedtime. If no syncope or other adverse effects occur 24 h later, use 1 mg twice daily. If a dose > 6 mg is used, a beta-blocker and diuretic are usually necessary.

Caution. Rare syncopal reactions after the first dose has been documented. Postural hypotension, dizziness, tachycardia, and palpitations with rare precipitation of angina have been documented. The drug is not recommended by the author.

Prazosin may cause

1. Increase in heart rate.
2. Increase in pulsative force and cardiac peak ejection velocity.
3. Increase in circulating norepinephrine levels.
4. Activation of the renin-angiotensin system.

Dosaxozin prazosin, terazosin, hydralazine, and vasodilators of this class do not significantly prevent or reduce LVH and have some nonbeneficial effects on the cardiovascular system.

Drug name:	*Terazosin*
Trade name:	Hytrin
Supplied:	1, 2, 5, 10 mg
Dosage:	Discontinue diuretics and give a daily 1-mg dose at bedtime, increasing if needed to 2 mg in a few weeks; maintenance 2-5 mg daily

The main advantage over prazosin is a half-life three to four times greater, which allows for once-daily dosing. The agent is widely used in men with benign prostatic hypertrophy. Caution is required in combination with calcium antagonists.

Drug name: *Phentolamine*

Trade name: Rogitine

Supplied: 10 mg/dL; as 1- or 5-mL ampules

Dosage: IV: 10-20 μg/kg/min; average 5-10-mg dose IV repeated as necessary

Infusion: 5-60 mg over 10-30 min at rate of 0.1-2 mg/min

Phentolamine and phenoxybenzamine block $alpha_1$-receptors and especially $alpha_2$- receptors. The drug is very expensive; action is rapid and lasts only minutes. Thus, nitroprus side is preferred for control of most hypertensive crises including pheochromocytoma.

Drug name: *Phenoxybenzamine*

Supplied: 10-mg capsules

Dosage: Pheochromocytoma: 10 mg every 12 h, increasing the dose every 2 days by 10 mg daily; usual dosage 1-2 mg/kg daily in divided doses; saline may be needed to prevent postural hypotension

The drug is given for 1-2 wk before surgical removal of pheochromocytoma. Postoperative hypotension may be avoided by discontinuing the drug several days before operation and limited use in selected patients. Beta-blockers may be added after 1 wk of alpha-blockade, but only if it is necessary to treat catecholamine-induced arrhythmias.

Contraindications

HF is a contraindication.

STEPS IN EMERGENCY

Hypertensive crisis is usually subclassified into either *hypertensive emergency* or *urgency. Hypertensive emergency* is defined as a severe sudden elevation in BP, generally diastolic > 120-130 mmHg; some clinicians add: and/or systolic BP > 220 mmHg. The systolic pressures should reflect a knowledge of previous BPs, such as a rise within days from 160-170 to >220 mmHg systolic; the rate of rise of BP in relation to previous BP is more important than the absolute BP. Most important, the sudden excessive elevation in BP should be associated with *acute* organ damage or dysfunction, which confers an immediate threat to the integrity of the cardiovascular system and to life.

In aortic dissection or acute pulmonary edema, a BP of 200/110 mmHg must be reduced. Emergencies require reduction in BP within minutes by intravenous (IV) therapy. For hypertensive emergencies, the goal is to produce an immediate but modest reduction in BP. The objective, in most patients, is to achieve no greater than a 20% reduction from baseline of the mean arterial pressure or to reduce the diastolic BP to 110 mmHg and no less than 100 mmHg over a period of several minutes to several hours depending on the clinical situation. The BP is maintained at this level for a further 12-24 h, at which time oral therapy should be instituted and a decision made as to the necessity for further lowering of BP. These guidelines do not apply in patients with aortic dissection, in whom BP must be reduced to a much lower level along with the use of a beta-blocking agent to decrease the rate of rise of aortic pressure.

Caution is necessary: nitroprusside and labetalol have caused precipitous reductions in BP in some patients, resulting in cerebral and myocardial ischemia and/or MI. A titrated reduction in BP is superior and safer than the widespread use of sublingual nifedipine, which has resulted in cerebral and myocardial ischemia and/or MI. Sublingual nifedipine is not approved by the FDA in the United States. In countries where sublingual or oral rapid-release nifedipine 5-10 mg is still used, caution is necessary to avoid the medication in patients who are suspected of having myocardial ischemia, MI, or cerebral ischemia. *Hypertensive urgencies* refer to other situations in which it is advisable to reduce markedly elevated BP within a day or two, rather than within minutes using oral drugs. These situations include

1. Upper level of stage 3 hypertension: BP commonly systolic > 220 mmHg, diastolic > 120 mmHg; decrease to 180/100 mmHg.
2. Presence of papilledema but without acute deterioration of specific organ systems.
3. Progressive but not acute target organ damage, as outlined earlier for hypertensive emergencies.
4. Rebound hypertension, for example, after withdrawal of methyldopa, clonidine, or DHPs. If stroke, decrease only to 180/105 mmHg.
5. Refractory hypertension: patients in this category, with diastolic BP > 120 mmHg and no acute complications, should be tried first on oral antihypertensive therapy and sedation plus furosemide 40-80 mg IV initially; this is especially necessary if the history suggests that volume expansion is present; with renal failure, volume overload is usually present.

Drug name: *Sodium nitroprusside*

Trade name: Nipride

Dosage

IV administration is given by infusion pump only. Take care to avoid extravasation. Wrap the infusion bottle in aluminum foil or other opaque material to protect it from light. The prepared solution must be used within 4 h.

One vial (50 mg) sodium nitroprusside in 500 mL 5% dextrose = 100 μg/mL.

Dose/kg	μg/kg/min	mL/kg/min
Average	3	0.03
Range	0.25-10	0.005-0.08
Dose for a 60-kg patient	mg/60 kg/min	mL/60 kg/min
Average	180	1.8
Range	30-300	0.30-3.0

Start the infusion at the lower dose range (0.5 μg/kg/min) and adjust in increments of 0.2 μg/kg/min, usually every 5 min until the desired BP reduction is obtained. The average dose is 3 μg/

kg/min (range, 0.5-8 μg/kg/min). Oral antihypertensive agents should be started immediately so that the patient can be weaned from nitroprusside as quickly as possible.

Clinical Role

Nitroprusside is a potent, rapidly acting, IV antihypertensive agent. The hypotensive effects are caused by peripheral vasodilation and reduction in peripheral resistance as a result of a direct action on vascular smooth muscle, partly through nitric oxide. There is also venous pooling. Because of vasodilation, variable reflex tachycardia occurs. There is a slight decrease in stroke volume and cardiac output. Myocardial oxygen consumption is reduced. The drug's effect on BP is almost immediate (within 0.5–2 min) and usually ends when the IV infusion has stopped. The brief duration of drug action is the result of its rapid biotransformation to thiocyanate. The ferrous ion in nitroprus side reacts with the sulfhydryl groups of red blood cells to produce cyanide ion, which is further reduced to thiocyanate in the liver, which, in turn, is excreted by the kidney.

Advice, Adverse Effects, and Interactions

Symptoms

1. Hepatic failure.
2. Compensatory hypertension, such as arteriovenous shunt or coarctation of the aorta; corrected hypovolemia or severe anemia.
3. Abnormalities of cyanide metabolism: Leber' s optic atrophy and tobacco amblyopia.
4. Malnutrition, vitamin B_{12} deficiency, and hypothyroidism.
5. Severe renal failure and inadequate cerebral circulation; caution is necessary in these patients.
6. Pregnancy.
7. Raised intracranial pressure.

Fatalities have resulted from cyanide poisoning. In the presence of liver disease, cyanide levels increase with evidence of metabolic acidosis, so it is necessary to measure cyanide levels. If kidney disease exists, thiocyanate levels must be monitored, especially if treatment is to be extended for more than 2 d. Acceleration of infusion from an accident or faulty equipment and failure to monitor the BP accurately have all been associated with hypotension and shock. Retrosternal chest pain and palpitations may also be experienced. Methemoglobinemia has been reported. Hydroxocobalamin decreases cyanide levels and may be useful to increase the margin of safety. Rebound hypertension can be a problem.

Acute poisoning

Amyl nitrite inhalations and IV sodium thiosulfate are used to treat acute cyanide poisoning. Nitrites form methemoglobin, which combines with cyanide ions to form relatively nontoxic cyanomethemoglobin.

Drug name: *Fenoldopam*

Trade name: Corlopam

Dosage: IV infusion: initial 0.1 μg/kg/min, titrate 0.05 to 0.1 μg/kg/min at intervals of 15 min to 1-1.6 μg/kg/min

Functions

Onset of action is within 5 min, and duration of action is 30 min, Unlike dopamine, fenoldopam (Corlopam) does not have alpha- or beta-adrenergic agonist activity. Fenoldopam is a peripherally acting selective D_1-receptor antagonist. The agent causes peripheral arterial dilation, and, importantly, significant renal, mesenteric, and coronary vasodilation occurs. Advantages over nitroprusside include the absence of significant rebound hypertension and an increase in renal blood flow and lack of coronary steal and thiocyanate and cyanide intoxication. Unlike other dopaminergic agonists, fenoldopam does not cross the blood-brain barrier.

Caution. Fenoldopam contains sodium metabisulfite ME and may cause allergic-type reactions and may precipitate asthmatic episodes in sensitive asthmatic patients. Avoid in glaucoma; the drug increases intraocular pressure. Also, avoid in patients with IHD because the drug may cause significant tachycardia that can worsen acute coronary syndrome, and the tachycardia must not be treated with a beta-blocker. This drug is approved for short-term use, <48 h.

Drug name:	*Labetalol*
Trade names:	Normodyne, Trandate
Supplied:	Ampules: 20 mL, 5 mg/mL
Dosage:	20-80-mg bolus every 10–15 min; IV infusion: 0.5-2 mg/min

Dosage

IV infusion of 20-160 mg/h (2 mg/min) under close and continuous supervision is given slowly to obtain the desired BP reduction. The onset of action is 5-10 min, and the duration is 3-6 h. The patient must be recumbent throughout the infusion and for at least 4 h afterward. The hypotensive effect may last 1-8 h after cessation of the infusion. Thus, labetalol is not as predictable as nitroprusside. Alternatively, bolus injections are used, starting with a 20-mg dose and gradually increasing the dosage every 10 min to a maximum of 80 mg. Labetalol is a very useful drug for the management of hypertensive emergencies.

The drug is especially useful for crises associated with dissecting aneurysm, renal failure, hypertensive encephalopathy, eclampsia, and clonidine withdrawal, as well as in malignant hypertension and in some patients with pheochromocytoma. It is useful perioperatively and postoperatively, such as during neurosurgery for clipping aneurysms and in ear surgery. Adverse effects include bronchospasm, nausea, vomiting, orthostatic hypotension, and, rarely, hepatic necrosis.

Drug name:	*Hydralazine*
Trade name:	Apresoline
Dosage:	10–20 mg

Dosage

Hydralazine intramuscularly, but preferably by IV infusion, is indicated for hypertensive emergencies, particularly if the condition is associated with renal failure or preeclampsia. The recommended test dose is 10 mg followed in 30 min by an IV infusion of 10-20 mg/h, depending on the response. A maintenance dose of 5-10 mg/h is recommended with continuous monitoring of heart rate and BP.

Oral hydralazine is commenced within 24 h, 100-200 mg daily. If there is no contraindication to beta-blockers, propranolol is given IV 1-4 mg and then orally 120-240 mg daily in addition (or the equivalent dose of another beta-blocker). Furosemide 40 mg IV followed by oral HCTZ or furosemide greatly improves the control of BP. Hydralazine is contraindicated in patients with IHD and aneurysms, in particular dissecting aneurysm, because the drug increases cardiac ejection velocity and the rate of rise of aortic pressure. Other agents have relegated the use of hydralazine to a role in patients with renal failure and eclampsia.

Drug name: Diazoxide

Trade names: Hyperstat, Eudemine (UK)

Dosage: 50-150 mg IV bolus injected undiluted and within 30 s directly into a peripheral vein

Diazoxide has a small role in the management of hypertensive emergency associated with severe renal dysfunction if infusion pumps and intensive monitoring are unavailable to allow the use of agents. The agents have rendered diazoxide *obsolete* for hypertensive emergencies.

Drug name: *Urapidil* (Europe)

Dosage: 12.5-mg bolus followed by infusion of 5-40 mg/h

Urapadil, an alpha-blocker with serotonin-agonist activity, has a role in selected patients with hypertensive emergencies. The dose is 12.5-25 mg by IV bolus and then infusion at 5-40 mg/h. The drug does not increase heart rate significantly.

The time to onset of action is 3-5 min, and the duration of action is 4-6 h; adverse effects include hypotension, headache, and dizziness. The drug is a useful addition to therapy for accelerated malignant hypertension, hypertensive encephalopathy, and eclampsia, and, unlike fenoldopam and nitroprus side, it does not cause tachycardia or coronary steal and thus may be useful for LVF.

Other Agents

Nimodipine is indicated for hypertensive emergencies associated with subarachnoid hemorrhage. *Enalaprilat* is advocated by some for LVF but must be avoided in acute MI, ischemia, renal vascular hypertension, and pregnancy. The drug's onset of action is 20 min and duration is 6 h. The drug may cause a precipitous fall in BP; nitrogycerin IV is preferred for the management of LVF; the addition of IV furosemide and sedation provide salutary effects.

4

Diseases of Circulatory System

The prevalence of cardiovascular diseases will undoubtedly increase as the mean age of the populations throughout the world increases. Heart disease is generally complex and multifactorial. Despite considerable advances in therapy, the understanding of the underlying mechanisms are still rather poor. Efforts are being directed towards the development of new drugs through the modification of the physicochemical properties of drug molecules, and the design and synthesis of new drugs with particular emphasis on drug delivery systems, pharmaceutical materials research, controlled drug release and targeting of drugs to the site of action to enhance the therapeutic effects. The perspectives of the user enables the pharmacist and clinicians in understanding the factors leading to the suboptimal use of drugs in society.

Identifying, solving and preventing drug-related problems in order to improve health and quality of life through pharmaceutical care in the community are therefore also important. Equally, the optimal utilization of pharmacist and clinician know-how will bring tremendous benefits to the patients from the viewpoint of social pharmacy in the promotion of safe and efficient use of drugs. In this chapter, key research developments in cardiovascular medicine covering the area of congestive heart failure, cardiac arrhythmias, hypertension, ischemic heart disease and a variety of related disorders of the cardiovascular system are critically reviewed. The physical, chemical, and biological aspects of structure–activity relationships and pharmaceutics is also briefly presented to highlight the importance of drug delivery systems and the importance of clinical pharmacy from a pharmacological and pharmaceutical perspective.

BLOOD VASCULAR SYSTEM

The heart acts as a muscular pump and beats about 70 times every minute. The heart pumps blood around the body at a rate of 5 L/min, that is, about 180 million gallons during a lifetime's pumping. The blood pressure is dependent upon the heart's constant pumping of both blood and the size of all its blood vessels through which the blood passes. The pumping of the heart or the heartbeat is caused by alternating contractions and relaxations of the myocardium. These contractions are stimulated by electrical impulses from the sinoatrial or SA node located in the muscle of the right auricle. An impulse from the SA node causes the two auricles to contract, forcing blood into the ventricles.

Contraction of the ventricles is controlled by impulses from the atrioventricular or AV node located at the junction of the two auricles. Following contraction, the ventricles relax and allow the pressure to fall. Blood again flows into the auricles and an impulse from the SA node repeats the cycle again; this process is called the cardiac cycle. The period of relaxation is called a diastole and the period of contraction is called a systole. Diastole is the longer of the two phases so that the heart can rest between the contractions. Hence, systolic pressure and diastolic pressure are observed. The rate of SA impulse is subconsciously regulated by the autonomic nervous system. In this way the heartbeat is either accelerated or slowed down in response to physical activity and other factors. The force of blood pumped through the arteries exerts its pressure on the arterial walls.

The body monitors blood pressure by means of receptors present in the main arteries of the heart and then controls it through changes in heartbeat and the flow of blood. Vasomotor effectors increase or decrease the diameter of the blood vessels and also regulate blood pressure. Physical exertion and digestion of food places a heavy demand on the heart's blood output, which can increase as much as ten-fold to meet the special needs. There are two main systems which operate to maintain arterial blood pressure. These are the sympathetic nervous system and the renin-angiontensin-aldosterone system. One of the processes that tends to worsen hypertension is renal glomerular sclerosis. The control system fails because narrowing of the renal vessels upsets the normal relationship between renal blood flow and arterial pressure. Blood pressure is governed by the following five elements:

1. Cardiac output
2. Volume of blood
3. Viscosity of blood
4. Resistance of the arterioles
5. Elasticity of the arterial walls.

Abnormality of any of these elements can cause blood pressure to be too high or too low.

CLINICAL THERAPY

Arrhythmia

Despite considerable progress in management over the recent years, coronary artery disease (CAD) remains the leading cause of death in the industrialized world. It is estimated that CAD is responsible for causing 152,000 deaths per year in the United Kingdom. 400,000 deaths per year in the United State of America, and one in eight deaths worldwide. Many of these deaths are attributed to the development of cardiac arrhythmias, especially ventricular fibrillation during periods of myocardial ischemia or infarction. Myocardial ischemia is characterized by ionic and biochemical alterations, creating an unstable electrical substrate capable of initiating and sustaining arrhythmias, and infarction creates areas of electrical inactivity and blocks conduction, which also promotes arrhythmogenesis. Arrhythmias are dysfunctions which cause abnormalities in impulse formation (automaticity), conduction (conductivity), or both in the myocardium.

Most disturbance of cardiac rhythm arises when a small group of cardiac cells competes with or replaces the normal pacemaker. Any circumscribed region that acts in such a way is called an ectopic focus, meaning that it is not located in the usual pacemaker region, the sinoatrial node. The result can be abnormal ectopic beats and rhythms. The three basic mechanisms or causes responsible for ectopic beats and rhythms (arrhythmias) are enhanced automaticity, reentry, and triggered activity.

Enhanced automaticity

The sinoatrial node shows the fastest rate of phase 4 depolarization and therefore, exhibits a higher rate of discharge than that occurring in other pacemaker cells exhibiting automaticity. The SA node normally sets the pace of contraction for the myocardium, and latent pacemakers are depolarized by impulses coming from the SA node. So, if cardiac sites other than the SA node show *enhanced automaticity*, they may generate competing stimuli, and arrhythmias may arise. Common causes of enhanced automaticity are an increase in chatecholamines, digitalis toxicity, hypoxia, potassium imbalance, and myocardial ischemia or infarction.

Abnormal conduction

A common mechanism by which abnormal conduction causes arrhythmias is called reentry excitation, which occurs if a unidirectional block is caused by myocardial injury or prolonged refractory period. With normal conduction, the electrical impulse moves freely down the conduction system until it reaches recently excited tissue that is refractory to stimulation. This causes the impulse to be extinguished. The SA node then recovers and fires spontaneously, and the conduction process starts all over again. Reentry excitation means that an impulse continues to reenter an area of the heart rather than coming to an end. For this to occur, the impulse must encounter an obstacle in the normal conducting pathway. The obstacle is usually an area which allows conduction in only one direction and causes a circular movement of the impulse. Re-entry is the most common cause of arrhythmias and this can occur in any level of the cardiac conduction system.

Triggered activity

Triggered activity is an abnormal conduction of latent pacemaker and myocardial cells in which the cells may depolarize more than once following stimulation by a single electrical impulse. The level of membrane action potential spontaneously and rhythmically increases after depolarization until it reaches threshold potential, causing the cells to depolarize. This phenomenon, called after depolarization, can occur immediately following repolarization early, in phase 3 (early after depolarization), or late in phase 4 (delayed after depolarization).

Triggered activity can result in atrial or ventricular ectopic beats occurring singly, in groups of two (paired or coupled beats), or in bursts of three or more beats (paroxysms of beats or tachycardia). Common cause of triggered activity include an increase of catecholamines, digitalis toxicity, hypoxia, and myocardial ischemia.

Nature of Arrhythmias

Sinus tachycardia

Sinus tachycardia is an arrhythmia originating in the SA node characterized by rate of over 100 beats per minute. Sinus tachycardia may caused by numerous conditions such as fever, hypotension, heart failure, thyrotoxicosis, stimulation of the sympathetic nervous system, and life style drugs such as alcohol, caffeine, and nicotine. Treatment of sinus tachycardia should be directed towards identifying the underlying cause of the arrhythmia. The P, QRS, and T deflection are all normal, but cardiac cycle duration (the PP interval) is altered.

Sinus bradycardia

Sinus bradycardia is an arrhythmia originating in the SA node, characterized by a rate of less than 60 beats per minute. Sinus bradycardia may occur with excessive vagal stimulation, deficient sympathetic tone, and sinus node dysfunction. It often occurs in healthy young adults, especially athletes and during sleep. Other conditions associated with sinus bradycardia include hypothyroidism, hypothermia, and drugs such as beta-blocker agents, amiodarone, diltiazem, and verapamil. A symptomatic sinus bradycardia does not require treatment. Acute, symptomatic sinus bradycardia can be treated with atropine or a temporary transvenous pacemaker. Chronic symptomatic sinus bradycardia requires insertion of permanent pacemaker. The P, QRS, and T deflection are all normal, but cardiac cycle duration (the PP interval) is altered.

Sinus arrest and sinoatrial exit block

Sinus arrest is an arrhythmia caused by episodes of failure in the automaticity of the SA node, resulting in bradycardia, asystole, or both. Sinoatrial exit block is an arrhythmia caused by a block in the conduction of the electrical impulse from the SA node to the atria, resulting in bradycardia, a systole, or both. Sinus arrest and SA exit block may be precipitated by an increase in vagal (para-sympathetic) tone on the SA node, hypoxia, hyperkalemia, or excessive dose of digitalis or propranolol. SA exit block may also result from quinidine toxicity.

Abnormal atrial contractions

Is an extra atrial contraction consisting of an abnormal (sometimes normal) P wave followed by a normal or abnormal QRS complex, occurring earlier than the next expected beat of the underlying rhythm. Common cause of premature atrial contractions (PACs) include an increase in catecholamines and sympathetic tone, infections, emotion, stimulants (alcohol, caffeine, and tobacco), lack of sleep, digitalis toxicity. Hypoxia. The electrophysiological mechanism responsible for PACs is either enhanced automaticity or reentry. The P wave superimpose QRS complex on PACs.

Atrial tachycardia

Atrial tachycardia is an arrhythmia originating in an ectopic pacemaker in the atria or the site of a rapid reentry circuit in the AV node with a rate between 160 and 240 beats per minute. It includes non-paroxymal atrial tachycardia and paroxymal atrial tachycardia (PAT). PAT is often called paroxymal supraventricualr tachycardia when the site of origin (either atria or AV junction) cannot be determined with certainty. Atrial tachycardia commonly starts and ends abruptly, occurring in paroxysms (paroxymal atrial tachycardia).

When atrial tachycardia does not start and end abruptly, it is called (nonparoxymal atrial tachycardia). By definition three or more consecutive premature atrial contractions are considered to be atrial tachycardia. The causes of atrial tachycardia are essentially the same as those of premature atrial contractions. Non-paroxymal atrial tachycardia is most commonly caused by digitalis toxicity. The electrophysiological mechanism which is most likely to be responsible for non-paroxymal atrial tachycardia is enhanced automaticity, for paroxymal atrial tachycardia, it is a reentry mechanism.

Atrial fibrillation

Atrial fibrillation is an arrhythmia arising in multiple ectopic pacemakers or sites of rapid reentry circuits in the atria characterized by an irregular, often rapid ventricular response. So it is disorganized, tremor-like movements of the atria. This lack of effective atrial contraction impairs ventricular filling, decreases cardiac output and may lead to the formation of atrial thrombi. Another characteristic is a very rapid atrial rate (400-600 beats per minute). Atrial fibrillation is a major cause of morbidity in the aging population. Initially preponderant in men over 60, it ultimately becomes a disease of women in part perhaps because of their greater longevity. Atrial fibrillation is commonly associated with advanced rheumatic heart disease, hypertension, or coronary heart disease. The electrophysiological mechanism responsible for atrial fibrillation is either enhanced automaticity or reentry. P waves do not appear in the ECG; they are replaced by continuous irregular fluctuations of potential, called F waves, and QRS complexes is normal.

Atrial flutter

Atrial flutter is an arrhythmia arising in ectopic pacemaker or sites of a rapid reentry circuits in the atria, characterized by rapid abnormal atrial flutter F waves and a slower, regular ventricular response. Atrial flutter is characterized by a rapid (270-230 atrial beats per minute). Chronic atrial flutter is most commonly seen in middle-age or elderly people with advanced rheumatic heart disease, particularly if mitral or tricuspid valvular disease is present, and in those with coronary or hypertensive heart disease. The electrophysiological mechanism responsible for atrial flutter is either enhanced automaticity or reentry. Atrial flutter is characterized by two or three P waves (atrial contraction) for every QRS complex (ventricular contraction).

Mechanism of PJC

A *premature junctional contraction* (PJC) is an extra ventricular contraction that originates in an ectopic pacemaker in the AV junction, occurring before the next expected beat of the underlying rhythm. Premature junctional contractions are also called premature junctional beats or complexes. PJCs are a result of digitalis toxicity, increase in parasympathetic tone on the SA node, an excessive dose of certain drugs (quinidine and procainamide) or sympathomimetic drugs, hypoxia and congestive heart failure. The electrophysiological mechanism responsible for premature junctional contractions is either enhanced automaticity or reentry. PJC consists of a normal or abnormal QRS complex with or without an abnormal P wave.

Junctional escape rhythm (JER)

Junctional escape rhythm is an arrhythmia originating in an escape pacemaker in the AV junction with a rate of 40-60 beats per minute. Junctional escape rhythm is a normal response of the AV junction when the rate of impulse formation in the SA node become less than that of the escape pacemaker in the AV junction, or when the electrical impulse from the SA node or atria fail to reach the AV junction because of a sinus arrest, sinoatrial exit block, or third-degree AV block. Generally, when an electrical impulse fails to arrive at the AV junction within 1.0-1.5 s, the escape pacemaker in the AV junction begins to generate electrical impulses at its inherent firing rate of 40-60 beats per minute.

Non-paroxymal junctional tachycardia

Nonparoxymal junctional tachycardia (accelerated junctional rhythm) is an arrhythmia originating in an ectopic pacemaker in the AV junction with a regular rhythm and a rate of 60-150 beats per minute. Non-paroxymal junctional tachycardia is most commonly a result of digitalis toxicity. Other common causes are excessive administration of catecholamines and damage to the AV junction from myocardial infarction or rheumatic fever. The electrophysiological mechanism responsible for non-paroxymal junctional tachycardia is enhanced automaticity.

Paroxymal junctional tachycardia

Paroxymal junctional tachycardia (PJT) is an arrhythmia originating in an ectopic pacemaker or the site of a rapid reentry circuit in the AV junction with a rate between 160 and 240 beats per minute. PJT is often called paroxymal supraventricular tachycardia when the site of origin cannot be determined with certainty. PJT may occur without apparent cause in healthy persons of any age with no apparent underlying heart disease. It may precipitate by an increase in catecholamines and sympathetic tone, stimulants or emotional stress. The electrophysiological mechanism responsible for PJT is a reentry.

First-degree AV block

First-degree AV block is an arrhythmia in which there is a constant delay in the conduction of the electrical impulse usually through the AV node. It is characterized by abnormally prolonged PR intervals that are greater than 0.20 s usually about 0.28 s and constant. First-degree AV block occurs commonly in acute inferior myocardial infarction because of the effect of an increase in parasympathetic tone, ischemia on the AV node and inflammation of the AV bundle.

Second-degree AV block

Second-degree AV block is an arrhythmia in which a block of conduction of the electrical impulse occurs. The sites of block may be located above or below the His bundle. A block below the bundle is usually more serious than one above the bundle, because the former is more likely to evolve to third-degree AV block. It is characterized by abnormally prolonged PR interval 0.25–0.45 s; some P waves trigger QRS complexes and others do not; 2:1, 3:2 P wave/ QRS complex ratios may occur. Second-degree AV block occurs commonly due to excessive vagal stimulation.

Third-degree AV block (complete AV block)

Third-degree AV block is the complete absence of conduction of the electrical impulses through the AV node, bundle of His, or bundle branches characterized by independent beating of the atria and ventricles. Atrial rhythm is usually approximately 100 beats per minute and ventricular rhythm is less than 40 beats per minute. Third-degree AV block occurs commonly due to ischemia of AV nodal fibres associated with myocardial infarction, increased vagal tone, electrolyte imbalance and digitalis toxicity.

Premature ventricular contractions

Premature ventricular contraction (PVCs) is an extra ventricular contraction consisting of an ectopic pacemaker in the ventricles. It occurs earlier than the expected beat of the underlying rhythm and is usually followed by a compensatory pause. PVCs is also defined as discrete and identifiable premature QRS complexes. PVCs may occur in healthy persons with apparently healthy hearts and without apparent cause. PVCs frequently occur in acute myocardial infarction; they may also be caused by an increase in sympathetic or parasympathetic tone lack of sleep and with coronary thrombosis. The electrophysiological mechanism responsible for PVCs is either enhanced automaticity or reentry. PVCs usually characterized by prolonged QRS complex and inverted T wave.

Ventricular tachycardia

Ventricular tachycardia is an arrhythmia originating in an ectopic pacemaker on the ventricles with a rate between 110 and 250 beats per minute. It is also defined as a run of four or more consecutive ventricular premature beats. Ventricular tachycardia may occur in paroxysms of three or more PVCs separated by the underlying rhythm (non-sustained ventricular tachycardia or paroxymal ventricular tachycardia) or persist for a long period of time (sustained ventricular tachycardia). Ventricular tachycardia usually occurs in the presence of significant cardiac disease. Most commonly it occurs in coronary artery disease like in acute myocardial infarction. Digitalis toxicity is also a common cause of ventricular tachycardia. The electro-physiological mechanism responsible for ventricular tachycardia is usually characterized by abnormally wide and bizarre QRS complexes.

Trosade de pointes

Trosade de pointes, a French term meaning twisting of points, is a syndrome characterized by clinically serious ventricular arrhythmia similar to fibrillation, with rapid asynchronous complexes, which is always spontaneously reversible (unlike ventricular fibrillation which is usually not reversible). It is usually associated with a prolonged QT interval in the ECG, which may be congenital (in congenital long QT syndrome) or drug induced (by classes IA and III antiarrhythmics).

Ventricular fibrillation

Ventricular fibrillation is an arrhythmia arising in numerous ectopic pacemakers in the ventricles, produces ineffective myocardial contraction so that there is no cardiac output. It is defined as a signal for which individual QRS deflections can no longer be distinguished from one another, and for which a rate can no longer be measured. Ventricular fibrillation is a lethal arrhythmia if it lasts for more than few minutes and it is the main mechanism of sudden cardiac death. Ventricular fibrillation usually occurs in the presence of significant cardiac disease, most commonly in coronary artery disease, myocardial ischemia, acute myocardial infarction, and in third-degree AV block. The electrophysiological mechanism responsible for ventricular fibrillation is either enhanced automaticity or reentry. The PR intervals and the QRS complexes are absent in ventricular fibrillation.

Accelerated idioventricular rhythm

Accelerated idioventricular rhythm is an arrhythmia originating in an ectopic pacemaker in the ventricles with heart rate less than 120 beats per minute. Accelerated idioventricular rhythm 1 is relatively common in acute myocardial infarction although it may also result from digitalis toxicity. The electrophysiological mechanism responsible for accelerated idioventricular rhythm is probably enhanced automaticity.

Ventricular escape rhythm

Ventricular escape rhythm is an arrhythmia originating in an escape pacemaker in the ventricles with a rate of less than 40 beats per minute. Ventricular escape rhythm usually occurs as a result of sinus arrest, sinoatrial block, or third-degree AV block. The PR intervals are absent in ventricular escape rhythm, and the QRS complexes exceed 0.12 s and are irregular.

Ventricular a systole

Ventricular a systole is the absence of all electrical activity within the ventricles, and it is one of the common causes of cardiac arrest. Ventricular a systole usually occurs in third-degree AV block. In the dying heart, ventricular a systole is usually the final arrhythmia following ventricular tachycardia, ventricular fibrillation, and ventricular escape rhythm. The P waves may be present or absent in ventricular a systole, the PR intervals and the QRS complexes are absent.

OBJECTIVES OF ARRHYTHMIAS

The objective of treating an arrhythmia is to provide an effective drug at an appropriate concentration that can be tolerated by the patient and is free of adverse effects. The concentration needed will depend on the patient, in particular, on their particular arrhythmia profile.

Class IA: Sodium Channel Blockers

Quinidine

Quinidine is the prototype class IA drug. At high doses, it can precipitate arrhythmias, which can lead to fatal ventricular fibrillation. Because of quinidine's toxic potential, calcium antagonists, such as verapamil, are increasingly replacing this drug in clinical use. In addition to class IA, quinidine depresses the contractility of the myocardium (negative inotropic effect), and reduces vagus nerve activity on the heart (antimuscarinic effect). Quinidine is most commonly used in the United States.

Mechanism of action. Quinidine binds to open and inactivated sodium channels and prevents sodium influx, thus slowing the rapid upstroke during phase 0. It also decreases the slope of phase 4 spontaneous depolarization.

Therapeutic uses. Quinidine is useful in the treatment of a wide variety of arrhythmias including atrial, AVjunctional and ventricular tachyarrhythmias.

Pharamacokinetics. Quinidine sulfate is rapidly and almost completely absorbed after oral administration. The remainder of the drug is excreted unchanged by the kidneys.

Adverse effects. Quinidine must never be used alone to treat atrial fibrillation or flutter as its antimuscarinic action enhances AV conduction and the heart rate may accelerate. The negative inotropic action of quinidine may result in hypotension and cardiac failure. Cardiotoxic effects are exacerbated by hyperkalemia. Plasma digoxin concentration is raised by quinidine (displacement from tissue binding and impairment of renal excretion). A potential adverse effect of quinidine (or any antiarrhythmic drug) is exacerbation of the arrhythmia.

Procainamide

Procainamide is a derivative of the local anesthetic procaine. The mechanism of action and the therapeutic uses of procainamide are similar to those of quinidine.

Pharmacokinetics. Procainamide is more than 75% bioavailable after oral administration. The intravenous preparation is relatively frequently used but can cause hypotension if rapidly administered. Procainamide has a relatively short half-life of 2-3 h. Available portion of the drug is acetylated in the liver to N-acetyl procainamide (NAPA). NAPA unlike the parent drug has properties of a class III drug. NAPA is eliminated via the kidneys.

Adverse effects. The most common problem with procainamide is lupus-like syndrome. Sixty-five percent of patients will develop antibodies within 12 months; only 12% will show symptoms that are reversible when the drug is stopped. Rare central nervous system side effects include depression, hallucinations, and psychosis, but gastrointestinal intolerance is less frequent than with quinidine.

Disopyramide

Disopyramide is most commonly used in Europe. The mechanism therapeutic uses of disopyramide are similar to those of quinidine.

Pharmacokinetics. About one half of the orally ingested drug is excreted unchanged by the kidneys. Approximately 30% of the drug is converted by the liver (5,17,18) to the less active mono-N dealkylated metabolite. Drug half-life is 5–8 h (6).

Adverse effects. Most of the side effects seen with disopyramide relate to its anticholinergic activity (dry mouth, blurred vision, urinary retention, and constipation).

Class IB: Sodium Channel Blockers

Lidocaine

Lidocaine, a local anesthetic, is the prototype of class IB.

Mechanism of action. Lidocaine, shorten phase 3 repolarization and decrease the duration of action potential. Unlike quinidine, which suppresses arrhythmias caused by increased normal automaticity, lidocaine suppresses arrhythmia caused by abnormal automaticity.

Therapeutic uses. It is the drug of choice for treating serious ventricular arrhythmias associated with acute myocardial infarction, cardiac surgery and electrical cardioversion. Lidocaine is also the drug of choice for emergency treatment of cardiac arrhythmias.

Pharmacokinetics: Lidocaine is given intravenously because of extensive first-pass transformation by the liver, which precludes oral administration. The drug is dealkylated and eliminated almost entirely by the liver, so dosage adjustments are necessary in the presence of hepatic dysfunction.

Adverse effects. Central nervous system side effects predominate; drowsiness, slurred speech, agitation, convulsions, and confusion.

Mexiletine and tocainide

Mexiletine and tocainide are oral analogues of lidocaine with similar pharmacological actions. They are used to suppress ventricular fibrillation or ventricular tachycardia. They are well absorbed from the gastrointestinal tract. Adverse effects are similar to those of lidocaine. Tocainide has pulmonary toxicity which may lead to pulmonary fibrosis.

Phenytoin

Phenytoin, an anticonvulsant, may be used to treat arrhythmias produced by digoxin intoxication. The antiarrhythmic properties of phenytoin generally resemble those of lidocaine. Phenytoin decreases automaticity and improves conduction through the AV node. Decreased automaticity helps control arrhythmias, whereas enhanced conduction may improve cardiac function, further, because heart block may result from digoxin, quinidine, or procainamide. Phenytoin may relieve arrhythmias without intensifying heart block. Phenytoin is given orally.

Class IC: Sodium Channel Blockers

These drugs are approved only for refractory ventricular arrhythmias. However, recent data have cast serious doubts on the safety of class IC drugs.

Flecainide

Mechanism of action. Flecainide suppresses phase 0 upstroke in Purkinje and myocardial fibres. This causes marked slowing of conduction in all cardiac tissues, with a minor effect on the duration of the action potential and refractoriness. Automaticity is reduced by an increase in the threshold potential rather than a decrease in the slope of phase 4 depolarization.

Therapeutic uses. Flecainide is useful in treating refractory ventricular arrhythmias. It is particularly useful in suppressing premature ventricular contractions. Flecainide has a negative inotropic effect and can cause worsening of congestive heart failure. The cardiac arrhythmia suppression trial investigators concluded that antiarrhythmic therapy with flecainide, in post myocardial infarction patients increased the risk of sudden death although the agent was able to suppress premature ventricular complexes.

Pharmacokinetics: Flecainide is absorbed orally, undergoes minimal biotrans-formation, and has a half-life of 16-20 h.

Adverse effects. Flecainide can cause dizziness, blurred vision, headache, and nausea. Like other class IC drugs, flecainide can aggravate preexisting arrhythmias or induce life-threatening ventricular tachycardia that is resistant to treatment.

Propafenone

Propafenone has similar actions and therapeutic uses to flecainide.

Pharmacokinetics. Propafenone undergoes extensive first-pass transformation to hydroxylated metabolite with reduced electrophysiological uses.

Adverse effects. Propafenone can cause nausea, weakness, and metallic taste. It may also have severe pro arrhythmic effects, similar to those of the other agents with class IC properties.

Class II: Beta-Adrenegic Blockers

These agents exert antiarrhythymic effects by blocking sympathetic nervous system stimulation of beta receptors in the heart and decrease risks of ventricular fibrillation. These drugs diminish phase 4 depolarization, thus depressing automaticity, prolonging AV conduction, and decreasing heart rate and contractility.

Blockage of receptors in the SA node and ectopic pacemaker decrease automaticity, and blockage of receptors in the AV node increase the refractory period. These drugs are effective for the treatment of supraventricular arrhythmias and those resulting from excessive sympathetic activity. Thus they are most often used to slow the ventricular rate of contraction in supraventricular tachyarrhythmias. They are also used for atrial flutter and fibrillation.

As a class, beta-blockers are being used more extensively because of their effectiveness and their ability to reduce mortality in a variety of clinical settings, including post-myocardial infarction and heart failure. With the exception of beta-blockers, none of the available antiarrhythmic drugs have significantly improved survival after myocardial infarction.

These drugs include propranolol which is a β-adrenegic antagonist that is most widely used in the treatment of cardiac arrhythmias. Propranolol reduces the incidence of sudden arrhythmic death after myocardial infarction. Esmolol is a very short-acting beta blocker used for intravenous administration in acute arrhythmias occurring during surgery or in emergency situations. Other useful drugs are pindolol, metoprolol, acebutolol, sotalol, atenolol, and timolol and nadolol.

Pharmacokinetics of class II. For long-term use, any of the oral preparations of beta-blockers is suitable. In emergencies propranolol or esmolol may be given intravenously.

Adverse effects. Adverse cardiac effects from over dose include heart block or even cardiac arrest. Heart failure may be precipitated when a patient is dependent on sympathetic drive to maintain output.

Class III: Potassium Channel Blockers

Potassium channel blockers (sotalol, amiodarone, bretylium, ibutilide, and dofetilide) block the outward-flowing potassium channel, and decrease potassium conductance which prevent or delay repolarization (prolongation of action potential duration and end resting potential).

Sotalol

Sotalol has both beta-adrenergic blocking and potassium channel blocking activity. Beta-blocking effects predominate at lower doses and class III effects predominate at higher doses.

Mechanism of action. Sotalol blocks rapid outward potassium current; this blockade prolongs both repolarization and the duration of the action potential, thus lengthening the effective refractory period.

Therapeutic uses. Sotalol is approved for treatment of ventricular tachycardia and fibrillation. It has also been used in smaller doses, to prevent atrial fibrillation. Sotalol is more effective in preventing arrhythmia recurrence and in decreasing mortality than mexiletine, procainamide, propafenone, and quinidine in patients with sustained ventricular-tachycardia.

Pharmacokinetics: Sotalol has a half-life of about 12 h after oral administration. It is eliminated by the kidneys.

Adverse effects. Sotalol had the lowest rate of acute or long-term adverse effects. The symptoms of trosade de pointes presents serious potential adverse effects typically seen in 3-4 % of patients. Fatigue, dizziness, and insomnia can occur.

Amiodarone

Amiodarone contains iodine and is related structurally to thyroxine. It has complex effects of classes I-IV actions. Amiodarone has antianginal as well as antiarrhythmic activity.

Mechanism of action: Its dominant effect is prolongation of the action potential duration and refractory period.

Therapeutic uses. Amiodarone is effective in the treatment of severe refractory supraventricular and ventricular tachyarrhythmias. Its clinical usefulness is limited by its toxicity.

Pharmacokinetics. Amiodarone is incompletely absorbed after oral administration. The drug is extensively deethylated in the liver to N-desethylamiodarone, full clinical effects may not be achieved for up to 6 weeks after initiation of treatment.

Adverse effects. Amiodarone shows a variety of toxic effects. Some of the more common effects include gastrointestinal tract intolerance, tremor, dizziness, hyper, or hypo thyroidism, liver toxicity, muscle weakness, and blue skin discoloration caused by iodine accumulation in the skin. More than one half of the patients show side effects sufficiently severe to prompt its discontinuation.

Bretylium

Mechanism of action. Bretylium has a number of direct and indirect electro-physiological actions, the most prominent of which are prolongation of the refractory period and raising of the intensity of the electrical current necessary to induce ventricular fibrillation in the His-Purkinje system.

Therapeutic uses. Bretylium is reserved for life-threatening ventricular arrhythmias, especially recurrent ventricular fibrillation or tachycardia.

Pharmacokinetics. Bretylium is poorly absorbed from the gastrointestinal tract and is therefore generally administrated parenterally. The drug is excreted unchanged in the urine.

Adverse effects. Bretylium can cause severe postural hypotension. Nausea and vomiting may occur after rapid intravenous administration.

Class IV: Calcium Channel Blockers

Calcium channel blockers block the movement of calcium into conductile and contractile myocardial cells. As antiarrhythmic agents, they act primarily against tachycardias at SA and AV nodes because the cardiac cells and slow channels that depend on calcium influx are found mainly at these sites. Thus, reduce automaticity of the SA and AV nodes, slow conduction, and prolong the refractory period in the AV node.

Verapamil and diltiazem

Verapamil shows greater action on the heart than on vascular smooth muscle, whereas nifidipine, a calcium channel blocker is used to treat hypertension exerts a stronger effect on vascular smooth muscle than on the heart. Diltiazem is intermediate in its action.

Mechanism of action. Calcium channel blockers, such as verapamil and diltiazem, are more effective against the voltage-sensitive channels, causing a decrease in the slow inward current that triggers cardiac contraction. Verapamil and diltiazem bind only to open, depolarized channels, thus preventing repolarization until the drug dissociates from the channel. By decreasing the inward current carried by calcium, verapamil and diltiazem slow conduction and prolong the effective refractory period in tissues dependent on calcium currents, such as the AV node.

Therapeutic uses. Verapamil and diltiazem are more effective against atrial than ventricular arrhythmias. They are useful in treating reentrant supraventricular tachycardia and reducing ventricular rate in atrial flutter and fibrillation.

Pharmacokinetics. Verapamil and diltiazem are absorbed after oral administration. They are metabolized by the liver, and the metabolites are primarily excreted by the kidneys.

Adverse effects. Both drugs may cause bradycardia and therefore may be contraindicated in patients with preexisting depressed cardiac function. These drugs can also cause a decrease in blood pressure caused by peripheral vasodialtion.

Class V: Other Antiarrhythmic Drugs

Digoxin

Digoxin shortens the refractory period in atrial and ventricular myocardial cells while prolonging the effective refractory period and diminishing conduction velocity in Purkinje fibres. Digoxin is used to control the ventricular response rate in atrial fibrillation and flutter. At toxic concentration, digoxin causes ectopic ventricular beats that may result in ventricular tachycardia and fibrillation.

Adenosine

Adenosine is a naturally occurring nucleoside, but at high doses the drug decreases conduction velocity, prolongs the refractory period, and decreases automaticity in the AV node. Intravenous adenosine is the drug of choice for abolishing acute supra-ventricular tachycardia. It has low toxicity but causes flushing, chest pain and hypotension.

Magnesium sulfate

Magnesium sulfate is given intravenously in the treatment of several arrhythmias, including prevention of recurrent episodes of trosades de pointes and treatment of digitalis induced arrhythmias. Its antiarrhythmic effects may derive from imbalances of magnesium, potassium, and calcium.

Non-Pharmacological Treatment of Arrhythmias

Non-pharmacologic treatment is preferred, at least initially for several arrhythmias. For example, sinus tachycardia usually results from such disorders as infection, dehydration, or hypotension and treatment should be aimed toward relieving the underlying causes. For ventricular fibrillation, immediate defibrillation by electrical counter shock (electroversion or cardioversion) is the initial treatment of choice. The technique consists of applying the shock, by means of a defibrillator to the chest wall of the patient, through two electrodes, one overlapping the lower border of the left scapula and the other over the third right intercostals space. The energy of the brief (2 ms) shocks used ranges from 100 to 400 joules. The impetus for non-pharmacologic treatments developed mainly from studies demonstrating that antiarrhythmic drugs could worsen existing arrhythmias, cause new arrhythmias, and cause higher mortality rates in clients receiving the drugs than clients not receiving the drugs.

Current technology allows clinicians to insert pacemakers and defibrillators, for example, the implantable cardioverter-defibrillators (ICD) have been useful to control brady arrhythmias or tachy arrhythmias. Similarly, radio waves have been used in radio frequency catheter ablation or surgery to deactivate ectopic foci. The ICD are designed to deliver an electric shock when ventricular fibrillation or tachycardia occurs. They are increasingly being recognized as most efficacious and cost-effective in initial therapy for the prevention of sudden cardiac death. The major goals of defibrillator therapy are detection and termination of ventricular tachyarrhythmias, prevention of sudden death, reduction in patient mortality and in improving the quality of life.

CLINICAL ASPECTS OF ANTIARRHYTHMIC DRUGS

The chemical structure of compounds found to possess antiarrhythmic activity covers a wide range, and because of this great structural heterogenicity, efforts to determine a definitive structure have not been too successful. Even some of the more recently discovered and currently used antiarrhythmic drugs were synthesized for quite different purposes, and routine pharmacological screening and clinical trials revealed their antiarrhythmic activity. The first extensive reviews on the SAR of antiarrhythmic drugs were by Szekeres and Papp and Conn.

These investigations indicated that most antiarrhythmic drugs possess a tertiary amine group that appears to be a key component for activity. The presence of aromatic ring system is important, and this moiety is usually connected to the tertiary amine via a hydroxy substituted alkyl chain ester or amide group.

The methoxy group on the aromatic ring may enhance activity. Since these early studies the number of compounds synthesized and tested for antiarrhythmic activity has increased dramatically. The structure activity relationships and approaches to new antiarrythmic agents of some class I antiarrhythmic agents are highlighted below.

Class lA: Antiarrhythmic Agents

Procainamide hydrochloride

Procainamide hydrochloride; p-Amino-N-[2-(diethylamino)ethyl]benzamide mono-hydrochloride. It was developed in the course of research for compounds structurally similar to procaine, which had limited effect as an antiarrhythmic agent because of its central nervous system side effects and short-lived action resulting from the rapid hydrolysis of its ester linkage by plasma esterases. Procainamide hydrochloride is also more stable in water than procaine because of its amide structure. Metabolism of procainamide hydrochloride occurs through the action of Nacetyltransferase. The product of enzymatic metabolism of procainamide hydrochloride is NAPA, which possesses only 25% of the activity of the parent compound. A study of the disposition of procainamide hydrochloride showed 50% of the drug was excreted unchanged in the urine, with 7% to 24% recovered as NAPA.

Disopyramide phosphate

Disopyramide phosphate. α-[2-(Diisopropylamino) ethyl]-α-phenyl-2-pyridineacetamide phosphate is an oral and intravenous class IA antiarrhythmic agent. Oral administration of the drug produces peak plasma levels within 2 h. the drug is approximately 50% bound to plasma protein and has a half-life of 6.7 h in humans. More than 50% is excreted unchanged in the urine. Disopyramide phosphate commonly exhibits side effects of dry mouth, constipation, urinary retention, and other cholinergic-blocking actions because of its structural similarity to anticholinergic drugs.

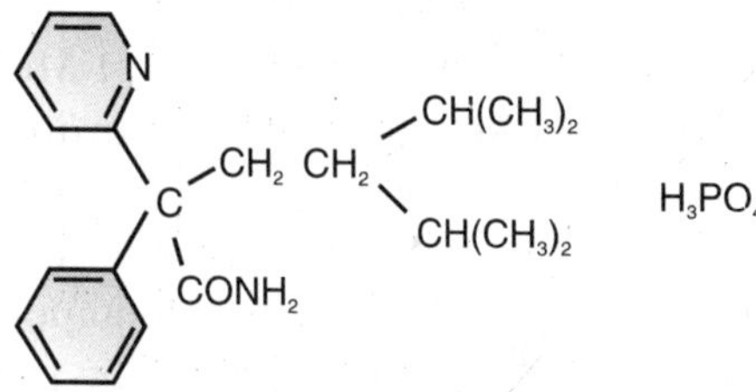

Fig. 4.1. Disopyramide phosphate.

Class IB: Antiarrhythmic Agents

Lidocaine hydrochloride

Lidocaine hydrochloride. 2-(Diethylamino)-2′,6′-acetoxylidide monohydrochloride. This drug was conceived as a derivative of gramine (3-dimethylaminomethylindole) and introduced as a local anesthetic, is now being used intravenously as a standard parenteral agent for suppression of arrhythmias associated with acute myocardial infarction and cardiac surgery. Lidocaine is the drug of choice for the parenteral treatment of premature ventricular contractions. Lidocaine hydrochloride administration is limited to the parenteral route and is usually given intravenously, though adequate plasma levels are achieved after intramuscular injections.

Lidocaine hydrochloride is not bound to any extent to plasma proteins and is concentrated in the tissues. It is metabolized rapidly by the liver. The first step is deethylation, with the formation of monoethylglycinexylidide, followed by hydrolysis of the amide. Metabolism is rapid, the half-life of a single injection ranging from 15 to 30 min. Lidocaine hydrochloride is a popular drug because of its rapid action and its relative freedom from toxic effects on the heart, especially in the absence of hepatic disease. Monoethylglycinexylidide, the initial metabolite of lidocaine, is an effective antiarrhythmic agent; however, its rapid hydrolysis by microsomal amidases prevents its use in humans.

Tocainide hydrochloride

2-Amino-2′,6′-propionoxylidide hydrochloride. Tocainide hydrochloride (pKa 7.7) is an analogue of lidocaine. It is orally active and has electrophysiological properties similar to lidocaine. Total body clearance of tocainide hydrochloride is only 166 mL/min, suggesting that hepatic clearance is not large. Because of low hepatic clearance, the hepatic extraction ratio must be small; therefore, tocainide hydrochloride is unlikely to be subject to a substantial first-pass effect. The drug differs from lidocaine in that it lacks two ethyl groups, which provides tocainide hydrochloride some protection from

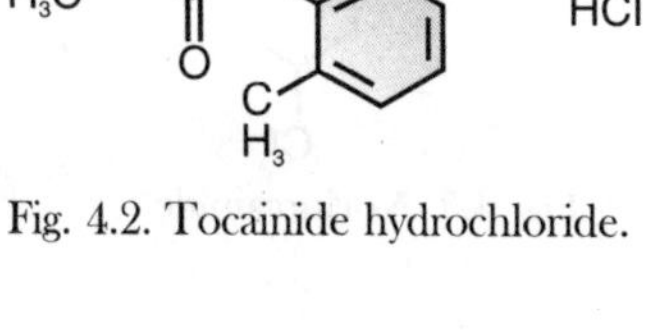

Fig. 4.2. Tocainide hydrochloride.

first-pass hepatic elimination after oral ingestion. Tocainide hydrochloride is hydrolyzed in a manner similar to lidocaine. None of its metabolites are active.

Class IC: Antiarrhythmic Agents

Encainide hydrochloride

Encainide hydrochloride. 4-Methoxy-N-[2-(1 -methyl-2-piperidinyl) phenyl] benzamide monohydrochloride. Encainide hydrochloride is a benzanilide derivative that has local anesthetic properties in addition to its class le antiarrhythmic action. The drug is metabolized extensively, producing products that also have antiarrhythmic properties. The metabolite, 3-methoxy-O-demethylencainide, is about equipotent to encainide hydrochloride. O-Demethylencainide is considerably more potent than the parent drug. The half-life of encainide hydrochloride is 2-4 h. The active metabolites have longer half-lives, estimated as being up to 12 h, and may play an important part in the use of this drug in long-term therapy. Encainide hydrochloride also undergoes N-demethylation to form N-demethylencainide.

Flecainide acetate

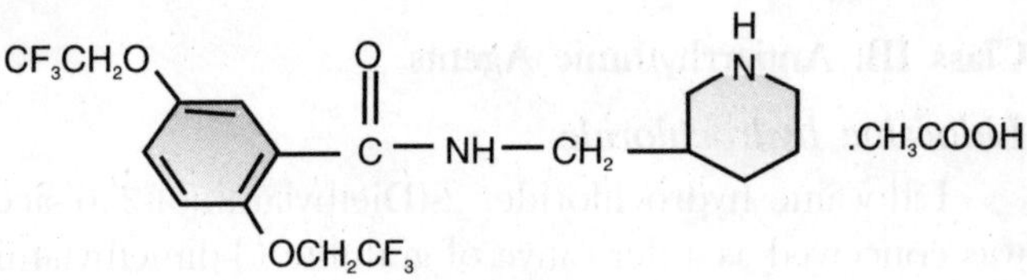

Fig. 4.3. Flecainide acetate.

Flecainide acetate. N-(2-Piperidinyl-methyl)-2,5-bis(2,2,2-trifluoroethoxy) benzamide monoacetate. It is a chemical derivative of benzamide. The drug undergoes bio-transformation, forming a meta-O-dealkylated compound, the antiarrhythmic properties of which are one-half as potent as those of the parent drug, and meta-O-dealkylated lactam of flecainide with little pharmacological activity. Flecainide acetate has some limitations because of central nervous system side effects.

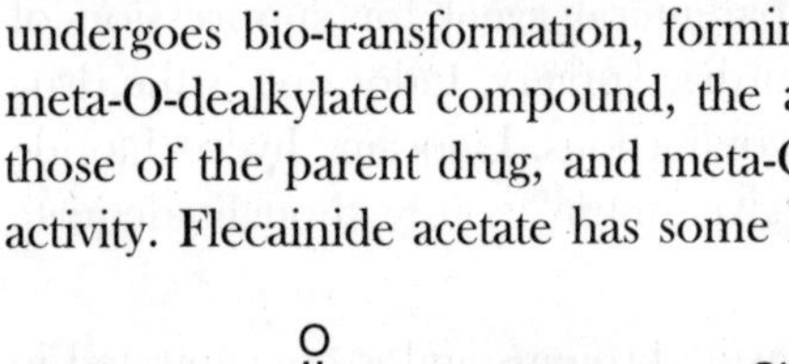

Fig. 4.4. Lorcainide hydrochloride.

Lorcainide hydrochloride

Lorcainide hydrochloride. N-(4-Chlorophenyl)-N-[1-(1 -methylethyl)-4-piperidinyl] benzeneacetamide monohydrochloride. Lorcainide hydrochloride undergoes metabolic N-dealkylation to produce norlorcainide. Metabolism is the product of the first-pass clearance after oral administration. The basis for this observation is that norlorcainide is not produced in significant amounts in the body following intravenous administration. Norlorcainide is an important metabolite of the parent drug, as it is cleared slowly from the body and has a half-life that is approximately three times longer. Accumulation of norlorcainide is of considerable clinical importance because the metabolite is equipotent to the original drug.

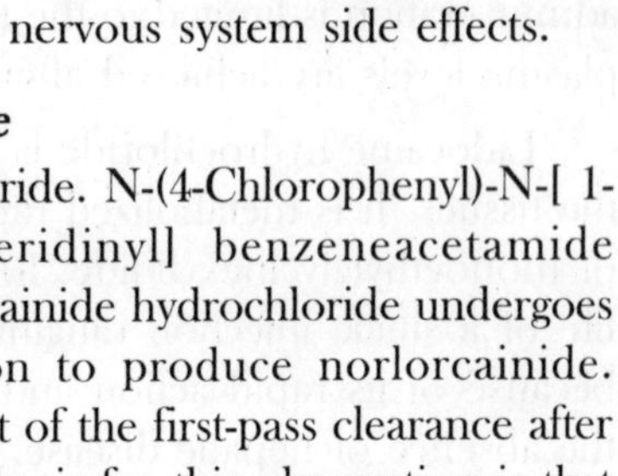

Fig. 4.5. Norlorcainide.

Moricizine

Moricizine, ethyl 10-(3-morpholinopropionyl)phenothiazine-2-carbamate, is a phenothiazine derivative used for the treatment of malignant ventricular arrhythmias.

New Chemical Entities

SUN-1165

SUN-1165 contains the highly basic pyrolizidinyl moiety of the less basic diethylamine group of lidocaine. Electrophysiological studies in vitro revealed that SUN- 1165 is a class IB drug. Compared to lidocaine and disopyramide, the compound had less central nervous system (CNS) toxicity and anticholinergic activity respectively.

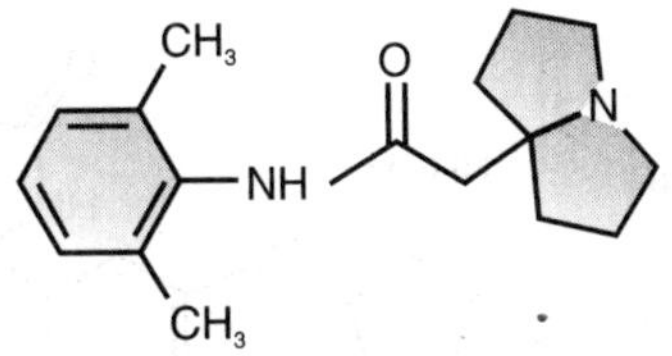

Fig. 4.6. SUN.

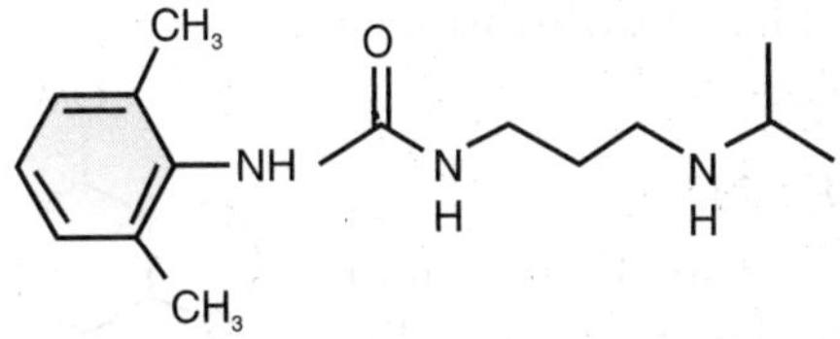

Fig. 4.7. Recainam (WY-423262)

Recainam (WY-423262)

Recainam (WY-423262) is class IB lidocaine-like drug. Therapeutic intravenous doses failed to produce CNS or negative inotropic side effects in dogs, and the drug safely and effectively reduced premature ventricular contractions in man.

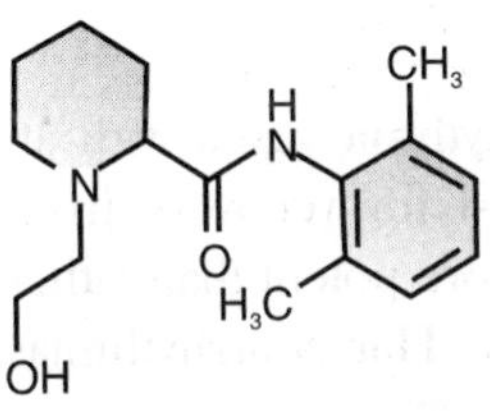

Fig. 4.8. Droxicainide

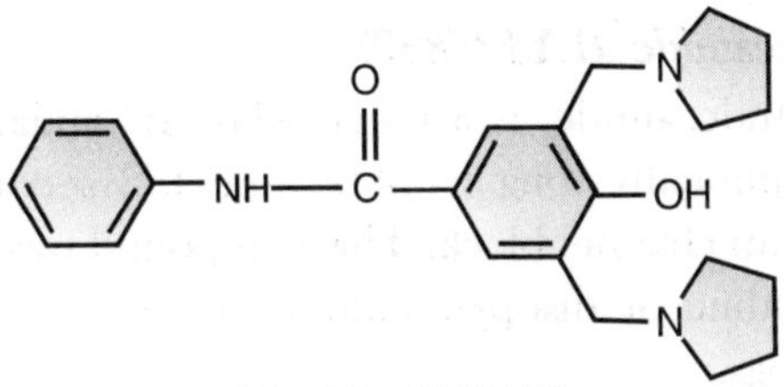

Fig. 4.9. ACC-9358

Droxicainide (ALS-1249)

Droxicainide (ALS-1249) was reported to have antiarrhythmic and local anesthetic properties that are qualitatively similar to lidocaine, but with a better therapeutic index over 24 hour period in infarct dogs than lidocaine. The piperidine and azepine analogues of droxicainide which lack the hydroxyethyl moiety are also active.

ACC-9358

Electrophysiological studies showed that ACC-9358 is a class I agent.

Propisomide (CM-7857)

Disopyramide continues to serve as a prototype for new antiarrhythmics. A series in which the phenyl group of disopyramide was replaced with alkyl moieties was synthesized. One of the compounds in this series, propisomide (CM-7857), was reported to possess a longer duration of action, fewer neurologic and gastrointestinal and less anticholinergic activity than disopyramide. As with disopyramide, the major metabolite in both plasma and urine was the mono-N-dealkylated analogue. A single dose of propisomide was compared to a variety of other antiarrhythmics in 10 patients. The compound was effective in 6; it increased PQ intervals, and had no effects on QRS or QT intervals.

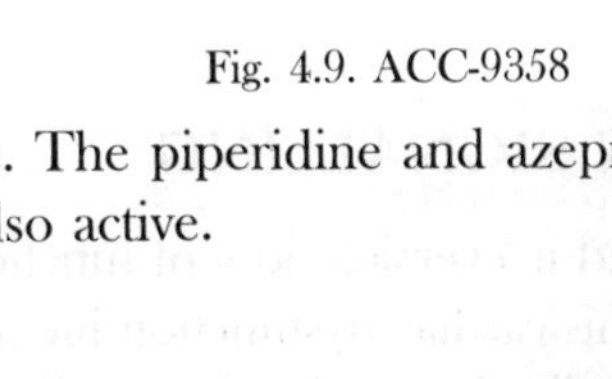

Fig. 4.10. Propisomide

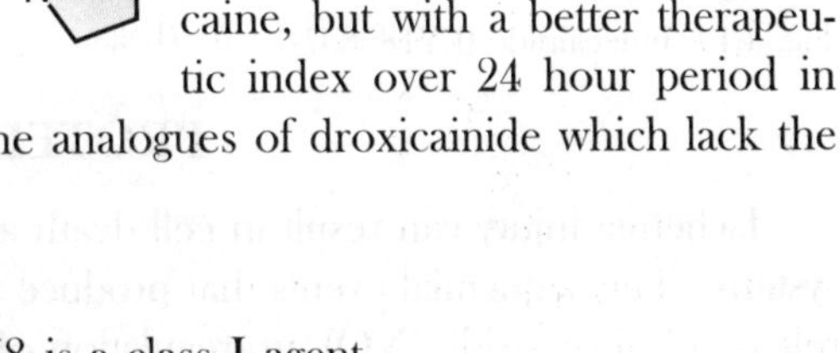

Fig. 4.11. AHR-10718

AHR-10718

AHR-10718 is the optimal compound in a series of aminoethylureas that suppressed arrhythmias in the ouabain-intoxicated and Harris dog models. The drug caused a use-dependent decrease in action potential upstroke velocity (V_{max}), Purkinje fibre conduction on velocity and APD.

Carocainide (MD77020)

Carocainide, a benzofuran antiarrhythrnic, decreased V_{max} in isolated papillary muscle and Purkinje fibres: it decreased the plateau amplitude and APD. The antagonized digitalis and infarction induced arrhythrnias in dogs.

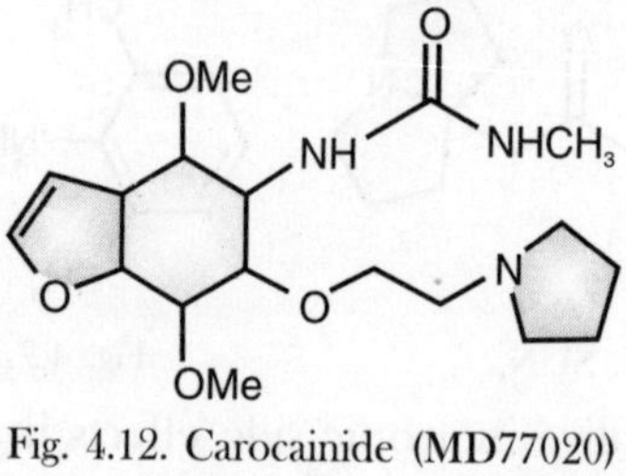

Fig. 4.12. Carocainide (MD77020)

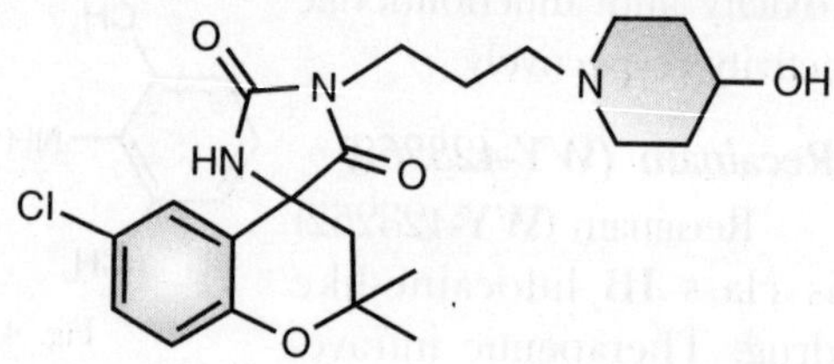

Fig. 4.13. E-0747.

E-0747

E-0747 was more potent than quinidine, disopyramide, lidocaine, or phenytoin. Moreover, at therapeutic doses, the compound appeared to have low potential for cardio-depression.

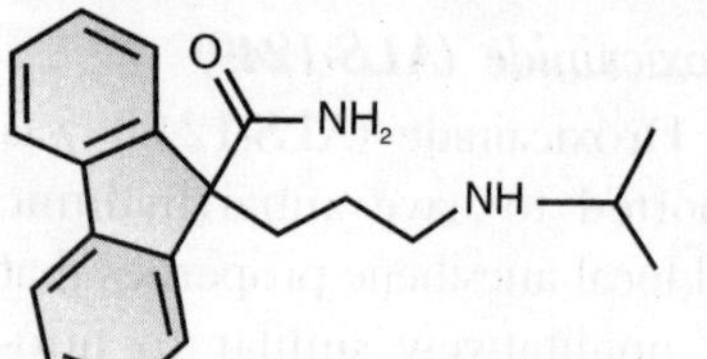

Fig. 4.14. Indecainide (LY185837).

Indecainide (LY135837)

Indecainide, is a potent class IC antiarrhythmic agent with an exceptionally long half-life time (52 seconds) for recovery from sodium channel block. This compound was more potent than either aprindine or disopyramide against ouabain or Harris arrhythmias in dogs.

PROTEIN SIGNALLING

Ischemia injury can result in cell death and irreversible loss of function in a variety of biological systems. The sequential events that produce this cardiac dysfunction include a decreased endothelial release of nitric oxide (NO), up-regulation of adhesion molecules on the endothelial surface leading to enhanced leukocytes-endothelium interaction and release of superoxide radicals. These radicals are largely responsible for producing cardiac dysfunction and enhanced necrosis. The time course of these events starts at 2.5-5 min post-reperfusion.

Signalling Pathways

An understanding of the intracellular signalling mechanisms by which cells protect themselves against ischemia-induced damage bears great clinical significance with respect to the treatment and prevention of tissue injury. Cells start to generate agonists especially those that signal via $G\alpha_1$, such as adenosine, acetylcholine, opioids and bradykinin, which bind to G protein coupled receptors (GPCR) and initiate a signalling cascade that involves activation of phosphoinositide-3-kinase (PI3K), endothelial NO synthase, tyrosine kinase, protein kinase c, glycogen synthase, mitogen-activated protein kinases, and other signalling pathways. Activation of these signalling pathways along with generation of reactive species leads to alterations in the activity of key mitochondrial proteins such as mitochondrial ATP-sensitive K^+ channels and the mitochondrial permeability transition pore.

Alterations on these mitochondrial proteins results in altered metabolism and inhibition of cell death, thus resulting in cardioprotection. Activation of GPCR leads to signalling via $G\alpha_1$ and

Gβγ and initiates a signalling cascade. Acetylcholine-induced protection leads to activation of PI3K pathways via GPCR transactivation of the epidermal growth factor, which also signals through Gβγ and lead to activation of PI3K. At the same time activation of GPCR lead to activation of mitogen-activated protein kinase (MAPK) pathway via Gα-dependant signalling, by trans-activation of the epidermal growth factor; MAPK pathway operate through sequential phosphorylation events to phosphorylate transcription factors and regulate gene expression.

PI3K generates phosphoinositides that localize kinases, such as phosphoinositide-dependant kinase (PDK), with substrates, leading to activation of downstream kinases such as protein kinase B (PKB, also known as Akt), mammalian target of rapamycin (mTOR) and p70S6-kinase. Both PDK1 and PKB play an important role in the activation of p70S6 kinase. PDK1 phosphorylation of thr 229 is required for activation of p70S6K. Prior phosphorylation of thr 389 by mammalian target of rapamycin (mTor) is necessary before PDK1 can phosphorylate thr 229. At the same time PKB phosphorylates and activates endothelial NO synthase (eNOS) and phosphorylates and inactivates GSK3β and the proapoptic BAD.

PI3K also play a role in the activation of protein kinase C (PKC), but the precise role of PI3K has not been determined; some studies suggested that PI3K activation of PKCε occurs via an eNOS-mediated mechanism. However, another study suggested that the reactive oxygen species (ROS), is generated by preconditioning is involved in activation of PKC. With regards to PKC there is increasing evidence for functional coupling of PKC to tyrosine kinase in the heart. Both PKC and tyrosine kinase show a cardioprotective effect via activation of transcription regulatory protein nuclear factor ($NF\text{-}_{K}B$) which occur through both tyrosine and serine phosphorylation of $I_{K}B\alpha$. Furthermore PKC activation has been shown to be important in activation of the mitoKATP channel, at the same time NO has been shown to activate mitoKATP channel.

PKCε also forms a complex with the components of the mitochondrial permeability transition (MPT) pore. However, the association of PKCε with components of the MPT pore does not demonstrate that this association is important in modulating MPT or cardioprotection. This study also reported that ERK and PKCε are contained in a multimeric mitochondrial signalling complex and that PKCε may lead to activation of ERK in this complex. One study reported that KATP channel opening can lead to activation of ERK, which is likely to be secondary to ROS generated opening of the mitoKATP channel.

CHEMOTHERAPY

Ischemic Heart Disease

This disease is usually caused by atherosclerosis which narrows the coronary arteries. Common risk factors include hypertension, genetic disposition, smoking, diabetes, and hyperlipidemia amongst a variety of other conditions. The manifestations of ischemic heart disease frequently results in sudden death, myocardial infarction or angina pectoris.

Nitrates

The nitrates in the body dilate blood vessels in three areas: viz. (i) venous circulation, which decreases venous return and preload on the heart, (ii) the arterioles, which reduce peripheral resistance and afterload. Both reduce stress on the myocardial wall and lower oxygen demand, (iii) the coronary arteries, especially in the presence of coronary spasm to improve the oxygen supply. A large number of patients tend to be tolerant to the anti-anginal effects of the nitrates as a result of depletion of the essential thio (-SH) groups especially if they are taken for prolonged periods (usually for more than 24 h).

Whenever it is necessary to prescribe, a nitrate-free period must be built-in into the treatment regimen on an individual basis for each patient. The nitrates are converted in the body to NO. This in combination with sulphydryl (-SH) groups forms nitrosothiols which thus activate the enzyme guanylyl cyclase to produce the secondary messenger, cyclic GMP. This causes the smooth muscle to relax with vasodilatation.

Specific instance of nitrates. Glyceryl trinitrate is well absorbed sublingually but, if taken orally, it is broken down by the liver by first pass metabolism. Isosorbide dinitrate (can be taken sublingually, orally or intravenously although, the absorption rate is slow, its duration of action is much longer). Isosorbide mononitrate may be taken orally but in general, its use is limited for practical reasons.

Beta-blockers

These types of drugs block the beta-adrenoceptors, that is, both beta-1 and beta-2, which are present in the heart to reduce the production of cyclic AMP. The beta-blockers may be either non-selective and block both types of beta-receptors, or sufficiently selective to bind to beta-1 receptors. The beta-blockers slow the heart and thereby reduce its force of contraction and decrease oxygen demand. On the other hand, they may also increase oxygen supply in response to a diastole when most of the blood flow in the coronary arteries is extended. Typical examples of beta-blockers: Atenolol: is a hydrophobic beta-1-selective blocker. Propranolol, is lipophilic and a non-selective beta-blocker and Metaprolol, is a lipophilic beta-1-selective blocker. Clinical uses include:

1. Hypertension,
2. Angina pectoris,
3. Following myocardial infarction, beta-1-selective blockers cause a reduction in risk of recurrence,
4. Useful in anxiety and manifestations of anxiety such as tremor and depression,
5. Arrhythmia.

Calcium-channel blockers

The calcium-channel blockers (antagonists) act by antagonising the movement of calcium through channels in the cell membranes. In cardiac muscle or vascular smooth muscle, any change in membrane potential causes calcium to enter the cell through the calcium channels that are voltage-dependent. The calcium channel blockers reduce the force of contraction of the heart and thus slow the heart rate in vitro. They also reduce smooth muscle contraction, and act as vasodilators of coronary and systemic arteries and alter cardiac rhythm.

Antagonisms

Calcium antagonists inhibit the slow inward current caused by entry of extra cellular calcium through the cell membrane of excited cells, particularly arteriolar smooth muscles, and cardiac arterial cells. Calcium ions play an essential role in the regulation of skeletal and smooth muscle contractility in the performance of the normal and diseased heart. The treatment of hypertension with the introduction of calcium channel blockers (CCBs), was a major breakthrough was infact approached when it was discovered that vascular smooth muscle is linked to the movement of calcium ions with the muscle cells from extra-corporeal circulation.

The channels in the cell membrane are normally occupied by calcium ions bound to receptor storage sites and function as voltage-dependant ionic gates. Their opening is controlled by the electronic gradient across the cell membrane. The gates open and allow the calcium ions to enter the vascular muscle cells upon stimulation by noradrenaline, which is released from the nerve endings. This causes the muscle cell system to contract in order to maintain blood pressure. Excessive stimulation, increases the heart rate and myocardial oxygen demand resulting in chronic hypertension and coronary spasm, causing a severe chest pain (angina).

Physiological control of calcium channel blockers

There are three major types of calcium channels blockers, viz. (i) voltage-dependent, (ii) receptor-operated, and (iii) stretch operated.

1. Voltage (potential)-dependent Ca^{2+}-channels (homologous to Na^+ and K^+ channels, consist of at least three types in the body, L, T, and N). L-type channels:
   (a) have large sustained conductance, become slowly inactive and are widespread in the cardiovascular system;
   (b) are responsible for the plateau phase (slow inward current) of action potential;
   (c) may trigger release of internal Ca^{2+};
   (d) are sensitive to Ca^{2+} channel blockers.

   Cardiac L-channels are regulated by cAMP-dependent protein kinase.

   T-type channels:
   (a) are structurally similar to L-type channels;
   (b) become inactive rapidly;
   (c) are involved in cardiac pacemaker activity, growth regulation and triggering contraction of vascular smooth muscles.

   T-type channels are not very sensitive to most of the L-type Ca^{2+}-channel blockers.

   N-type channel:

   They are found only in neuronal cells and are not very sensitive to Ca^{2+}- channel blockers used for treating cardiovascular disorders.
2. Receptor-operated Ca^{2+}-channels (e.g. alpha-adrenergic receptors): not very sensitive to Ca^{2+}-channel blockers.
3. "Stretch" operated Ca^{2+} channels are not particularly important in maintaining vascular smooth muscle tone as they do not appear to be sensitive to Ca^{2+}-channel blockers.

Two types of calcium channel blockers are used in clinical situations: those which are selective for L-type (long-lasting, large-currents, or slow) and those that are non-selective. In clinical practice, selective agents are primarily used. The selective calcium channel blockers share a similar antihypertensive mechanism of action: they inhibit the influx of extracellular calcium through L-type channels, resulting in relaxation of vascular smooth muscle and reduction in vascular resistance. The reason why calcium antagonists are useful as drugs is probably because they are a heterogeneous group of compounds which have marked differences in chemical structure, binding sites, tissue selectivity, clinical activity, and therapeutic effect.

Calcium antagonists are chemically and pharmacologically a diverse groups of drugs. These include verapamil and its analogues, the benzo-1,5-thiazepine related to diltiazem and the 1,4-dihydropyridines, for example, nifedipine. These were briefly discussed earlier. Typical examples of calcium-channel blockers: There are generally three types of calcium-channel blockers. They bind to different although related receptor sites on the calcium channel. Their effects vary with different tissues.

1. Verapamil is less effective on peripheral circulation but more active on the heart. It slows the heart and reduces the force of contraction and can be taken orally or intravenously.
2. Diltiazem is more effective on the heart compared with nefedipine and is more active on peripheral circulation than verapamil. It is metabolised by the liver.
3. *Dihydropyridines*: Examples are nifedipine, nitrendipine, nimodipine, nislodipine and felodipine. They mainly cause arterial vasodilatation and reduce blood pressure and afterload on the heart i.e. the load against which the heart ejects blood. It increases heart rate and the force of contraction. Nefedipine is orally well absorbed and is metabolised by the liver. It can be used in combination with beta-blockers in angina or hypertension. Although rare, combination therapy may precipitate heart failure. Verapamil in combination with beta-blockers has been reported to be dangerous especially when administered intravenously. It can cause severe hypotension and bradycardia.

The pharmacological effects of the newer types of dihydropyridine drugs show that they can improve efficacy and vascular selectivity with a longer duration of action.

Clinical uses include:

1. Hypertension;
2. Angina, especially due to coronary artery spasm;
3. Raynaud's phenomenon (nefedipine);
4. Supraventricular tachycardia (verapamil).

The binding sites for all three chemical types of calcium channel blockers are present in many tissues, including myocardium, smooth muscle, skeletal muscle, and glandular tissue. Each of the three selective calcium channel blockers interact with a specific receptor domain found on a large membrane-spanning protein that constitute a substantial portion of the L-type, voltage dependent calcium channel. These receptor sites are all located on the surface on the alpha subunit of the channel. The 1,4-dihydropyridine receptor is most accessible and is located on the surface of the channel. This receptor has been the most widely studied of the three groups, and therefore a relatively greater number of dihydropyridines have been designed to bind and inhibit at that site.

Congestive Heart Failure

Heart failure usually results from damage to the myocardium typically from ischemic heart disease, cardiomyopathy or myocarditis. Hypertension also contributes to cardiac failure especially if it is poorly controlled or results from damage to the heart valves and fluid overload. Apart from drugs, metabolic diseases such as, for example, thyrotoxicosis also makes a contribution. The overall result is that cardiac failure results in poor perfusion of vital organs and tissues and causes congestion of the lungs. Treatment of Congestive Heart Failure is mainly carried out by:

1. Diuretics,
2. Cardiac stimulants,
3. Angiotension converting enzyme inhibitors (ACE) inhibitors,
4. Angiotensin II receptor antagonists,
5. Cardiac glycosides, digoxin and digitalis.

Diabetes insipides

These are still the mainstay of treatment in cardiac failure, the most potent diuretics are loop diuretics such as frusemide and bumetanide. They act by promoting sodium and chloride excretion. That is, they promote water loss to relieve systemic and pulmonary congestion. They can act at different sites in the nephron in different ways. Other than diuresis, diuretics may work through different mechanisms. For example, thiazide diuretics which are less potent and more appropriate for elderly patients for maintenance therapy only partially effect lowering of blood pressure via diuresis. This effect is to enlarge the result of vasodilatation. In acute cardiac failure diuretics are used intravenously. Some examples of these types of drugs include: spironolactone, amiloride, trimterene, thiazides, frusemide, butetanide, ethacrynic acid, mercurials, mannitol, and xanthines. In general there are four major groups:

1. Thiazides,
2. Loop diuretics,
3. Potassium-sparing diuretics,
4. Osmotic diuretics.

Thiazides and related drugs

Thiazides inhibit the absorption of sodium ions and chloride ions at the distal convoluted tubule. Water is lost with the sodium ions. More sodium ions then reach the distal tubule, where it is exchanged with potassium ions. Thiazides also cause potassium loss. Magnesium is also lost while Ca^{2+} is retained. The principal sites of action are the distal nephron, distal tubule, Loop of Henle, and proximal tubule. The maximal hypotensive response to thiazides can be reached at relatively low doses. They are ineffective in the presence of renal impairment. At one time, thiazides were used in extremely high doses with a high incidence of adverse effects. Nowadays, thiazides are used in low doses as they are equally effective in hypertension and result in fewer problems.

Typical examples are: Bendrofluazide which is well absorbed after oral administration. It causes diuresis within approximately 1-2 h, with lasting effect for up to 12 h and is metabolized by the liver.

Pharmaceutical applications

1. Congestive heart failure,
2. Hypertension,
3. Congestive cardiac failure.

A number of adverse reactions can occur such as dehydration; this is a common risk in the use of all diuretics especially in the elderly patients. Other effects include: hypokalemia (usually mild), hyperuricemia/gout, impaired glucose tolerance/diabetes mellitus, insulin resistance (significance is uncertain), Hypercholesterolemia (usually in short-term therapy rather than in long-term therapy), impotence, and hypercalcemia. Contraindications are gout, diabetes mellitus, and renal disease.

Loop diuretics

Loop Diuretics inhibit the resorption of chloride ions in the ascending limb of the loop of Henle with loss of chloride ions and sodium ions. Water loss with electrolytes then follows. The responsive action of loop diuretics is dose related, and is mainly effective even in renal impairment, unlike thiazides. They have a much greater diuretic effect compared with the thiazides. Their use in acute pulmonary oedema causes venodilatation when administered intravenously. They also reduce preload before diuresis starts. Typically, like the thiazides, the loop diuretics cause potassium loss although the unlike effect is that they cause Ca^{2+} loss. Typical examples are: frusemide or bumetamide which act within 10 minutes of an IV dose. After oral administration, their effects tend to peak at about 1 hour with lasting effects of up to 4-6 hours.

Pharmaceutical applications

1. Edema, and
2. Congestive heart failure.

Typical adverse effects include: hypokalemia, renal impairment (which results from dehydration and pre-renal failure and direct toxic effects on the kidneys), and ototoxicity (usually deafness with frusemide). Typically, drugs such as probenecid and the non-steroidal anti- inflammatory drugs interfere with the action of diuretics. Loop diuretics can usually be combined with potassium sparing diuretics in order to reduce potassium loss. Similarly, they can be combined with ACE inhibitors in hypertension or heart failure. However, diuretics can reduce excretion of lithium ions which can potentially lead to serious toxicity.

Potassium-sparing diuretics

These types of diuretics are chemically a diverse group that, unlike other diuretics, retain potassium ions. They tend to act in the distal convoluted tubule where potassium and sodium ions are exchanged, and are thus considered to be weak diuretics. Their main use is to treat or prevent hypokalemia which is caused by thiazides or loop diuretics. In the past, potassium supplements were widely recommended, although very high doses are needed and this apparently reduces patient compliance.

It is not normally necessary to treat or prevent a diuretic-induced hypokalemia, but if needed, potassium sparing diuretics have been indicated for clinical use. Spironolactone is an aldosterone antagonist. Aldosterone itself causes sodium ion retention with potassium loss. In contrast, spironolactone displays the opposite effect. Amiloride and Triamterene both interfere with sodium and/ or potassium ion exchange. All these diuretics are administered orally.

Pharmaceutical applications

Use of spironolactone diuretic is limited due to its carcinogenicity, which can cause:

1. Ascites (from liver disease);
2. Resistant congestive heart failure (with loop diuretics);
3. Primary hyperaldosteronism (Conn's syndrome).

Amiloride and Triamterene are useful in:

1. Diuretic-induced hypokalemia,
2. Oedema.

Typical adverse effects are: hyperkalemia and gynecomastia (with spironolactone). Again, combinations with potassium supplements or ACE inhibitors may lead to serious hyperkalemia.

Osmotic diuretics

These drugs are only employed under restricted circumstances. The compounds are filtered in the glomerulus and excreted. Due to their osmotic effects water is retained within the tubule. This causes diuresis of water with some loss of electrolytes.

Pharmaceutical applications

1. Mannitol is given by IV for the treatment of oliguria in incipient renal failure in order to lower intracranial pressure. It is also used in glaucoma.
2. Generally, hypernatremia, hyperkalemia, or dehydration are common side-effects. Plasma volume is increased which may precipitate cardiac failure.

Angiotension converting enzyme inhibitors

The renin–angiotensin–aldosterone is involved in the regulation of vascular tone. Although, their precise mode of action is still not entirely clear it is thought that they exert their clinical effect at least in part by vasodilation. ACE inhibitors are widely distributed in the body; although the systemic effects of the inhibitors and their effects on local angiotensin production in the tissues is unclear. They appear to prevent the formation of angiotensin Il and thus block its vasoconstrictor effects and decrease the production of aldosterone and fluid retention. They breakdown inflammatory peptides (typically bradykinin), this can contribute to its anti-hypertensive effect. Typical examples are: captopril which was the first ACE-inhibitor is given orally. It is cleared by the kidneys. Enalapril is a prodrug which is also given orally and is activated to the active, enalaprilat by the liver. Clearance of the drug occurs via the kidneys. Enalapril is usually given once daily.

Pharmaceutical applications

1. *Cardiac failure.* Angiotension converting enzyme inhibitors reduce mortality.
2. *Hypertension* (*widely employed*). ACE inhibitors may be especially useful in the treatment of hypertension in diabetes patients, due to a delay in the onset of microalbuminuria. Both captopril and enalapril appear to be well tolerated.
3. *Myocardial infarctian.* Particularly apparent in patients who experience heart failure as a complication. The mechanism is somewhat unclear but appears to involve a reduction in left ventricular enlargement.

A number of diverse side-effects have been observed. Some examples include:

1. *Severe hypotension.* Especially in cardiac failure, on first dose, or patients on high-dose diuretics. The dose is usually reduced as much as possible prior to taking ACE inhibitors.
2. *Hyperkalemia.* Potassium losing effects of aldosterone are blocked.
3. Impairment of renal function through decrease in renal perfusion and intra renal hemodynamics.
4. *Cough.* Although very common but not serious, probably as a result of potentiation of bradykinin.
5. *Angioedema.* This is more serious and may be caused by potentiation of bradykinin.
6. Other adverse effects, which usually resolve when treatment is stopped, includes rashes, loss of taste and gastro-intestinal disturbances.

Contraindications include:

1. Renal artery stenosis is often asymptomatic but should be suspected in patients with hypertension and peripheral vascular disease.
2. ACE inhibitors should be used with caution in other forms of renal disease, as they have been reported to precipitate a reduction in renal function.

Coadministering with potassium sparing diuretics or potassium supplements presents a risk of serious hyperkalemia. ACE inhibitors are most effective in hypertension when used with a diuretic, although they increase dependence on the renin-angiotensin system.

Angiotensin II receptor antagonists

These are a group of drugs which block the receptor at which angiotensin II acts. Their effects are generally the same as ACE inhibitors. Other benefits, include maintenance of renal blood flow, although this has not been proven. Losartan was the first drug of this type and is currently used for hypertension.

Cardiac glycosides, digoxin and digitalis

At the cellular level, digitalis compounds generally block the exchange of intracellular sodium ions for extracellular potassium by inhibiting Na^+/K^+-ATPase in the cell membrane; This increases intracellular sodium ions and encourages exchange for $Ca^{2þ}$ in order to raise its intracellular Ca^{2+} levels to increase the force of cardiac contraction (i.e. a positive inotropic action). These effects cause characteristic changes in the ECG. In the body, digitalis and related compounds stimulate the vagus and increase vagal tone. These oral inotropes are most used in clinical practice. Digoxin is one of the most commonly prescribed.

The principle effect of digoxin is to increase myocardial contractility. Some evidence suggests that in patients with a large dilated heart, digoxin reduces wall tension, improves oxygen usage, educes the end-diastolic volume and improves cardiac function. In chronic failure, digoxin is of modest value and is useful to control heart rate in the patient with atrial fibrillation. Digoxin is well absorbed and can be taken orally or parenterally. It has a long pharmacological half-life of 36-38 h and is excreted unchanged by the kidneys with a very narrow therapeutic ratio (the relationship between clinical and toxic effects) and therefore, it is useful to monitor the plasma levels of digoxin. For rapid effects the dose can be loaded initially and followed by a maintenance dose. Digoxin is metabolised by the liver.

Pharmaceutical applications

1. Atrial fibrillation (this is to slow the ventricular response)
2. Congestive heart failure. This is probably useful in sinus rhythm, but there is some uncertainty).

About 15% of patients on long term treatment will experience toxicity at some time. Severe overdoses can lead to hyponatremia and hyperkalemia due to cellular effects. Hypokalemia or hypercalcemia both increase the risk of digitalis toxicity. Old age or Hypothyroidism also increases sensitivity to standard doses. The first manifestation of digoxin toxicity is usually gastro-intestinal disturbance, that is, anorexia, nausea, vomiting, anorexia, and occasionally diarrhoea. Thereafter any type of cardiac arrhythmia or conduction problem may occur.

Neurological effects include: fatigue, depression, malaise, insomnia altered colour vision (green/ yellow are predominant), and confusion. Cardiovascular effects include any form of cardiac arrhythmia especially ventricular bigeminy and bradycardias and complete heart block. Electrocardiogram may show first-degree heart block at an early stage, although serious arrhythmias may occur without prior warning. Gynecomastia is also another unpleasant side effect. Contraindications associated with digitalis compounds include:

1. Myocardial ischemia, which may increase oxygen demand;
2. Renal impairment, which may occur especially with digoxin;
3. Precipitation of heart block;
4. Disease of the AV node which may cause heart block.

Interaction with diuretics may worsen digoxin toxicity and cause hypokalemia. Amiodarone, quinidine and verapamil all increase the levels of plasma digoxin levels by causing a reduction in its clearance. Additionally, interaction with cholestyramine may block absorption of digoxin.

Clinical treatment of digoxin toxicity mainly depends on severity but generally, involves the following measures:

1. Withdrawal of drug;
2. Intravenously administer correct low serum Kþ;
3. Use lignocaine or phenytoin to treat ventricular arrhythmias;

4. In life-threatening situations, digoxin-specific can be used since antibodies will bind to and inactivate digoxin.

Cardiac Arrhythmias

The cardiac arrhythmias have already been well covered, some of which, do not require treatment; for example, atrial or ventricular ectopics rarely need treatment. Some of the others are more serious and require treatment for prophylaxis or termination including correction of the underlying cause. Antiarrhythmic drugs are usually classified according to the Williams system. This is based on the changes in the action potential that is produced by drugs in isolated cardiac cells. A number of drugs may have mixed properties and do not fit neatly into any particular class. The classification is not especially useful and is of limited clinical value.

Hypertension

Hypertension (high blood pressure) is considered to be one of the major medical problems of the world. In the United States nearly 25 million cases are diagnosed each year. Research shows that effective control of high blood pressure can reduce associated morbidity (notably stroke, kidney, and congestive heart failure) and prolong life. It may be argued whether hypertension is a disease or merely a risk factor towards disease. Nevertheless, hypertension increases the chance of developing organ damage and is a major cause of death and disability. Diagnosis of hypertension must be made carefully and with consideration since once made, it carries life-long implications in terms of medical care and management. Blood pressure may be initially high due to patient apprehension and unfamiliarity with the surroundings. Repeat readings usually show a fall although, subsequently steady values are obtained. This is the basal blood pressure upon which the diagnosis of hypertension is based and decisions are made as to what treatment should be adopted.

In hemodynamic terms:

$$\text{Blood pressure} = \text{cardiac output} \times \text{peripheral resistance}$$

Hypertension can therefore be a consequence of either a raised cardiac output and /or increased systemic vascular resistance. Whilst no single causative mechanism can be identified, the following factors are all thought to play a part in the development of hypertension. It has been suggested that this mechanism is upset in hypertension due to increased resistance of blood vessels in the kidneys, which in turn upsets the filtration of water and increases the absorption of sodium (Na^+). Thus, blood volume expands and blood pressure increases to set up a new steady-state in terms of sodium. The blood pressure at this point is significantly high. In order to influence the regulatory mechanisms a variety of nervous and chemical stimuli are required. However, it should be remembered that the stimulus from nerve endings is chemical in nature and nervous control of blood pressure occurs through the autonomic nervous system.

Neural impulses in parasympathetic or sympathetic fibres achieve their effects by discharge of specific neurotransmitter substances, acetylcholine and noradrenaline, respectively. The beta-receptors play a key role in this mechanistic pathway. Baroreceptors and chemoreceptors are also involved in nervous control of the heart and have an inverse relationship between arterial blood pressure and heart rate. The World Health Organization has suggested that blood pressure levels which are consistently above 160/95 mmHg should be regarded as "*hypertension*". However, many clinicians now feel that these levels are too high and that the threshold for diagnosis should be lower. In general, levels persistently above 140/90 mmHg may be regarded as being consistent with established hypertension.

In this respect, the use of the word persistent is important as some patients have occasional high readings which are interspersed with normal values commonly known as "labile" hypertension. Around 25% of patients with this type of hypertension are likely to develop "established" hypertension. The prevalence of hypertension is often asymptomatic and carries both morbidity and mortality. It is associated with increased risks from ischemic heart disease and cerebrovascular disturbances, and in severe cases, with heart and renal failure. The therapeutic advances in the treatment of hypertension and related conditions are now described.

Blood pressure

Blood pressure is determined by a combination of genetic and environmental factors. The chances of being hypertensive are greater within a family with a history of hypertension. The most constant environmental influence on hypertension is obesity. The mechanisms by which obesity appears to cause high blood pressure are not fully understood. However, it is not certain whether it is due to the high intake of nutrients although blood pressure is generally found to be lower in vegetarians. A decrease in the daily intake of salt may also reduce high blood pressure, but there is still much conjecture on this subject. Other factors such as alcohol consumption or stress also increase blood pressure.

Hypertension presents a unique problem in therapeutics. It is usually a life long disease that causes few symptoms until it reaches an advance stage. Pharmacological treatment varies depending on the severity of hypertension. In mild cases a single drug is usually sufficient such as a beta-blocker or a diuretic. The definition of hypertension remains controversial. Unfortunately, there is no natural dividing line between "*hypertension*" and "*normotension.*" Blood pressure varies widely over 24 h, both in those who have high and normal blood pressure.

Effects of hypertension

In the United Kingdom, high blood pressure is one of the most common medical condition that requires long-term treatment. About one sixth of all men and women have high blood pressure. With increasing age, the blood pressure levels rise considerably. Annual deaths in the United Kingdom from hypertensive disease have been reported to be approximately 500 per million of the population. Since high blood pressure has no warning signs and no clear symptoms, it is often called a silent killer. People may not be aware of it until they have trouble with their heart, brain, or kidneys. If high blood pressure is untreated, it may cause:

1. The heart to enlarge and lead to heart failure.
2. Aortic aneurysm (consistently high levels of blood pressure accompanied by degenerative changes in the vessel wall leading to a dilation of the aorta. The dilation may form an aneurysm, that is, like a bump or swelling protruding from the vessel resulting in a thin endothelial wall. Persistent hypertension if left untreated.
3. May precipitate rupture of the aorta and cause death.
4. The blood vessels in the kidneys to narrow and cause kidney failure.

5. The arteries in the body to "harden" faster, especially those associated with the heart, brain and the kidneys and lead to a heart attack or stroke.
6. Peripheral arterial disease (hypertension increases the incidence of atheroma of the legs. Intermittent claudication and ischemia of the feet are common clinical manifestations.

Etiology of hypertension

Hypertension is classified according to the causes responsible for rise in blood pressure. Hypertension is often called essential or primary hypertension if there is no obvious underlying cause (i.e. idiopathic). More than 90% of all high blood pressure cases are considered to be "*essential* or *primary hypertension.*" Conversely where there is an apparent causal link, for example to an endocrine disorder or kidney disease, it is then classed as secondary hypertension. There are several factors which contribute to high blood pressure such as:

1. Genetic factors which are often associated with a family history of high blood pressure.
2. Environmental factors: Numerous factors have been related to the development of hypertension, but only the following appear to be important:
   (a) *Obesity.* Blood pressure rises with increasing obesity.
   (b) *Alcohol intake.* Consumption of alcohol acutely raises blood pressure.
   (c) *Salt intake.* Recent studies have shown that high-sodium intake is directly associated with higher blood pressure.
   (d) *Smoking.* Smoking increases blood pressure leading to damage of the blood vessels and the heart.
   (e) Stress and lack of regular physical exercise.

Several factors are often involved in elevating blood pressure. Apart from family history, changes in the above factors can help to treat blood pressure.

Effective methods of hypertension

Treatment with drugs is one of the effective methods for attaining larger reduction in blood pressure. Using drugs to control hypertension without producing unacceptable side effects is important. In many countries management of hypertension is now based on the "stepped care" approach i.e. a drug regimen which aims to start the patient off with as low a dose as possible with, for example, a diuretic or beta-blocking agent. If this single drug therapy fails to control blood pressure (i.e. mild hypertension: blood pressure 140-160/90-105 mmHg) adequately then two types of drugs are given together to reduce blood pressure (i.e. moderate hypertension: blood pressure 160-200/105-115 mmHg). Should this combination fail to give acceptable clinical results then a third drug is added to the regimen to reduce blood pressure [i.e. severe, accelerated hypertension: blood pressure greater than 115 mmHg (diastolic)].

With the introduction of other drugs such as angiotension converting enzyme inhibitors and calcium antagonists, this traditional approach is now being challenged. However, as a general framework of the understanding of the overall approach to the treatment of hypertension, "stepped care" represents a good starting point for treatment of hypertension.

Physiological role of antihypertensive drugs

The history of high blood pressure drugs goes back to 1950s when the first ganglionic blocking drugs were introduced. These drugs interfered with the constriction of blood vessels by nervous control and were the first successful treatment for malignant hypertension. A new generation of drugs have been discovered recently. Many patients can be satisfactorily treated with a single antihypertensive drug; the choice of the drug therapy is determined by safety convenience and freedom form side effects.

Another group of patients will require a combination of two or three antihypertensive agents to give a good control with fewer side effects as previously described. The principal drugs used in single drug treatments of hypertension are diuretics, beta-adrenoceptors, ACE inhibitors and calcium antagonists. Some of the most commonly used drugs to treat high blood pressure are now briefly reviewed:

1. *Diuretics.* Sometimes called "water pills", these drugs flush excess water and sodium from the body by increasing urination. This reduces the amount of fluid in the blood and flushes sodium from blood vessels so that they can open wider, with increased blood flow and reduction of pressure against the vessels. Diuretics are often used in combination with other blood pressure drugs.
2. *Beta-blockers*: These drugs slow the heartbeat by blocking the effects of nerve impulses to the heart and blood vessels, thereby lessening the burden on the heart.
3. *ACE (angiotensin-converting enzyme) inhibitors.* They inhibit formation of the hormone angiotensin II and cause the blood vessels to narrow and thereby increase the blood pressure.
4. *Calcium channel blockers.* They prevent calcium from entering into the muscle cells of the heart and blood vessels, in order to relax the blood vessels and decrease the blood pressure.
5. *Alpha-beta blockers.* These combine the actions of alpha-blockers (which relax the blood vessels) and the beta-blockers (which slow the heartbeat). Their dual effect reduces the amount and pressure of the blood which passes through the blood vessels.
6. *Vasodilators.* These agents directly dilate the blood vessels by relaxing the smooth muscle.

Special treatments

Lowering of blood pressure urgently is only indicated in patients with hypertensive encephalopathy, or hypertension-induced renal or heart failure although, these conditions are rare. High blood pressure in the absence of the above conditions do not warrant urgent blood pressure reduction, as it may be dangerous to reduce cerebral or coronary perfusion. In cases where urgent treatment is required, the patient should be admitted to hospital for rest and very carefully monitored. Usually oral therapy with beta-blockers, calcium antagonists or ACE inhibitors may be adequate. In very rare cases where parenteral therapy is necessary, sodium nitroprusside is used by intravenous infusion. Sodium nitroprusside is a powerful arterial and venous vasodilator. It acts very rapidly with a very short duration of action.

Blood pressure rises within a minute of the infusion being stopped. It allows control of blood pressure while other slower-acting drugs are being introduced. Prolonged use for periods of more than 24 h. causes the production of cyanide (a toxic metabolite). Labetalol, a combined alpha- and beta-blocker may also be administered intravenously. Resistant hypertension can occur with a failure to respond to triple therapy with standard drugs. This must be fully investigated to exclude the possibility of secondary hypertension.

Minoxidil may be used in resistant hypertension. It is a vasodilator which causes fluid retention and tachycardia. It should only be used with diuretics and a beta-blocker. A unusual side effect of the drug is the stimulation of male pattern hair growth. Therefore, it is generally best to avoid the use of this drug in women.

Angina Pectoris

Angina can be managed by treating the acute attack by:

1. Resting the patient.
2. Administering an oral dose of glyceryl trinitrate tablets (GTN) or spray of medication. Patients who do not respond to attacks within minutes of taking the first or second tablet must seek medical attention without delay as this could indicate the early stages of a myocardial infarction (heart attack).

Prophylactic therapy must be taken regularly. There are generally three types of drugs employed: beta-blockers, calcium-channel blockers, and oral nitrates. Each type of drug differs in its mode of action, although a combination of drugs can be used for the same patient. Aspirin is frequently used as long as there are no contraindications. Aspirin reduces the risk of threat of any serious thrombotic complications that may arise. For patients who previously had myocardial infarction, there is a good chance that aspirin may reduce the risks of further infarctions.

Myocardial Infarction (Heart Attack)

This usually occurs when a blood clot blocks a coronary artery and causes coronary thrombosis in a ruptured atherosclerotic plaque, occluding a coronary artery, so that part of the myocardium ceases to function. The patient will feel a severe chest pain. In addition to loss of myocardium, the patient is likely to be at risk from potentially fatal arrhythmias. Therefore, prompt action should be taken to reduce further loss of myocardium and to improve survival of the patient. About one in ten patients who have unstable angina may develop acute myocardial infarction. Management of myocardial infarction often requires alternative drugs in order to deal more effectively with further complications. For example, the following measures have been indicated:

1. Use of sublingual glyceryl trinitrate (if a severe attack of angina occurs, as opposed to myocardial infarction);
2. Use of aspirin (oral): providing there is no contraindication. This can improve survival of the patient.

Other treatments involve:

1. Pain relief by IV administration of opiates usually with an anti-emetic drug.
2. Use of thrombolytic drugs if no contraindication exists.
3. Other types of treatments may also be employed to decrease mortality although they are generally less widely adopted, including the use of atenolol by IV or the use of nitrates.

CONCLUDING REMARKS

The cardiovascular system is a complex network of blood vessels with the heart at its center. The function of the system is to circulate oxygenated blood and nutrients to the various tissues in the body and return deoxygenated blood to the lungs for reoxygenation. The pharmacology of the sympathetic nervous system is important in cardiovascular medicine in view of the stimulating action of the sympathetic receptors. The beta-blockers are among some of the most commonly used drugs. They cause the heart rate to slow down and decrease cardiac contractility. Systemic hypertension is a major cause of death and disability because it is prevalent, often asymptomatic, and carries morbidity and mortality.

The problem of identifying and treating hypertension particularly in the mild to moderate category arises from the fact that uncomplicated hypertension is largely considered to be a symptomless condition. However, hypertension is usually a consequence of either raised cardiac output and/or increased systemic vascular resistance. Complications of raised blood pressure, arrhythmia, and atrial fibrillation can lead to further impairment and cause congestive heart failure. Although clinical manifestations of coronary heart disease do not differ between hypertensive and normotensive patients, the degree of severity of atherosclerosis is generally greater.

The classical clinical feature of coronary heart disease is angina pectoris, the course of which can result in myocardial infarction. There is now an increasing interest in the clinical pharmacology of drugs in the treatment cardiovascular diseases. Recognition of the central role and the pharmacological action of some of the most commonly used drugs in cardiovascular therapy, such as the antihypertensives, calcium channel blockers, beta-blockers, alpha-blockers, *angiotension converting enzyme* (ACE) inhibitors, ganglion and adrenergic neurone blockers, vasodilators, diuretics, cardiac glycosides, and drugs for combating arrhythmias, have all been well covered herein to demonstrate their pharmacological importance and widespread use in clinical practice as drug therapies in cardiac and advanced congestive failure.

5

Cardiac Drugs

Several cardiac drugs manifest some form of interaction with each other or with non- cardiac drugs. It is of paramount importance for the physician to be aware of these interactions, which may be potentially harmful to patients or may negate salutary effects. In addition, the prescribing physician must be conscious of these interactions so as to avoid errors that may provoke medicolegal action. The number of cardiac drugs available to the clinician has increased by more than 100% over the past 20 yr. Also, the drug armamentarium for the treatment of psychiatric, rheumatic, neurologic, infectious, and gastrointestinal (GI) diseases has expanded greatly. Consequently, drug interactions have increased. Although drug interactions may have been briefly described for drug groups in earlier chapters, this chapter gives a structured summary of interactions. This chapter deals with:

1. The adverse cardiovascular effects of noncardiac drugs.
2. The interaction of cardiovascular drugs with agents used for treating diseases of other system and other agents with pharmacologic properties, such as caffeine, alcohol, and tobacco.
3. The interactions that occur between cardiac drugs.

Drug interactions are usually:

1. *Pharmacokinetic.* Occurring when one drug alters the absorption, distribution, metabolism, or excretion of another.
2. *Pharmacodynamic.* Occurring between drugs that have similar or opposite pharmacologic effects or adverse effects; for example, these effects may occur because of competition at receptor sites or action of the drugs on the same physiologic system. In some instances, the hemodynamic effects of one agent increase or decrease the hemodynamic effects of another agent.

CO-OPERATIVE EFFECT OF CARDIOVASCULAR DRUGS

Each cardiac drug, (or drug group) co-operative effect is discussed under the following categories:

1. Cholinergic effects.

2. Cardiovascular effects:
   (a) Vasodilator (arterial or venous, afterload or preload reduction), hypotension, presyncope.
   (b) Inotropic: Positive or negative; alteration of ejection fraction (EF); propensity to precipitate congestive heart failure (CHF).
   (c) On the sinoatrial (SA) or atrioventricular (AV) node.
   (d) On the conduction system.
   (e) Proarrhythmic.
3. Hematologic, including anticoagulant activity and/or immune effects.
4. Plasma levels.
5. Renal clearance of the drug; effect on renal function and serum potassium (K^+) concentration.
6. Hepatic and GI.

ACE Inhibitors/ARBs

Effect on blood vascular system

When angiotensin-converting enzyme (ACE) inhibitors are used in combination with vasodilators, hypotension may occur. Preload-reducing agents, such as nitrates or prazosin, may cause syncope or presyncope.

Diuretics stimulate the renin-angiotensin system (RAS) and enhance the antihypertensive effects, producing a salutary interaction in hypertensive individuals, but hypotension may ensue in patients with CHF. The hypotensive effects are longer lasting with long-acting ACE inhibitors.

Effect on nephric tubules

1. *Probenecid* inhibits the tubular excretion of ACE inhibitors and increases the blood levels of these agents.
2. ACE inhibitors may cause a 20-30% decrease in digoxin clearance and an increase in digoxin levels.
3. ACE inhibitors and ARBs may decrease glomerular filtration rate (GFR) in some patients with renal dysfunction and may result in decreased clearance of other cardiac drugs that are eliminated by the kidney. Interaction may occur with renally excreted beta-blockers, nadolol, atenolol, sotalol, and antiarrhythmic agents.
4. Serum K+ may increase to dangerous levels if potassium-sparing diuretics (amiloride, eplerenone, triamterene, spironolactone, Moduretic, Moduret, Dyazide) are used concurrently. This occurrence is enhanced in patients with renal dysfunction.
5. A normal serum creatinine level in patients with a lean body mass may not reflect underlying renal dysfunction, and *caution* is necessary, particularly in the elderly.
6. Interaction with cyclosporine may cause hyperkalemia.
7. These agents may decrease the diuretic effect of furosemide.

Table 5.1 Angiotensin-converting enzyme inhibitors: potential interactions

1. Aspirin: Appears to decrease salutary effects
2. Potassium-sparing diuretics: amiloride, Moduretic, Moduret, triamterene, Dyazide, Dytide, spironolactone (Aldactone), Aldactazide
3. Preload-reducing agents: Diuretics, Nitrates, Prazosin
4. Renally excreted cardioactive agents
   Antiarrhythmics; beta-blockers (atenolol, sotalol, nadolol)
5. Digoxin
6. Agents that may alter immune status: Allopurinol, Acebutolol, Hydralazine, Pindolol, Procainamide
7. Tocainide
8. Furosemide
9. Probenecid
10. NSAIDs
11. Lithium
12. Cyclosporine
13. Phenothiazis
14. Imipramine

Immune adverse effects

The risk of immune adverse effects may increase with concurrent use of agents that alter the immune status: assess antinuclear antibody levels and also assess for neutropenia with concurrent use of acebutolol, allopurinol, hydralazine, pindolol, procainamide, and tocainide.

Alimentary canal

The absorption of fosinopril is reduced by antacids.

ANTIARRHYTHMIC AGENTS

Adenosine

The electrophysiologic effects of adenosine are competitively antagonized by methylxanthines (theophylline) and are potentiated by dipyridamole, which inhibits its cellular uptake, and severe hypotension may occur.

Amiodarone

Effects on blood vascular system

1. The proarrhythmic effect of amiodarone is low, but interaction may occur and may increase the risk of torsades de pointes when amiodarone is used concurrently with the following agents: diuretics, which cause hypokalemia, class IA agents (quinidine, disopyramide, procainamide), sotalol, tricyclics, phenothiazines, and erythromycin.
2. Amiodarone has a mild vasodilator effect, and interaction may occur with anti-hypertensive agents, causing hypotension. This interaction is more prominentwith intravenous amiodarone.

3. Because of a mild inotropic effect, CHF may be precipitated, especially when amiodarone is combined with agents that possess negative inotropism (e.g., beta-blocking agents, verapamil, or diltiazem).
4. Combination with verapamil or diltiazem may cause sinus arrest or AV block.

Table 5.2 Amiodarone: potential interactions

1. Antihypertensive agents
2. Drugs that are negatively inotropic: Verapamil, diltiazem, beta-blockers
3. Agents that inhibit SA and AV node conduction: Diltiazem, verapamil
4. Agents that ↑ the QT interval Class IA agents: quinidine, disopyramide, procainamide Sotalol Tricyclic antidepressants Phenothiazines Erythromycin (other macrolides) Pentamidine Zidovudine
5. Agents that decrease serum K^+ concentration (diuretics)
6. Agents that are renally eliminated: digoxin, flecainide, procainamide
7. Anticoagulants
8. Cimetidine

Oral anticoagulants

Their activity is increased.

Plasma levels

Digoxin levels may double because amiodarone decreases renal clearance of digoxin. Flecainide and procainamide levels increase.

Disopyramide

Effects on blood vascular system

The negative inotropic effect of disopyramide is increased by beta-blocking agents, verapamil, diltiazem, and flecainide. Proarrhythmic effects, in particular torsades de pointes, are increased with agents that prolong the QT interval or cause hypokalemia. Sinus and AV node effects may occur when disopyramide is used concurrently with drugs that depress the sinus node (verapamil, diltiazem, digitalis).

Cholinergic effects

Because disopyramide inhibits muscarinic receptors, the anticholinergic activity may cause constipation. Thus, the drug should not be combined with verapamil. Disopyramide may precipitate glaucoma, and this effect may be counteracted by timolol or betaxolol.

Flecainide

Effects on blood vascular system

The powerful, negative, inotropic effect of flecainide is worsened by negative inotropic agents; interaction occurs with verapamil, diltiazem, beta-blockers, and disopyramide; intraventricular conduction defects may occur when the drug is combined with class IA agents.

Plasma levels

Plasma levels increase with concomitant *amiodarone* administration; thus, the flecainide dose should be decreased. Flecainide causes an increase in *digoxin* levels.

Lidocaine; Lignocaine

Plasma levels

1. Lidocaine plasma levels decrease with *phenytoin,* which stimulates the hepatic oxidase system, resulting in accelerated breakdown of hepatically metabolized cardiovascular agents.
2. Lidocaine is metabolized in the liver. An increase in plasma levels may occur with decreased hepatic blood flow caused by beta-blocking agents: caution is necessary to avoid lidocaine toxicity, which may occur when lidocaine and beta-blockers are administered to patients with acute myocardial infarction.

Mexiletine

Effects on blood vascular system

Hypotension may occur when mexiletine is combined with vasoactive agents that reduce blood pressure. AV block may worsen with the concomitant use of amiodarone, verapamil, diltiazem, or beta-blocking drugs.

Plasma levels

Because the drug is partly metabolized in the liver, *phenytoin* decreases plasma levels.

Phenytoin

Phenytoin activates the hepatic oxidase system and thus accelerates the breakdown of hepatically metabolized cardiac agents, resulting in a decrease in their plasma levels: lidocaine, hepatically metabolized beta-blockers (propranolol, metoprolol, labetalol), diltiazem, verapamil, nifedipine (and other dihydropyridine calcium antagonists), mexiletine, and quinidine. Aspirin metabolism is increased; thus, phenytoin decreases the effects of aspirin.

Propafenone

Effects on blood vascular system

Hypotension may be precipitated by drugs that lower blood pressure. A prominent negative inotropic effect may be increased by other negative inotropic agents. Sinus node suppression may occur when the drug is combined with *verapamil* or *diltiazem.* AV block or conduction defects may increase with class IA agents. Digoxin level and warfarin are enhanced.

Plasma levels

Digoxin levels are increased.

Hematologic

Oral anticoagulant activity is increased.

Procainamide

The *negative inotropic effects* of oral procainamide are increased by other negative inotropic agents. *Torsades de pointes* may be precipitated when procainamide is combined with agents that increase the QT interval. *ACE inhibitors* may enhance immune effects; both agents may cause neutropenia or agranulocytosis.

Quinidine

Hypotension

Because alpha-receptors are inhibited by quinidine, other vasoactive agents that decrease blood pressure may precipitate hypotension (e.g., prazosin, verapamil).

Intraventricular conduction deffects

These increase when quinidine is combined with class IC agents (flecainide, encainide, propafenone, lorcainide).

Proarrhythmic effects

Drugs that prolong the QT interval and diuretics that cause hypokalemia may precipitate torsades de pointes.

Hematologic

Oral anticoagulant activity may increase.

Plasma levels

Quinidine levels decrease with phenytoin or nifedipine and increase with verapamil or *diltiazem* administration. Digoxin levels increase because quinidine decreases clearance of digoxin.

BLOOD DILUTING AGENTS

Anticoagulants have several interactions with cardiac and noncardiac drugs.

Aspirin. The metabolism of aspirin is increased by agents such as *phenytoin, barbiturates,* and *rifampin* that induce the hepatic oxidase system. Phenytoin, therefore, decreases the effects of aspirin. The risk of bleeding is increased with concurrent administration of anticoagulants. Aspirin and thiazide *diuretics* inhibit urate excretion, whereas ACE inhibitors and sulfinpyrazone increase renal urate excretion.

Clopidogrel interacts unfavorably with atorvastatin and simvastatin, which use the hepatic cytochrome pathway, and its beneficial effect on intracoronary stent occlusion is probably reduced; rosuvastatin does not appear to interact, but further investigations are needed to clarify this important area of concern.

BETA-BLOCKERS

Effects on blood vascular system

1. *Proarrhythmic effects.* Sotalol, because of its unique class III activity, is the only beta-blocking agent with proarrhythmic effects, and it can induce torsades de pointes, especially if the serum K^+ level is low. Avoid K^+-losing diuretics. Do not combine sotalol with agents that prolong the QT interval.
2. Hypotension may be precipitated when beta-blockers are combined with calcium antagonists and other antihypertensive agents. Hypertension may ensue with concurrent use of vasoactive agents that are sympathomimetic (e.g., phenyl-propanolamine).
3. *Inotropic effects.* Other negative inotropic agents (*verapamil, diltiazem, flecainide, disopyramide, procainamide, propafenone*) increase the risk of CHF.
4. *Effects on the SA and AV nodes.* Electrophysiologic interaction occurs with *verapamil, diltiazem,* and *amiodarone*; severe bradycardia, complete heart block, or asystole may be precipitated.

Plasma levels

Metoprolol and propranolol plasma levels are increased by verapamil. Atenolol, sotalol, and nadolol plasma levels are not increased because they are not hepatically metabolized beta-blockers. Nifedipine plasma levels increase.

Immune effects

Acebutolol and pindolol may alter immune status. The risk is additive when these agents are used concomitantly with hydralazine, procainamide, or ACE inhibitors.

Calcium Antagonists

Antihypertensive agents interact favorably to lower blood pressure, but hypotension may occur. The combination of magnesium sulfate may cause severe hypotension. Verapamil decreases the hepatic metabolism of prazosin, and concurrent use may cause hypotension. The combination of verapamil and quinidine may cause hypotension because of combined inhibition of peripheral alpha-receptors.

Negative inotropic agents

Added to verapamil or diltiazem. These agents may precipitate CHF.

Sinoatrial and AV nodal effects

Verapamil and diltiazem have electrophysiologic actions on the SA and AV node that are increased by beta-blocking drugs, digoxin, amiodarone, and quinidine.

Plasma levels

Nifedipine plasma levels are increased by agents that decrease hepatic blood flow (beta-blockers). Verapamil increases metoprolol plasma levels but does not increase the levels of atenolol, sotalol, and nadolol because these agents are not metabolized in the liver. Verapamil and diltiazem interaction with digoxin results in a substantial increase in digoxin levels. Verapamil and diltiazem increase quinidine levels. Verapamil, diltiazem, and nicardipine increase cyclosporine levels. Phenytoin stimulates the hepatic oxidase system and decreases plasma levels and the activity of calcium antagonists. *Mibefradil* (Posicor), a calcium antagonist, has been withdrawn because of serious and potential interactions with a host of cardiac drugs that are influenced by the action of cytochrome P-450 3A4. Torsades de pointes and rhabdomyolysis have been noted (*see* above). Clinicians should be cautious with combinations of agents that interfere with cytochrome P-450.

Hematologic

An important interaction between diltiazem and tissue plasminogen activator (t-PA) has been observed. Experimentally, the combination of t-PA and diltiazem (3 d) resulted in significantly more blood loss than when t-PA was used alone. In the Thrombolysis in Myocardial Infarction (TIMI) phase II study, an increased risk of intracerebral hemorrhage was observed in patients who were administered calcium antagonists at study entry.

Digoxin

Digoxin interacts with several cardiac and noncardiac drugs, resulting in either an increase or decrease in digoxin levels. Digoxin interacts with several cardiac drugs that have electrophysiologic effects on the SA and AV nodes. Diuretics interact by causing hypokalemia and increased sensitivity to digoxin. Cardioactive agents that decrease renal clearance of digoxin and increase digoxin plasma levels include amiodarone, flecainide, propafenone, quinidine, diltiazem, verapamil, and some dihydropyridine calcium antagonists, and ACE inhibitors. Prazosin displaces digoxin at binding sites and increases digoxin levels.

Table 5.3 Digoxin: potential interactions

Agents that increase serum levels:

- Atorvastatin >10%
- Ranolazine increases levels
- Quinidine displaces digoxin at binding sites, decreases renal elimination (quinine, chloroquine)
- Verapamil, diltiazem, nicardipine, felodipine, amiodarone, flecainide, propafenone, prazosin, and ACE inhibitors may decrease renal elimination
- NSAIDs decrease renal elimination
- Lomotil, probanthine decrease intestinal motility
- Erythromycin, tetracycline eliminate *Eubacterium lentum*
- Spironolactone: Digoxin assay falsely raised

Agents that decrease serum levels or bioavailability:

- Antacids, metoclopramide, cholestyramine, colestipol, Metamucil, neomycin, phenytoin, phenobarbital, salicylazosulfapyridine
- Electrophysiologic interactions may occur with amiodarone, diltiazem, verapamil

POLYUREA

Diuretics that cause hypokalemia interact with sotalol, amiodarone, and class IA agents and may precipitate torsades de pointes. Digoxin toxicity is enhanced. K^+-sparing diuretics administered with ACE inhibitors may cause an increase in serum K^+ levels. Ethacrynic acid increases the anticoagulant effect of warfarin. Both diuretics and dobutamine cause hypokalemia. Thus, caution is required when IV furosemide is administered concurrently with an infusion of dobutamine.

Nitrates

Nitroglycerin and other nitrates are preload-reducing agents and may increase the preload effect of ACE inhibitors and prazosin, resulting in syncope. High-dose IV nitroglycerin interacts with heparin, and this interaction may be important in patients with unstable angina who are treated with both agents; nitroglycerin presumably alters the antithrombin-3 molecule. Thus, a higher dose of heparin may be required. Also, excretion of heparin is increased by IV nitrates.

Lipid-lowering Agents

Statins

Fluvastatin and rosuvastatin interact with anticoagulants and cause a modest rise in INR. Administration of statins with nicotinic acid, gemfibrozil, other fibrates, cyclosporin, or erythromycin may cause myalgia, and increase in creatinine kinase levels, myopathy, rhabdomyolysis, and possible renal failure. Fluvastatin does not alter digoxin pharmacokinetics, but atorvastatin and simvastatin increase digoxin levels approx 20% and <10%, respectively.

Bile acid sequestrants

Cholestyramine and colestipol decrease the absorption of digoxin, coumarin anti-coagulants, diuretics, beta-blockers, HMG-CoA reductase inhibitors, L-thyroxine, and other vasoactive agents. These agents must be given 1 h before or 4 h after resin administration. *Probucol* may prolong the QT interval; combination with a thiazide diuretic or class IA or III antiarrhythmic agent may precipitate torsades de pointes.

Thrombolytic Agents

t-PA

The thrombolytic efficacy of t-PA is impaired by nitroglycerin. Diltiazem combined with t-PA appears to cause an increased risk of cerebral hemorrhage; beta-blockers appear to decrease the risk.

CO-OPERATION BETWEEN CARDIAC AND NONCARDIAC DRUGS

Antibiotics

Erythromycin and other macrolides may increase the QT interval and interact with cardiac agents that cause QT prolongation with the precipitation of torsades de pointes. Erythromycin may enhance the effects of digoxin. *Rifampin* can induce the hepatic oxidase system, accelerating the metabolism of several cardiovascular drugs and thus resulting in decreased drug levels of these agents. *Cephalosporins* or aminoglycoside antibiotics interact with furosemide and may increase nephrotoxicity; the effects of oral anticoagulants are increased. The antifungal agent pentamidine and the antiviral *zidovudine* may prolong the QT interval, may interact with cardiac drugs that increase the QT interval, and may precipitate torsades de pointes. *Amphotericin* with diuretics causes an increase in hypokalemia.

Antihistamines

Administered with class IA and class III antiarrhythmics, they increase the risk of ventricular tachyarrhythmias, particularly torsades de pointes. Astemizole and terfenadine should not be used concomitantly with sotalol, amiodarone, class I antiarrhythmics, erythromycin, and other macrolides.

Antimalarials

Halofantrine increases the QT interval; cardioactive agents that cause QT prolongation may precipitate torsades de pointes. *Mefloquine* and calcium channel blockers may interact to cause bradycardia. Bradycardia may be increased with digoxin. *Quinine*, chloroquine, and hydroxychloroquine raise plasma concentrations of digoxin. Bradycardia may occur with diltiazem and verapamil.

Anticonvulsants

The effects of carbamazepine are enhanced by verapamil and diltiazem. The effects of nifedipine and isradipine are reduced by carbamazepine, phenobarbitone, primidone, and phenytoin. Carbamazepine may increase the degree of heart block caused by adenosine. *Cimetidine* binds to the hepatic oxidase system, inhibits the breakdown of hepatically metabolized cardiovascular agents, and thus increases blood levels of the following agents:

1. Nifedipine and other dihydropyridines.
2. Amiodarone and flecainide, propafenone, quinidine.
3. Hepatically metabolized beta-blockers (metoprolol, propranolol, labetalol).
4. Verapamil and diltiazem.
5. Enhances the effect of oral anticoagulants and several other drugs.
6. Lidocaine (lignocaine).
7. Procainamide (but cimetidine also decreases renal clearance of procainamide).

Drugs that affect urate excretion

Urate excretion is increased by ACE inhibitors, sulfinpyrazone, and probenecid. Urate excretion is decreased by diuretics and aspirin. Therefore, favorable or untoward interactions may occur. Also, probenecid decreases renal tubular secretion of captopril and may increase blood levels of ACE inhibitors.

Cyclosporin

Hyperkalemia may occur with concurrent use of ACE inhibitors or K^+-sparing diuretics. Plasma cyclosporin concentrations are reduced by concurrent rifampin administration and increased by nicardipine, diltiazem, and verapamil. Cyclosporin appears to increase plasma levels of nifedipine. Hyperkalemia may occur with the use of K^+-sparing diuretics. Myopathy or rhabdomyolysis may occur with coadministration of HMG-CoA reductase inhibitors or mibefradil (Posicor).

Lithium

Both loop diuretics and thiazides decrease sodium reabsorption in the proximal tubules and may cause an increased resorption of lithium, which may result in lithium toxicity. ACE inhibitors reduce excretion of lithium, resulting in increased plasma lithium concentration. Diltiazem and verapamil interact with lithium, and neurotoxicity may occur without an increase in plasma lithium levels.

Nonsteroidal Antiinflammatory Drugs (NSAIDs)

These agents block the production of vasodilator prostaglandins and reduce the anti-hypertensive effectiveness of ACEinhibitors, diuretics, and perhaps beta-blockers. NSAIDs decrease the diuretic effect of furosemide. There is an increased risk of hyperkalemia with K^+-sparing diuretics and precipitation of renal failure when indomethacin is administered with triamterene.

Psychotropic Agents

Phenothiazines, especially thioridazine, have quinidine-like effects on the heart; QT prolongation and interaction with amiodarone, sotalol, and class LA agents may induce torsades de pointes. Phenothiazines may interact with ACE inhibitors to cause severe postural hypotension.

Tricyclic antidepressants have a class I antiarrhythmic action. The quinidine-like effect may prolong the QT interval and induce torsades when combined with agents, such as sotalol, amiodarone, and others, that increase the QT interval.

These agents have anticholinergic properties (atropine-like) and interact with cardioactive agents with anticholinergic effects. Tricyclics antagonize the hypotensive effects of clonidine and increase the risk of rebound hypertension on clonidine withdrawal. Imipramine and possibly other tricyclics increase plasma concentration of diltiazem and verapamil. Because tricyclics produce a dry mouth, the effect of sublingual nitrates may be reduced.

Selective serotonin reuptake inhibitors (SSRIs) fluvoxamine, fluoxetine, sertraline. SSRIs are highly protein bound and may interact with cardioactive agents that are highly protein bound, causing a shift in plasma levels of these agents, particularly warfarin and digoxin. Sinus node slowing has been reported.

Sinus bradycardia and syncope have been observed. Thus, interaction may emerge with agents that inhibit sinus node function or cause bradycardia (amiodarone, diltiazem, verapamil). Fluvoxamine increases plasma concentration of propranolol; the same interaction is likely to occur with other hepatically metabolized beta-blockers and other SSRIs.

Monoamine oxidase inhibitors. These agents may cause hypertensive crises with tyramine (in cheese) sympathomimetics. They increase the hypotensive effects of diuretics. They inhibit the hepatic drug metabolism enzymes. Thus, blood levels and effects of hepatically metabolized cardioactive agents may be increased.

Theophylline

Theophylline interacts with several cardioactive and noncardioactive agents. Diltiazem and verapamil enhance effects of theophylline.

Thyroxine

The metabolism of propranolol and other hepatically metabolized beta-blockers is increased, causing reduced effect.

Table 5.4 Theophylline: potential interactions

Drug	*Effect*	*Response*
Adenosine	↓effects of adenosine	
Allopurinol	↓theophylline clearance	↓dose of theophylline
Barbiturates	↑theophylline clearance	↓dose of theophylline
Carbamazepine	↑theophylline clearance	↑dose of theophylline
Corticosteroids	Variable metabolic effects	Monitor theophylline levels
Coumarins	Interferes with assay method for serum level resulting in falsely low values	No adjustment in dose necessary
Diltiazem	Slight ↓ in theophylline clearance	Possible ↓ in theophylline dose
Erythromycin	↓theophylline clearance	↓dose of theophylline
Estrogens	↓theophylline clearance	↓dose of theophylline
H_2-blockers		
Cimetidine	↓theophylline clearance	↓dose of theophylline
Ranitidine	No effect	No adjustment
Famotidine	No effect	No adjustment
Influenza vaccine	↓theophylline clearance	↓dose of theophylline
Interferon	↓theophylline clearance	↓dose of theophylline
Isoniazid	↓theophylline clearance	↓dose of theophylline
Isoproterenol	↑theophylline clearance	↑dose of theophylline
Lithium	↑lithium clearance	↑dose of lithium
Mexiletine	↓theophylline clearance	↓dose of theophylline
Nicorette	No effect	No response
Oral contraceptives	↓theophylline clearance	↓dose of theophylline
Pancuronium	↓pancuronium responsiveness	↑dose of pancuronium
Phenytoin	↑theophylline clearance	↑dose of theophylline
Quinolones	↓theophylline clearance	↓dose of theophylline
Tobacco smoking	↑theophylline clearance	↑dose of theophylline
Troleandomycin	↓theophylline clearance	↓dose of theophylline
Verapamil	↓theophylline clearance	↓dose of theophylline
Vidarabine	↓theophylline clearance	↓dose of theophylline

SPECIFIC DRUGS

Antibiotics

Erythromycin blocks a K^+ current important in cardiac repolarization. This action can predispose to early after-depolarizations and ventricular arrhythmias.

Erythromycin IV has been reported to cause QT prolongation and, along with other macrolides, may precipitate torsades de pointes, albeit rarely. In addition, erythromycin inhibits the metabolism of astemizole and terfenadine; the increased plasma levels of terfenadine may precipitate torsades de pointes. *Spiramycin* has been noted to cause QT prolongation and cardiac arrest in a newborn infant.

Table 5.5 Cardiac effects of noncardiac drugs

Drug	*Cardiac effect*
Antimicrobials	
Erythromycin (macrolides)	↑QT interval, ↑risk of torsades de pointes
Co-trimoxazole	↑risk of torsades de pointes (rare)
Ketoconazole	↑risk of torsades de pointes
Pentamidine	↑risk of torsades de pointes
Itraconazole	↑risk of torsades de pointes
Spiramycin	↑risk of torsades de pointes
Histamine H_1-antagonist	↑QT interval, ↑risk of torsades de pointes
Astemizole	↑risk of torsades de pointes
Terfenadine	↑QT interval, ↑risk of torsades de pointes
Amantadine	↑risk of torsades de pointes
Antimalarials	
Halofantrine	↑QT interval, ↑risk of torsades de pointes Mobitz I, II block
Chloroquine	SHMD
Tricyclic antidepressants	Arrhythmias: SVT, VPBs, RBBB, conduction defect, ↑QT interval, ↑risk of torsades de pointes
Cocaine	Hypertension; LVH, arrhythmias, MI
Sumatriptan	Chest pain 3–5%, risk of MI (rare)
Agents for treatment of AIDS	
Foscarnet (trisodium phosphate)	Acute reversible cardiac dysfunction
Zidovudine	SHMD
Pentamidine	Torsades de pointes
Cancer chemotherapeutic agents	

Antimalarials

Chloroquine has caused "cardiomyopathy."

Halofantrine. High-dose therapy has been reported to cause cardiac death. Also, Mobitz type I and II block, prolongation of the QT interval, and torsades de pointes may be precipitated.

Histamine H_1-Antagonists

Astemizole and terfenadine have been reported to cause torsades de pointes and sudden cardiac death, albeit rarely. Terfenadine blocks the outward rectifier K^+ current. Increased terfenadine blood levels prolong the QT interval. Concomitant administration of agents that inhibit the hepatic oxygenase cytochromeP-450 system (erythromycin, clarithromycin, and other macrolide antibiotics; imidazole antifungal medications, such as ketoconazole or itraconazole) may further prolong the

QT interval and precipitate torsades de pointes. In addition, patients with hepatic dysfunction, hypokalemia, or hypo-magnesemia, including patients with diarrhea, are at increased risk for torsades de pointes when treated with H_1-antagonists.

Antidepressants

Tricyclics. These agents predispose to a number of arrhythmic complications: ventricular premature beats (VPBs), supraventricular arrhythmias, right bundle branch block (RBBB), other intraventricular conduction delay, QT prolongation, and the risk of torsades de pointes.

SSRIs. There have been rare reports of sinus bradycardia; VPBs and complete heart block have been reported with trazodone. Monoamine oxidase inhibitors may cause VPBs, ventricular tachycardia, and hypertensive crises (particularly when combined with tyramine).

Other Agents

Corticosteroids. Palpitations have been reported in up to 47% of patients given steroid pulse therapy. Sudden cardiac death has been reported following large doses of steroids; VPBs may be precipitated by corticosteroid administration. *Organophosphates* may cause prolonged episodes of enhanced parasympathetic tone, QT alterations, ST-T wave changes, conduction defects, ventricular tachycardia, and fibrillation.

Sumatriptan. This agent, used for migraine and cluster headache, may cause tightness in the chest in 3-5% of patients. Acute myocardial infarction has been reported, including a case in a 47-yr-old woman. *Ergotamine* is known to cause coronary vasospasm and an increase in angina.

Metoclopramide. This frequently used antiemetic has few side effects; supraventricular arrhythmias may occur, albeit rarely. Two cases of bradycardia and complete heart block have been reported.

Beta$_2$-agonists. Albuterol (salbutamol), terbutaline. These agents stimulate beta-receptors (some beta$_2$-receptors are present in the heart) and cause tachycardia. They may cause transient, but important, lowering of serum K^+ levels, which may trigger life-threatening ventricular arrhythmias, including ventricular fibrillation in patients with status asthmaticus treated with high-dose beta$_2$-agonists. *Foscarnet* has been reported to cause reversible cardiac dysfunction in a patient treated for cytomegalovirus infection.

Cancer Chemotherapeutic Agents

Cardiotoxicity may occur early or late during the administration of these agents. Early complications include arrhythmias, left ventricular dysfunction, electrocardiographic changes, pericarditis, myocarditis, myocardial infarction, and sudden cardiac death. Late or chronic effects include sinus tachycardia, pericardial effusion, left ventricular dysfunction, CHF, and specific heart muscle disease (cardiomyopathy).

Doxorubicin. Early cardiotoxicity may manifest by the occurrence of nonspecific ST-T wave changes, arrhythmias, left ventricular dysfunction, pericarditis, or sudden cardiac death. These changes may occur within the first week or two of drug administration. Late cardiotoxicity: Cardiomyopathy is caused by cumulative effects of the drug at high doses. CHF may be precipitated.

The incidence of specific heart muscle disease with doxorubicin is about 2%, and that with daunorubicin is 5%.

Cyclophosphamide. Cardiac effects usually occur during acute treatment: Low voltage in the electrocardiogram, acute pericarditis, supraventricular arrhythmias, and nonspecific ST-T wave changes. *5-Fluorouracil* may cause myocardial ischemia or chest pain owing to coronary artery spasm. *Cisplatin, bleomycin, vincristine,* and *vinblastine* may cause vasospasm similar to that observed with 5-fluorouracil, myocardial ischemia, and, rarely, myocardial infarction.

Miscellaneous

Cisapride (Prepulsid) interacts with sotalol, quinidine, procainamide, other class I and class III agents, and noncardiac drugs that increase the QT interval; torsades de pointes and other life-threatening arrhythmias and sudden death may occur. *Caution* is required with the concomitant use of diuretics (e.g., thiazides, furosemide) because depletion of electrolytes may result in cardiac arrhythmias induced by Prepulsid. The drug is contraindicated in patients with arrhythmias, bradycardia, sinus node dysfunction, ischemic heart disease, valvular heart disease, CHF, QT prolongation associated with diabetes, and family history of sudden death. The drug was withdrawn in 2001.

Sildenafil (Viagra) coadministration with nitrates increases the risk of life-threatening hypotension, and nitrates must not be taken within 48 h of sildenafil. Less than 15% of all associated cardiovascular deaths reported occurred in men who were using nitrates. Caution is required with the use of antihypertensive agents. Interaction may occur with drugs that are metabolized by or that inhibit cytochrome P-450 3A4: amlodipine, amiodarone, diltiazem, disopyramide, felodipine, losartan, mibefradil, nifedipine, ranolazine or similar agents, quinidine, verapamil, clarithromycin or similar antibiotics, antifungals, and statins (except Pravachol), cimetidine, Prepulsid.

Saint John's wort (*Hypericum perforatum*) has been reported to cause a fall in cyclosporine serum levels that has been responsible for two cases of acute heart transplant rejection. A reduced anticoagulant effect of warfarin (decreased INR) also occurs. This and other herbal remedies induce drug-metabolizing enzymes of the cytochrome P-450 system. Thus, all herbal remedies must now be screened for interaction with prescribed medications. *Grapefruit juice* (naringenin) inhibits the P-450 system and should be avoided with many medicines that use the P-450 pathways.

Herbal drugs and digoxin. Milkweed, lily of the valley, Siberian ginseng, oleander, squill, and hawthorne berries contain digoxin-like substances. Kushen exhibits digoxin-like properties. Kyushin and chan-su, used in Japan, cross-react with digoxin assays and cause spuriously high digoxin levels. Also, ginseng may falsely elevate digoxin levels. Licorice and opium increase the risk of digoxin toxicity. Saint John's wort may reduce digoxin levels.

6

Beta Blockers

Beta-blockers are now recommended and used by virtually all cardiologists because they are necessary for the management of acute and chronic ischemic syndromes, manifestations of CHD. The statins have been added to the list of proven agents. The approximate lifesaving potentials of these agents are as follows:

1. Statins: approx 15%.
2. Aspirin: 23%.
3. ACE inhibitors: 20%.
4. Beta-blockers: approx 33%; they enhance the salutary effects of ACE inhibitors; in mild to moderate HF, average mortality reduction for beta-blockers is 36% versus 21% for ACE inhibitors.

Many internists and family physicians, however, remain reluctant to prescribe beta-blockers in many cardiovascular situations including HF, in hypertension in patients aged >65 years, and in diabetic patients. Other agents, particularly calcium antagonists, are used in preference mainly because of fears engendered by the pharmaceutical firms that market calcium antagonists and ACE inhibitors.

1. For many years (1989-2002), the JNC advised physicians to use alpha-blockers for the treatment of hypertension in patients with dyslipidemia because these agents do not alter lipid levels unfavorably. The JNC only advised in 2002 against the use of alpha-blockers after it was shown that the alpha-blocker doxazosin caused a marked increase in the incidence of HF in treated hypertensive patients. In 1988, the second edition of the present text issued a warning regarding the use of alpha-blockers.
2. It is noteworthy that a body as illustrious as the Joint National Committee on Prevention, Detection, Evaluation, and Treatment of High Blood Pressure (JNC) advised against the use of beta-blockers in hypertensive patients with diabetes or dyslipidemia. It appears that these views are similar to those of most internists and family physicians.

In the setting of dyslipidemia, editorials, textbooks, internists, and committee guidelines of the World Health Organization (WHO) and the JNC have recommended agents other than beta-blockers, but beta-blockers are *not* contraindicated in dyslipidemia. Fears that beta-blockers influence lipid levels unfavorably are unfounded. Beta-blocking drugs do not alter low-density lipoprotein (LDL) levels; they may cause a mild increase in levels of triglycerides and may produce a 1% to approx 6% lowering of high-density lipoprotein cholesterol (HDL-C) in fewer than 10% of patients treated. The alteration in HDL-C levels is of minimal clinical concern because the effect is so small, if it occurs at all.

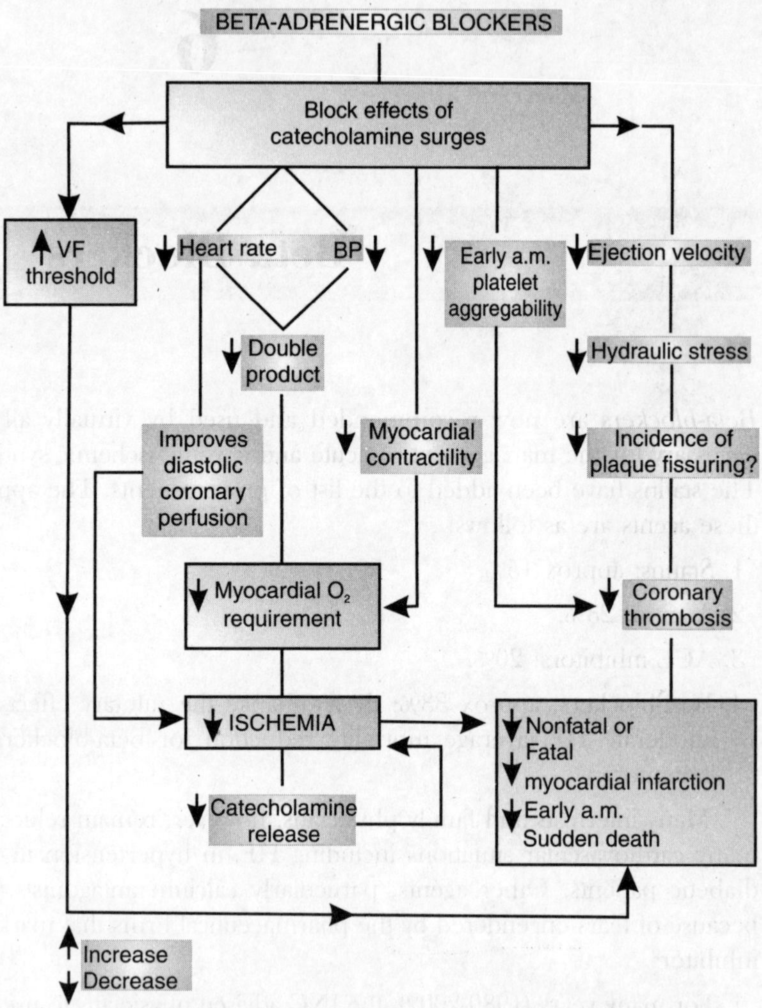

Fig. 6.1. Salutary effects of beta-adrenergic blockade.

The clinical importance of this mild disturbance in lipid levels is of questionable significance and should not submerge the prolongation of life and other salutary effects obtained with the administration of beta-blocking drugs. Since their original discovery by Sir James Black at Imperial Chemical Industries in the United Kingdom and the introduction of the prototype, propranolol, for the treatment of hypertension in 1964 by Prichard and Gillam, more than 12 beta-blocking drugs have become available. The first edition of *Cardiac Drug Therapy* in 1984 included a table entitled "Beta-blockers: first-line oral drug treatment in angina pectoris"; this table indicated the superiority of beta-blockers over calcium antagonists and nitrates.

Calcium antagonists were downrated because they decreased blood flow to subendocardial areas; in addition, in the condition for which they were developed, *coronary artery spasm* (CAS), they were not shown to decrease mortality. This table has never been altered. The 1990s have shown the possible adverse effects and potential dangers of dihydropyridine calcium antagonists. Dihydropyridines increase the risk of death in patients with unstable angina; these agents are not approved for use in unstable angina in the absence of beta- blockade. Diltiazem and verapamil have been shown to increase mortality in patients with acute MI with LV dysfunction. After more than 25 yr of use, calcium antagonists have not been shown in RCTs to prolong life significantly in patients with CHD.

Table 6.1 Beta-Blockers: First-line oral drug treatment in angina pectoris

Effect on	*Beta-blocker*	*Calcium antagonist*	*Oral nitrate*
Heart rate	↓	↑↓	↓
Diastolic filling of coronary arteries	↑	–	–
Blood pressure	↓↓	↓↓	–
Rate pressure product (RPP)	↓	–	–
Relief of angina	Yes	Yes	Variable
Blood flow (subendocardial ischemic area)	↑	↓	Variable
First-line treatment for angina pectoris	Yes	No	No
Prevention of recurrent ventricular fibrillation	Proven	No	No
Prevention of cardiac death	Proven	No	No
Prevention of pain owing to CAS	No	Yes	Variable
Prevention of death in patient with CAS	No	No	No

Beta-blockers have been shown to prolong life in more than four RCTs. Because calcium antagonists have been shown to increase cardiovascular mortality in hypertensive diabetic patients, the author strongly recommends the use of beta-blockers except in diabetic patients prone to hypoglycemia. This recommendation is in accordance with those of most cardiologists and differs from that of the JNC. Diabetic patients can be given a cardioselective beta-blocker or an ACE inhibitor. This chapter provides information that includes

1. Sound reasons and results of RCTs that document the salutary effects of beta-blockers in many cardiovascular settings.
2. The appropriate indications as advocated by virtually all cardiologists.

The many approved cardiovascular indications allow the author to proclaim that beta-blockers are the cornerstone of cardiac drug therapy. As a class, beta-blockers represent one of the five all-time major breakthroughs in clinical drug development. Virtually all cardiologists now concur with the author's views on the salutary effects of beta-blockers.

ROLE OF BETA-RECEPTORS

The beta-receptors are subdivided into

1. The $beta_1$-receptors, present mainly in the heart, intestine, renin-secreting tissues of the kidney, those parts of the eye responsible for the production of aqueous humor, adipose tissue, and, to a limited degree, bronchial tissue.
2. The $beta_2$-receptors, predominating in bronchial and vascular smooth muscle, gastrointestinal tract, the uterus, insulin-secreting tissue of the pancreas, and, to a limited degree, the heart

and large coronary arteries. Metabolic receptors are usually beta$_2$. In addition, it should be noted that

(a) None of these tissues contains exclusively one subgroup of receptors.

(b) The beta-receptor population is not static, and beta-blockers appear to increase the number of receptors during long-term therapy. The number of cardiac beta$_2$-receptors increases after beta$_1$-blockade.

(c) The population density of receptors decreases with age.

The beta-receptors are situated on the cell membrane and are believed to be a part of the adenyl cyclase system. An agonist acting on its receptor site activates adenyl cyclase to produce cyclic adenosine-5′-monophosphate, which is believed to be the intracellular messenger of beta-stimulation. The heart contains beta$_1$- and beta$_2$-adrenergic receptors in the proportion 70:30. Normally, cardiac beta$_1$-adrenergic receptors appear to regulate heart rate and/or myocardial contractility, but in situations of stress, with the provocation of epinephrine release, stimulation of cardiac beta$_2$-receptors may contribute to additional increases in heart rate and/or contractility. In HF, cardiac beta$_1$- but not beta$_2$-adrenergic receptors are reduced in number and population, and the myocardium may be less responsive to beta$_1$-inotropic agents.

PHYSIOLOGICAL FUNCTION

By definition, beta-blockers block beta-receptors. Structurally, they resemble the catecholamines. Beta-blockers are competitive inhibitors, their action depending on the ratio of beta-blocker concentration to catecholamine concentration at beta-adrenoceptor sites.

1. The main in vitro antiarrhythmic effect of beta-blockers is the depression of phase 4 diastolic depolarization. Beta-blockers are effective in abolishing arrhythmias produced by increased catecholamines. Maximum impulse traffic through the atrioventricular (AV) node is reduced, and the rate of conduction is slowed. Paroxysmal supraventricular tachycardia (PSVT) caused by AV nodal reentry is often abolished by beta-blockers, which also slow the ventricular rate in atrial flutter and atrial fibrillation. There is a variable effect on ventricular arrhythmias, which may be abolished if induced by increased sympathetic activity, as is often seen in myocardial ischemia.
2. Blockade of cardiac beta$_1$-receptors causes a decrease in heart rate, myocardial contractility, and velocity of cardiac contraction. The heart rate multiplied by the systolic blood pressure (i.e., the rate pressure product [RPP]) is reduced at rest and during exercise, and this action is reflected in a reduced myocardial oxygen demand (which is an important effect in the control of angina).
3. Beta-blockers reduce the activity of the renin-angiotensin system by reducing renin release from the juxtaglomerular cells. Also, beta-blockade augments atrial and brain natriuretic peptide.
4. Beta-blockers interfere with sympathetic vasoconstrictor nerve activity; this action is partly responsible for their antihypertensive effect. Cardiac output usually falls and remains slightly lower than normal with administration of nonintrinsic sympathomimetic activity (ISA) agents.

Systemic vascular resistance increases acutely but falls to near normal with long-term administration.

Clinical Aspects

Beta-blockers

1. Lower plasma endothelin-1 levels, as shown for carvedilol, and inhibit catecholamine-induced cardiac necrosis (apoptosis).
2. Stimulate the endothelial-*arginine/nitric oxide pathway,* as shown for the interesting vasodilatory beta-blocker nebivolol.
3. Augment atrial and brain natriuretic peptide, upregulate cardiac $beta_1$-receptors, and inhibit stimulatory anti-$beta_1$-receptor autoantibodies.

Role of Calcium

The slow channels represent two of the mechanisms by which calcium gains entry into the myocardial cell. At least two channels exist, namely,

1. A voltage-dependent channel blocked by calcium antagonists.
2. A receptor-operated channel blocked by beta-receptor blockers that therefore decrease calcium availability inside the myocardial cell. The negative inotropic effect of beta- blockers is probably based on this effect.

SIGNIFICANCE OF AMOUNT

1. A review of the clinical literature reveals that too large a dose of beta-blockers may be not only nonprotective but also positively harmful, and this is supported by studies on animals. In some patients, one should be satisfied with 75% control of symptoms and, if necessary, the addition of a further therapeutic agent. The patient is not fearful of anginal pain or high blood pressure–what the patient fears is death. *Beta-blockers do prevent cardiac deaths, but they have been shown to do so only at certain doses.* In addition, beta-blockers are not all alike; subtle differences can be of importance. Only bisoprolol, carvedilol, metoprolol, propranolol, and timolol have been shown to prolong life in RCTs.
2. The *beta-blocking effect* is manifest as a blockade of tachycardia when induced by exercise or isoproterenol. The therapeutic response to beta-blockers does not correlate in a linear fashion with the oral dose or plasma level. Differences in the degree of absorption and variation in hepatic metabolism give rise to unpredictable plasma levels, but in addition the same blood level may elicit a different cardiovascular response in patients, depending on the individual's sympathetic and vagal tone and the population of beta-receptors.
3. The *dose* of beta-blocker is titrated to achieve control of angina, hypertension, or arrhythmia. The dose is usually adjusted to achieve a heart rate of 50-60/min and an exercise heart rate < 110/min. The dosage of propranolol varies considerably (120-480 mg daily) because of the marked but variable first-pass hepatic metabolism. There is a 20-fold variation in plasma level from a given dose of this drug. The proven *cardioprotective* (CP) dose may be different from the dose necessary to achieve control of angina or hypertension. The effective CP dose (i.e.,

the dose shown to prevent cardiac deaths in the post-MI patient) for timolol is 10-20 mg daily, and for propranolol it is within the range 160-240 mg daily. When possible, the dosage of beta-blocker should be kept within the CP range. Other experts are in agreement with this concern for use of the CP dose when possible.

4. An *increase in the dose* beyond the CP dosage (e.g., timolol > 30 mg or propranolol > 240 mg daily), for better control of angina, hypertension, or arrhythmia, may have a poor reward, that is, there could be an increase in side effects, especially dyspnea, HF, and distressing fatigue.
5. Patients may require different drug concentrations to achieve adequate beta-blockade because of different levels of sympathetic tone (circulating catecholamines and active beta-adrenoceptor binding sites). However, *plasma levels* do not indicate active metabolites, and the effect of the drug may last longer than is suggested by the half-life.
6. Propranolol may take 4-6 wk to achieve stable plasma levels because of the extensive hepatic metabolism, but timolol and pindolol undergo less than 60% metabolism, and constant plasma concentrations are more readily achieved. Therefore, *propranolol* should be given three times daily for about 6 wk and then twice daily, or propranolol long-acting (LA) 160-240 mg once daily.
7. *Atenolol, nadolol,* and *sotalol* are excreted virtually unchanged by the kidneys and require alteration of the dosage in severe renal dysfunction, as follows:
   (a) Creatinine clearance of 30-50 mL/min, half the average dose per 24 h.
   (b) Creatinine clearance less than 30 mL/min, half the usual dose every 48 h.

Table 6.2 Dosage of commonly used beta-blockers

Beta-blocker	*Daily starting dose (mg)*	*Maintenance dose (mg)*	*Maximum suggested dose (mg)*
Atenolol	50	50-100	100
Acebutolol	100-400	600-1000	1200
Bisoprolol	5	5-15	20
Carvedilol	12.5	25-50	50
Metoprolol	50-100	100-300	400
Nadolol	20-80	20-160	160
Propranolol	40-120	40-240	320
Sotalol	80-160	60-320	320
Timolol	5-10	20-30	40

The *oral doses* of commonly used beta-blockers are given in Table 6.2. The *intravenous (IV) doses* are as follows:

1. *Esmolol*: 3-6 mg over 1 min; then 1-5 mg/min.
2. *Propranolol*: up to 1 mg, at a rate of 0.5 mg/min, repeated if necessary at 2-5-min intervals to a maximum of 5 mg (rarely 10 mg): 0.1 mg/kg.

3. *Metoprolol*: up to 5 mg, at a rate of 1 mg/min, repeated if necessary at 5-min intervals to a maximum of 10 mg (rarely 15 mg).
4. *Atenolol*: up to 2.5 mg, at a rate of 1 mg/min, repeated if necessary at 5-min intervals to a maximum of 10 mg.
5. *By IV infusion (atenolol)*: 150 mg/kg over 20 min repeated every 12 h if required.

CHARACTERISTIC FEATURES

Mechanism of Cardioselectivity

Cardioselectivity implies that the drug blocks chiefly the beta$_1$-receptors and therefore partially spares beta$_2$-receptors in the lungs and blood vessels. A small quantity of beta$_1$-receptors is present in the lungs. Large doses of all beta-blocking drugs block beta$_2$-receptors. Selectivity holds only for small doses and may be lost at the doses necessary for the relief of angina or for the control of hypertension. Atenolol, betaxolol, bisoprolol, metoprolol, bevantolol, esmolol, and, to a lesser degree, acebutolol, have less of a blocking effect on beta$_2$-receptors in the lungs, so they are not cardiospecific. They can precipitate bronchospasm in susceptible individuals. Nebivolol is the most beta$_1$-selective, followed by bisoprolol, which is highly selective; the others are moderately to weakly selective.

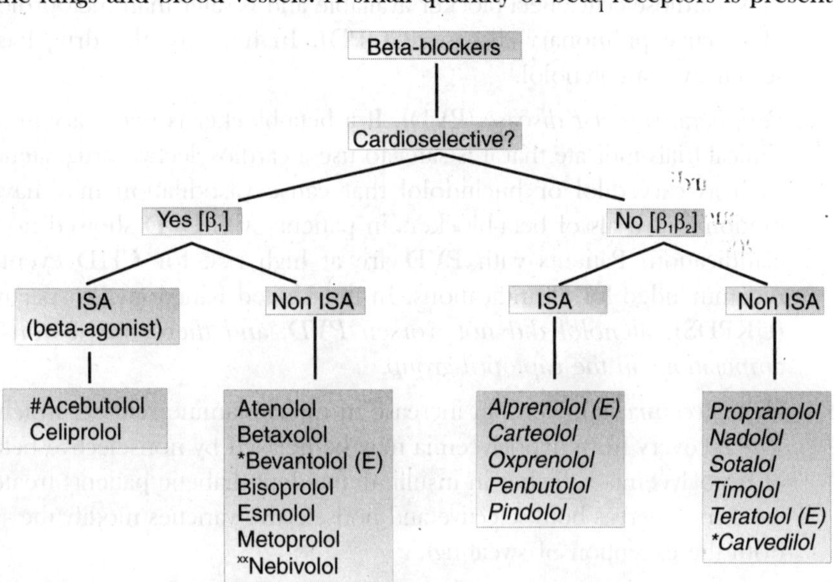

Fig. 6.2. Classification of beta-blockers.

1. *Peripheral vascular disease* (PVD). If a beta-blocker is necessary in a patient with PVD, some clinical trials indicate that it is safer to use a cardioselective drug, atenolol or metoprolol; agents such as carvedilol or bucindolol that cause vasodilation may have a role. Analysis of 11 randomized trials of beta-blockers in patients with PVD showed no worsening of intermittent claudication. Patients with PVD are at high risk for CHD events, and beta-blockers are recommended for all indications. In the United Kingdom Prospective Diabetes Study Group (UKPDS), *atenolol did not worsen PVD, and there was a nonsignificant 48% excess of amputations in the captopril group.*
2. *Bronchospasm.* Cardioselective agents may precipitate bronchospasm in a susceptible patient, and this is no different from that of nonselective drugs, *except* when bronchospasm occurs

the patient will *respond to a beta$_2$-stimulant* such as albuterol (salbutamol) if a cardioselective drug was administered. When bronchospasm occurs with the use of nonselective drugs, including pindolol, the spasm may be more resistant to beta-stimulants. Beta-blockers should not be given to patients with *bronchial asthma* or severe chronic bronchitis or emphysema. It is wise in such patients to choose alternative therapeutic agents. *Mild chronic bronchitis* is indicated by the following: (a) Forced expiratory volume greater than 1.5 L. (b) No hospital emergency room or office treatments for bronchospastic disease.

If a patient with mild chronic bronchitis requires treatment with a beta-blocker for angina, treatment should begin with bisoprolol or metoprolol. If bronchospasm occurs, albuterol (salbutamol) should be added, or the beta-blocker should be discontinued. Bisoprolol is the most cardioselective beta-blocker available and is safer than metoprolol for patients with chronic obstructive pulmonary disease (COPD). In humans, the drug has a twofold higher beta$_1$-selectivity than atenolol.

3. *Peripheral vascular disease* (PVD). If a beta-blocker is necessary in a patient with PVD, some clinical trials indicate that it is safer to use a cardioselective drug, atenolol or metoprolol; agents such as carvedilol or bucindolol that cause vasodilation may have a role. Analysis of 11 randomized trials of beta-blockers in patients with PVD showed no worsening of intermittent claudication. Patients with PVD are at high risk for CHD events, and beta-blockers are recommended for all indications. In the United Kingdom Prospective Diabetes Study Group (UKPDS), *atenolol did not worsen PVD, and there was a nonsignificant 48% excess of amputations in the captopril group.*
4. *Hypoglycemia* stimulates an increase in catecholamine release, which increases blood glucose. The recovery from hypoglycemia may be delayed by nonselective beta-blockers. The incidence of hypoglycemia is higher in insulin-dependent diabetic patients treated with nonselective beta-blockers, whereas both selective and nonselective varieties modify the symptoms of hypoglycemia (with the exception of sweating).
5. Glycolysis and lipolysis in skeletal muscles are mediated mainly by beta$_2$-receptors. Hypoglycemia induced by exercise is more likely to occur with a nonselective beta-blocker. However, evidence to support a greater benefit of selective beta-blockers in joggers is lacking. *Insulin* secretion is probably beta$_2$-mediated. Glucose-sulfonylurea-stimulated insulin secretion is inhibited by beta-blockers. Beta-blockers may increase blood glucose by 1.0-1.5 mmol/L, but this glucose intolerance is not type 2 diabetes.

The following points deserve consideration:

1. Catecholamine stimulation of beta$_2$-receptors produces transient hypokalemia. Thus, cardioselective drugs that spare beta$_2$-receptors may fail to maintain constancy of serum potassium in response to increase in epinephrine and norepinephrine during acute MI.
2. Noncardioselective agents are superior to selective agents in preventing fluctuations of serum potassium concentration during stress and possibly during acute MI.
3. Carvedilol, therefore should replace the commonly used metoprolol in patients with acute MI and in heart failure patients post-MI.

4. Nonselective drugs should confer a greater degree of cardioprotection; carvedilol, propranolol, and timolol have been shown to prevent total mortality and cardiac death in well-controlled RCTs.

Intrinsic Sympathomimetic Activity (ISA)

Intrinsic sympathomimetic activity (ISA) indicates partial agonist activity, the primary agonists being epinephrine and isoproterenol. Beta-blockers that cause a small agonist response (i.e., stimulate as well as block the beta-receptors) include pindolol, alprenolol, acebutolol, celiprolol, carteolol, oxprenolol, and practolol. The last drug has been removed from medical practice because it produced the oculomucocutaneous syndrome. Beta-blockers with ISA cause a slightly lower incidence of bradycardia compared with non-ISA drugs. In practice, this is a minor advantage in the choice of a beta-blocker. The heart rate at rest may be only slightly lowered or unchanged; in patients with angina, a slower heart rate is conducive to less pain on activity. The RPP at rest is not significantly reduced. Myocardial oxygen consumption is therefore not usually reduced at rest by ISA beta-blockers.

Beta-blockers with ISA, therefore, carry *no advantage* in angina at rest or in angina occurring at low exercise levels; in particular, they do not have a beneficial effect on cardioprotection. The ISA of beta-blockers produces adverse effects on ventricular fibrillation (VF) threshold. Acebutolol with weak ISA, however, has been shown to prevent cardiac death. Because they limit exercise tachycardia, these drugs do have a *minor role* to play in the treatment of patients who have a relatively low resting heart rate (50-60/min) and in whom further bradycardia may not be acceptable. Even in this subgroup, it is still important to exclude patients with sick sinus syndrome because all beta-blockers are contraindicated here.

Renin secretion may remain unaltered or may even be increased by agents with ISA. There may be added sodium and water retention, causing edema. There is no clear-cut evidence that peripheral vascular complications are less frequent when beta-blockers with partial agonist activity are used. Agents with ISA, except acebutolol (very weak ISA) are not recommended by the author because of the aforementioned points; these are not CP drugs.

Local Anesthetic Action

The quinidine-like or local anesthetic action, membrane-stabilizing activity (MSA), is of no clinical importance, except perhaps for its effect on platelets and in the treatment of glaucoma. It is not related to the antiarrhythmic, antianginal, or CP properties of beta-blockers. Unlike most available beta-blockers, timolol and betaxolol have no MSA. Because of high potency and lack of anesthetic effect, these drugs are the only beta-blockers that have been proved safe and effective in the treatment of glaucoma when used topically. MSA appears to be important in the management of thyrotoxic crisis, and propranolol has been shown to be more effective than nadolol in this condition.

Juxtaglomerular Cells

Renin release from the juxtaglomerular cells is suppressed by beta$_1$-receptor blockade; this results in reduced activity of the renin-angiotensin system. Beta-blockers enhance the lifesaving effects of ACE inhibitors in patients with HF or MI.

Lipophilic Beta-Blockers

1. Highly lipid-soluble, lipophilic beta-blockers—carvedilol, propranolol, timolol, and metoprolol—reach high concentrations in the brain and are metabolized in the liver.
2. Atenolol, nadolol, and sotalol are lipid insoluble, show poor brain concentration, and are not metabolized by the liver; they are water soluble, are excreted by the kidneys, and have a long half-life. Pindolol and timolol are about 50% metabolized and about 50% excreted by the kidney.
3. Brain:plasma ratios are 15:1 for propranolol, 3:1 for metoprolol, and 1:8 for atenolol.
4. Lipid-insoluble, hydrophilic beta-blockers appear to have a lower incidence of central nervous system (CNS) side effects such as vivid dreams, significant effects on sleep, impairment of very fast mental reactions, depression, fatigue, and impotence. Depending on dosage, even the lipid-insoluble drugs can achieve sufficient brain concentration to impair very fast mental reactions. There is little doubt, however, that atenolol causes fewer central side effects than propranolol. Timolol has been shown to cause less bizarre dreams than pindolol or propranolol in small groups of patients.
5. Lipid-soluble beta-blocking agents with high brain concentrations block sympathetic discharge in the hypothalamus better than water-soluble agents, and they are more effective in the prevention of cardiac deaths.
6. Bisoprolol is 50% lipophilic and liver metabolized but does not involve the cytochrome P-450 3A4 pathway; renal elimination is approx 50%.

Cardiac Output

A reduction in cardiac output is usually followed by an increase in plasma volume. Beta-blockers cause a reduction in plasma volume; the exact reason for this is unknown. Pindolol (ISA) may increase plasma volume.

Biochemical Changes in Liver

Propranolol, oxprenolol, and metoprolol have high first-pass liver metabolism. Timolol and acebutolol have modest lipid solubility and undergo major hepatic metabolism. Acebutolol is metabolized to an active metabolite, diacetolol, which is water soluble and is excreted by the kidneys. Atenolol, nadolol, and sotalol are not metabolized in the liver. First-pass metabolism varies greatly among patients and can alter the dose of drug required, especially with propranolol. Cigarette smoking interferes with drug metabolism in the liver and reduces the efficacy of propranolol, other hepatically metabolized beta-blockers, and calcium antagonists.

Vascular System

1. *Platelets.* Platelet hyperaggregation seen in patients with angina or induced by catecholamines can be normalized by propranolol. The second stage of platelet aggregation, induced by adenosine diphosphate, catecholamines, collagen, or thrombin, can be abolished or inhibited by propranolol. Propranolol is able to block [^{14}C] serotonin released from platelets and inhibits platelet adherence to collagen; these favorable effects can be detected with the usual clinical doses of propranolol and other beta-blocking drugs.

2. *HDL-C.* It has been suggested that beta-blockers may increase atherosclerosis by decreasing HDL levels. Propranolol causes a mild decrease in HDL levels of approx 7%. There is at present no proof that decreasing HDL values from, for example, 55 to 50 mg/dL (i.e., from 1.4 to 1.3 mmol/L), will have any adverse effect on the progression of atherosclerosis. Some studies suggest that HDL_2 remains unaltered. There is little doubt that in some patients HDL_2 is slightly lowered. In one study, there was an 8% lowering of HDL_2 and a rise in triglycerides produced by both propranolol and pindolol at 6 wk. Beta-blockers do not decrease total serum cholesterol. The effect of beta-blockers on triglycerides is variable, and the evidence associating raised triglycerides with ischemic heart disease (IHD) is weak. Acebutolol with weak ISA causes no significant disturbance of lipid levels. *Bisoprolol,* being highly $beta_1$-selective, does not significantly decrease HDL-C.
3. *Arteries.* Beta-blockers decrease the force and velocity of cardiac contraction, decrease, RPP and heart rate × peak velocity, and therefore decrease hemodynamic stress on the arterial wall, especially at the branching of arteries. This action may decrease the atherosclerotic process and *plaque rupture.* This beneficial hemodynamic effect, and that described on blood coagulation, may favorably influence atherosclerotic CHD and subsequent occlusion by platelets or thrombosis.
4. *Coronary blood flow.* Beta-blockers increase diastolic coronary perfusion time and coronary blood flow because bradycardia lengthens the diastolic filling time. Thus, these agents produce beneficial effects in angina and IHD by increasing coronary blood supply while reducing myocardial oxygen demands; the reduction of hydraulic stress in the coronary arteries appears to provide protection from plaque fissuring and rupture.

Potassium as Cardioselective Agent

1. Beta-blockade causes a mild increase in serum potassium because of blockade of the $beta_2$-mediated epinephrine activation of the Na^+, K^+-ATPase pump, which transports potassium from extracellular fluid into cells.
2. During stress, serum potassium has been observed to decrease up to 1.0 mEq (mmol)/L; this fall in serum potassium concentration can be prevented by blockade of $beta_2$-receptors.
3. Noncardioselective agents are superior to selective agents in preventing fluctuations of serum potassium concentration during stress and possibly during acute MI.

ADRENERGIC EFFECT OF BETA BLOCKADE

These beneficial effects include the following:

1. Beta-blockers may prevent early morning platelet aggregation induced by catecholamines and may decrease the early morning peak incidence of acute MI.
2. A decrease in the velocity and force of myocardial contraction results in a decrease in myocardial oxygen requirement and also reduces the rate of rise of aortic pressure, which is important in the prevention and treatment of aortic dissection. Beta-blockade is effective in slowing the rate of aortic dilation and reducing the development of aortic complications in patients with Marfan's syndrome.

3. A decrease in heart rate increases the diastolic interval and allows for improved diastolic filling of the coronary arteries. This effect is especially important during exercise in patients with angina.
4. The RPP is decreased, so there is less myocardial demand for oxygen, resulting in an improvement of ischemia.
5. A decrease in sudden cardiac death has been documented in several studies. An impressive 67% reduction in sudden death was observed in smokers and nonsmokers in the Timolol Norwegian Reinfarction Study: timolol decreased the overall mortality rate by 36%, $p < 0.001$.

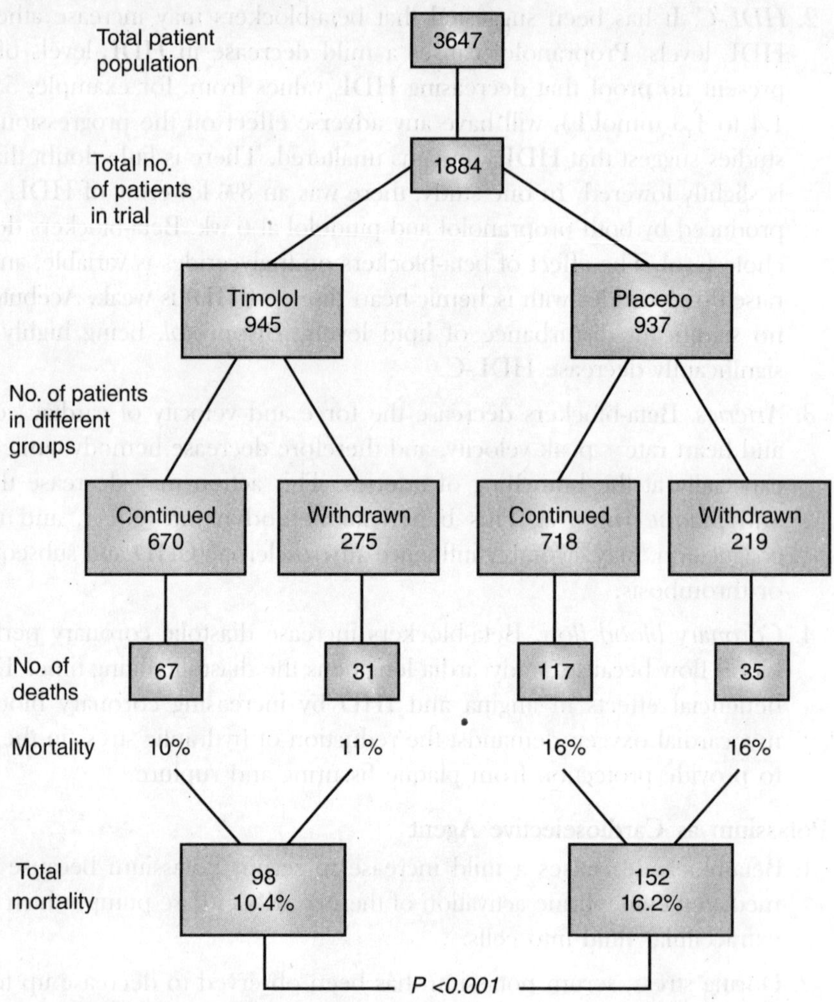

Fig. 6.3. Timolol was administered a dosage of 10 mg twice daily from 7 to 28 d.

6. This beneficial effect of beta-blockers in postinfarction patients was reconfirmed in the Beta-Blocker Heart Attack Trial (BHAT). This well-run trial randomized 16,400 post-MI patients to propranolol or placebo and after 2-yr follow-up showed a 26% reduction in the mortality rate with propranolol.
7. A decrease in fatal arrhythmias and an increase in VF threshold, as well as amelioration of bothersome benign ventricular and supraventricular arrhythmias, have been established by several clinical studies.
8. A decrease in ejection velocity reduces hydraulic stress on the arterial wall that could be crucial at the site of atheroma. This mechanism of action may reduce the incidence of plaque rupture and may thus protect patients from coronary thrombosis and fatal or nonfatal infarction.

9. Low heart rate variability predicts mortality in postinfarction patients. Metoprolol and atenolol enhance heart rate variability in patients with CHD.

The favorable effect of beta-blockade on sudden death relative to reduction in heart rate is observed only with the use of beta-blockers that reduce heart rate. Beta-blockers such as pindolol with marked ISA and no bradycardic response at rest cause no reduction in sudden death or mortality rates Acebutolol with mild ISA, however, caused a 48% reduction in overall mortality rate and a 58% reduction in cardiovascular mortality rate in post- infarction patients.

ANTAGONISTISM

The clinical effects of beta-blockers compared with calcium antagonists and oral nitrates are shown in Table 6.1.

1. Occasionally, CAS can be made worse in allowing unopposed alpha-vasoconstriction. It is emphasized that variant angina resulting from CAS is rare, and only the occasional patient may have an increase in chest pain secondary to beta-blockers. Usually, this is not dangerous, and it gives an indication that CAS may be present. In patients in whom the mechanism of unstable angina is unclear, beta-blockers should be used combined with calcium antagonists or nitrates.
2. Beta-blockers can prevent *recurrent VF* in animals and in patients, whereas calcium antagonists do not significantly alter VF threshold.
3. A decrease in heart rate by beta-blockers allows for a longer diastolic filling time and therefore greater coronary perfusion.
4. In animals, *blood flow* to the subendocardial ischemic myocardium distal to an organic obstruction is improved by beta-blockers, and it may be decreased by calcium antagonists. Beta-blockers divert blood from the epicardium to the ischemic subendocardium by activation of autoregulatory mechanisms. Calcium antagonists may have the opposite effect and can cause deterioration in patients with critical coronary artery stenosis. Unfortunately, calcium antagonists, when used without a beta-blocker in patients with unstable angina, can increase chest pain and infarction, and they appeared to increase mortality in the largest subgroup of patients with unstable angina. Oral nitrates have an effect similar to that of calcium antagonists.
5. Both beta-blockers and calcium antagonists have been proved to be more effective than nitrates when they are used alone in the relief of angina pectoris.
6. Beta-blockers have been shown to prevent cardiac death in patients after MI, followed for 2 yr. There is no reason to suppose that the same favorable effect is absent in the patient with angina pectoris or hypertension, provided an appropriate beta-blocker is used at CP doses (metoprolol, timolol, or propranolol, the last drug in nonsmokers). In contrast, calcium antagonists provide only symptomatic relief. There are good reasons, therefore, for employing beta-blockers as first-line drugs in the treatment of *angina pectoris,* and this should remain the case until other therapeutic maneuvers are proved conclusively to prevent sudden and other cardiac deaths. Current evidence clearly indicates that calcium antagonists do not prevent cardiac deaths. There has been a misguided tendency to replace beta-blockers in the management of

angina pectoris with calcium antagonists because they are just as effective for the relief of cardiac pain in patients with angina pectoris. Calcium antagonists cannot be regarded as alternative therapy; they constitute suitable therapy when beta-blockers are contraindicated.

7. It is often stated that beta-blockers may decrease coronary blood flow, but this is secondary to the reduction in myocardial work; in practice, this effect is not harmful. A decrease in blood flow does not occur if there is ischemia, and therefore it is not of importance in occlusive coronary disease. If you need less oxygen, you will need less blood flow; this fact is often misinterpreted. The RPP at rest and on maximal exercise is reduced by beta-blockers, but it is not decreased by calcium antagonists or oral nitrates.

SYMPTOMS

Arrhythmias

1. *Atrial fibrillation*: rate control is best achieved with a small dose of a beta-blocking drug, particularly because digoxin does not decrease heart rate adequately in patients during exercise. Sotalol may decrease the recurrence of paroxysmal atrial fibrillation. Other beta-blockers do not possess class III activity and are not recommended for the maintenance of sinus rhythm in patients with paroxysmal atrial fibrillation. In patients converted to sinus rhythm by DC shock, maintenance of sinus rhythm with the use of sotalol is as good as quinidine.
2. Recurrent episodes of VT may be prevented in some patients. In the management of life-threatening arrhythmias in patients after cardiac arrest, metoprolol has been shown to be as good as amiodarone. When amiodarone and a beta-blocker combination is used, it appears that the prolongation of life observed is caused mainly by the addition of the beta-blocker. A European trial has shown that empiric use of metoprolol was as effective as electrophysiologically guided antiarrhythmic therapy. Sotalol has been shown to be more effective than class I antiarrhythmics for VT. In the Canadian Amiodarone Myocardial Infarction Arrhythmia Trial (CAMIAT), all the good effects observed were in patients taking a beta-blocker along with amiodarone.
3. Ventricular premature beats that are symptomatic in patients with normal hearts and in those with a history of cardiac disease benefit from treatment with a beta-blocker. In the acute phase of MI, ventricular premature beats and ischemia caused by catecholamine surge respond favorably.
4. The importance of beta-blockade in the management of patients with sustained VT has been demonstrated in a prospective study.
5. AV nodal reentrant tachycardia may be aborted by beta-blocker therapy.
6. Esmolol IV may be used instead of diltiazem or verapamil, particularly in the perioperative period for most supraventricular tachycardias (SVTs) including atrial fibrillation.
7. Recurrent VF is unique among cardiac arrhythmias because the management is immediate countershock. Antifibrillatory drugs can be useful in the prevention of recurrent VF. Beta-blockers have long been known to have a role in the management of patients with persistently recurring VF. Beta-blockers decrease the incidence of VF in patients with acute

MI. Beta-blockers increase VF threshold and should be given by the IV route in patients with recurrent VF if lidocaine or bretylium fails. IV beta-blockers are given by some before the use of bretylium.

8. Nonsustained ventricular tachycardia (VT) is more appropriately treated with a beta-blocker than with antiarrhythmic agents. The use of most antiarrhythmics has dwindled because of the increased mortality attributed to these agents.

Hypertension

1. *Young patients.* A beta-blocker is the drug of choice in younger and older white patients with or without comorbid conditions, particularly IHD, MI, diabetes, and hyperlipidemia. Beta-blockers are generally known to be effective in white people aged less than 65 yr. In a study by Matterson and colleagues in African-American patients younger than 60 yr, atenolol was, surprisingly, the second most effective drug, after diltiazem, and it was more effective than hydrochlorothiazide (HCTZ).
2. *Elderly patients.* Contrary to common opinion, beta-blockers have been proved effective in older white patients. The statement that beta-blockers are not advisable in elderly hypertensive patients, as given in textbooks and excellent editorials, review articles, and JNC and WHO guidelines, is false because it is based on the results of the Medical Research Council (MRC) trial in the elderly. Unfortunately, of the 4396 elderly hypertensive patients in this trial, 25% were lost to follow-up, and the 63% randomized to a beta-blocker either withdrew or were lost to follow-up. More than half the patients were not taking their assigned therapy by the end of the study. Cruickshank' s review quotes these relevant statistics and yet states "certainly beta-blockers should not be first-line therapy for elderly hypertensives."
3. The misleading MRC trial rings out this statement: elderly hypertensive patients respond best to diuretic-based rather than beta-blocker-based therapy in terms of fewer heart attacks. If this poorly conducted MRC trial carries so much credibility, then diuretics should be used instead of beta-blockers to prevent heart attacks in all cardiac patients (hypertensive and normotensive) with high risk for CHD events. However, every practitioner knows that diuretics are not prescribed for prevention of CHD events in patients at risk for CHD events. Why prescribe them in place of beta-blockers in elderly hypertensive patients based on one unsound RCT? In the HANE (HCTZ, atenolol, nitrendipine, enalapril) study, at 8 wk the blood pressure response rate was significantly higher for atenolol (64%) than for enalapril (50%), HCTZ (46%), or nitrendipine (45%). Effectivity was maintained at the end of 1 yr. A beta-blocker was as effective in elderly patients as in younger ones.
4. In patients with hypertension or angina with LV dysfunction, a beta-blocker used in combination with an ACE inhibitor improves outcome.
5. For prevention and regression of LVH, a beta-blocker or ACE inhibitor is more effective than a diuretic or calcium antagonist.

Acute Myocardial Infarction

For the first 7 d, beta-blockers given within 4 h of an acute MI slightly reduce the 7-d mortality rate. There is evidence that beta-blockers can lower the incidence of acute MI, decrease the size of infarction, reduce the incidence of VF, and reduce the incidence of early cardiac rupture. Reduction of infarct size has been documented with the early use of timolol in acute MI. Beta-blockers are recommended from d 1 and for 2 yr or more; their combination with ACE inhibitors is superior to ACE inhibitors alone. During this period, these agents prevent sudden death and nonfatal and fatal MI and reduce overall mortality. In the CAPRICORN study, carvedilol produced an outstanding reduction in cardiac deaths and events in early post-MI patients.

Caution is needed because beta-blockers may increase cardiac mortality when used indiscriminately in patients in whom the drugs are contraindicated, particularly IV use in patients who are hemodynamically unstable, with impending cardiogenic shock, pulmonary edema, or acute MI with HF.

Heart Failure

The potentially deleterious effects of overactivation of the sympathetic nervous system in HF are now well established; beta-blockers curb this sympathetic overactivation, hence the rationale for the judicious use of titrated doses of beta-blockers in patients with impaired LV function.

1. Carvedilol has been shown in an RCT in patients with class II-III HF to decrease mortality and the risk of hospitalization for recurrent HF. Thus, approved beta-blockers have a new indication for the management of patients with LV dysfunction or class I-III HF.
2. In COPERNICUS, carvedilol produced a significant reduction in serious events in patients with compensated class IV HF, and the indication has been extended from class I to compensated class IV patients.
3. Most important, in MERIT-HF, only a small dose of a beta-blocking drug administered to patients in the low-dose subgroup achieved risk reduction similar to that observed in the high-dose subgroup.

Beta-Blocker Therapy

Beta-blocker therapy is associated with a marked long-term survival benefit among patients undergoing successful *elective percutaneous coronary intervention* (PCI); beta-blocker therapy was associated with a reduction from 6.0% to 3.9% at 1 yr (p = 0.00 14).

Congenital Syndromes

1. *Congenital syndromes* are rare, but they do respond to beta-blockers. These agents appear to restore an imbalance between the left and right stellate ganglia. If recurrent episodes of torsades de pointes develop in a patient with a prolonged QT interval, the condition usually responds to propranolol and acutely to pacing. Propranolol has been proved to be useful in the congenital long-QT syndromes.
2. Beta-blockers have a variable effect in the acquired idiopathic long-QT syndromes.

Effect on Aortic Pressure

A beta-blocker is the drug of choice to reduce the rate of rise of aortic pressure, which increases dissection. Even if the systolic blood pressure is 100-120 mmHg, propranolol or another IV preparation should be commenced. If the blood pressure exceeds 130 mmHg, a careful infusion of nitroprusside, or trimethaphan along with furosemide, is given to maintain systolic blood pressure at 100-120 mmHg. Labetalol has both beta- and alpha-blocking properties and has a role in therapy.

Note: The important effect of a beta-blocker in the treatment of dissecting aneurysm is not to lower blood pressure, but to decrease the force and velocity of myocardial contraction (*dp/dt*) and thus to arrest the progress of the dissection. The drug decreases the reflex tachycardia provoked by nitroprus side. The aforementioned beneficial effects of beta-blocking drugs can be translated to protection of arteries in patients with moderate to severe hypertension and in those with aneurysms. These agents are useful during the surgical clipping of intracranial aneurysms.

Effect on Cardiac Valves

Patients who have palpitations respond favorably to beta-blockers. These drugs should not be prescribed routinely, however, if the patient has only the occasional brief episode of palpitations. Asymptomatic atrial or ventricular ectopic beats require no medication. Chest pain in mitral valve prolapse syndrome is usually noncardiac. Care should be taken not to overtreat with medications and to avoid cardiac neurosis because this syndrome is usually benign. A beta-blocker can be tried if symptoms are distressing. Clinical trials with the use of carvedilol in patients with mitral regurgitation have shown salutary effects that include improved geometry of the left ventricle. Beta-blockers are the cornerstone of treatment for pregnant patients with mitral stenosis. These agents may prevent fulminant pulmonary edema.

Effect of Propranolol

Propranolol, by inhibiting right ventricular contractility, is of value in the acute treatment and prevention of prolonged hypoxic spells.

1. IV propranolol is used only for severe hypoxic spells.
2. Oral propranolol for the prevention of hypoxic episodes is useful in centers where surgical correction is not available or the surgical mortality rate is greater than 10%.

Hypertrophic Symptoms

Medical management remains poor and does not significantly alter the mortality rate. Propranolol is effective for the relief of symptoms such as dizziness, presyncope, angina, and dyspnea, but the dose may need to be as high as 160-480 mg daily. Beta-blockers possessing ISA do not have a role here. *Verapamil* is a useful alternative if beta-blockers are contraindicated or have failed to relieve symptoms. *Disopyramide* appears to be of value in patients with hypertrophic cardiomyopathy, but it needs further randomized studies for confirmation. *Amiodarone* is useful for the control of arrhythmias associated with this condition, but it must not be combined with verapamil.

Marfan's Syndrome

Progressive dilation of the aorta, culminating in aortic dissection, is the most dreaded complication in patients with Marfan's syndrome. Prophylactic beta-blockade slows the rate of aortic dilation and retards the development of aortic complications. Propranolol has been shown to increase cros slinking of collagen in animals known to have a higher rate of aortic dissection.

Subarachnoid Hemorrhage

Arrhythmias and other electrocardiographic abnormalities, LV dysfunction with reduced EF, and myocardial damage including subendocardial necrosis have been documented in patients with subarachnoid hemorrhage. These changes are believed to be caused by the release of sympathetic outflow that occurs with CNS damage and by local norepinephrine release from sympathetic nerves in the heart. Beta-blockers are indicated in individual patients with subarachnoid hemorrhage in conjunction with nimodipine therapy.

Perioperative Mortality

Optimization of cardiac medications in cardiac patients undergoing noncardiac surgery requires the addition of a beta-blocker. Beta-blockade allows safer induction of anesthesia and prevents the hypertensive response to endotracheal intubation; also, these agents prevent recurrent arrhythmias and have been shown to improve morbidity and mortality. Atenolol has been shown in an RCT to decrease morbidity and mortality when given before operation and for up to 1 wk postoperatively.

Neurocardiogenic Syncope (Vasovagal/Vasodepressor Syncope)

If syncopal episodes are bothersome, a beta-blocking drug is advisable. Cardiac sympathetic overstimulation, vigorous LV contraction, and stimulation of intramyocardial mechanoreceptors (C fibers) are underlying mechanisms in the causation of unexplained syncope in patients with a normal heart; a beta-blocker or disopyramide is the logical therapy and appears to be beneficial in patients with disabling syncope. Esmolol IV can be used to predict the outcome of oral beta-blocker therapy. Although a small trial has shown no beneficial effect for some beta-blockers, a noncardioselective agent with greater vasoconstrictive properties such as timolol should be tested in RCTs.

Diabetic Patients

Diabetic patients at risk are a new indication for beta-blockers. The UKPDS results confirm that in type 2 diabetes, beta-blockers significantly reduced all-cause mortality, risk for MI, stroke, PVD, and microvascular disease. In addition, over the 9-yr follow-up, the change in albuminuria and serum creatinine was the same in the captopril and beta-blocker groups. Also, in the SOLVD HF study, surprisingly, in contrast to enalapril, beta-blockers were renoprotective in both the ACE inhibitor and placebo groups.

PRECAUTIONS

Warnings

Beta-blockers are relatively safe if the warnings and contraindications are carefully respected.

1. Beta-blockers are not advisable in patients with decompensated class IV HF. (HF in the presence of acute MI with complete clearing of failure in a few days is not a contraindication.) RCTs have established that these agents save lives in patients with class I-III HF and are strongly indicated.
2. Severe cardiomegaly is a relative contraindication.
3. They are not advisable in patients with EF < 20%.

4. They are not advisable in symptomatic bradycardia or chronotropic incompetence.
5. They are not advisable in conduction defects, or second- or third-degree AV block.
6. Bronchial asthma is a contraindication. Chronic bronchitis and emphysema are relative contraindications depending on their severity and the necessity for beta-blockade.
7. Severe allergic rhinitis is a relative contraindication.
8. Avoid abrupt cessation of therapy. A worsening of angina or precipitation of acute MI has occurred on abrupt withdrawal of therapy. Although this happens only rarely, the patient must be warned. The incidence of this syndrome is said to be infrequent with pindolol because of ISA. Do not discontinue suddenly before surgery. When it is necessary to discontinue beta-blockers, the dosage should be reduced gradually over 2-3 wk, and the patient should be advised to minimize exertion during this period. Added therapy or nitrates and/or calcium antagonists are required during the withdrawal phase.
9. Insulin-dependent diabetes prone to hypoglycemia represents a relative contra-indication.
10. Severe PVD, including Raynaud's disease, is a contraindication.

Side Effects

Cardiovascular

Precipitation of HF, AV block, hypotension, severe bradycardia, intermittent claudication, cold extremities, Raynaud' s phenomenon, and dyspnea may occur.

Central nervous system

Depression may occur, especially with propranolol, and psychosis can occur. Dizziness, weakness, fatigue, vivid dreams, insomnia, and rare loss of hearing may occur.

Gastrointestinal

Nausea, vomiting, and epigastric distress are possible.

Respiratory

Bronchospasm, laryngospasm, respiratory distress, respiratory arrest (rare-with over-dose) may occur.

Skin, Genitourinary

Rashes, exacerbation of psoriasis, reduction of libido, and impotence (a clinical trial comparing propranolol with diuretics showed that impotence was significantly more common in the diuretic-treated group) are possible. Based on studies of 39,745 patients, 149 (0.4%) reported impotence. The true incidence of sexual dysfunction is in the 1-5% range. Reportedly, problems with erection occurred in 5.5% of patients administered a beta-blocker versus 19.6% for patients who received a diuretic. Very rare cases of retroperitoneal fibrosis have been reported with oxprenolol, atenolol, metoprolol, timolol, propranolol, sotalol, pindolol, and acebutolol.

The mucocutaneous syndrome observed with practolol has not been reported with other beta-blockers. Positive antinuclear factor has been reported, however, with acebutolol, pindolol, and labetalol. A lupus-like syndrome has been seen rarely with the use of practolol, and a case has been reported with the use of acebutolol. Labetalol can cause a lupus syndrome and hepatic necrosis. Large-scale clinical trials with timolol and propranolol in post-MI patients followed for 1-2 yr showed that these two drugs do not cause an increased incidence of HF. If patients are selected carefully, HF is not usually precipitated by beta-blockers.

Drug Interactions

Cimetidine, chlorpheniramine, hydralazine, and other drugs that decrease hepatic blood flow increase the bioavailability of hepatic metabolized beta-blockers. Systemic hypertension may be precipitated by combined beta-blocker and epinephrine or by timolol and phenylephrine ophthalmic therapy. Close monitoring of the blood pressure is therefore necessary when a beta-blocker is combined with phenylephrine or other alpha-adrenergic agonists.

Individual beta-blockers

Drug name:	*Acebutolol*
Trade names:	Monitan, Sectral
Supplied:	100, 200, 400 mg
Dosage:	100-400 mg twice daily, max. 1000 mg daily

Acebutolol is a relatively cardioselective and hydrophilic agent that possesses mild ISA and mild membrane-stabilizing activities. The very mild ISA activity prevents undue symptomatic bradycardia. Also, no significant decrease in HDL cholesterol occurs with long-term administration. In the Treatment of Mild Hypertension Study, acebutolol caused no significant change in HDL after 1 yr of therapy. Weak cardioselectivity and moderate lipid solubility are CP features. Mild ISA does not appear to negate the CP potential of acebutolol.

It is possible that strongly cardioselective, water-soluble agents render less cardioprotection, and moderate or strong ISA is counter-productive. The drug appears to have CP properties and caused about a 48% decrease in cardiac deaths in a randomized post-MI study. Clearly, a moderate degree of ISA is a disadvantage and destroys several salutary effects of beta-blockade, but mild ISA is an advantage in patients who require beta-blockade and demonstrate bothersome sinus bradycardia when they are given small doses of a non-ISA beta-blocker.

Drug name:	*Atenolol*
Trade name:	Tenormin
Supplied:	25, 50, 100 mg
Dosage:	Initial: 25-50 mg once daily

This drug was a favorite of trialists from 1980 to 2005 but has often proved ineffective in decreasing CVD outcomes compared with newer agents. One notable exception is the UKPDS study. In these diabetic patients, renal dysfunction may have increased the serum levels of the drug, thus increasing effectiveness. In this small but long-term follow-up study in high-risk patients, atenolol was as effective as captopril in reducing CVD outcomes. The drug is almost totally excreted by the kidney. There is only a fourfold variation in plasma level, and the main difference from nadolol is that it is a $beta_1$-cardioselective drug.

Atenolol is a long-acting preparation and can be used once daily. Atenolol has not caused the mucocutaneous syndrome; indeed, patients who developed the syndrome while taking practolol were cleared of the syndrome when treated with atenolol. Blood levels are maintained in heavy smokers, the opposite of the effect of propranolol and hepatically metabolized (HM) beta-blockers. In smokers who will not stop, bisoprolol, carvedilol, metoprolol, or timolol should be used. The author has discontinued use of atenolol since 2001 because the drug has lower CVD protective effects compared with other proven beta-blockers.

Drug name: *Bisoprolol*

Trade names: Zebeta, Monocor, Emcor

Supplied: 5-10 mg

Dosage: 5-10 mg daily, max. 20 mg

Ziac, Monozide: bisoprolol and hydrochlorothiazide 6.25 mg

This agent is highly $beta_1$-selective, and it is more cardioselective than metoprolol; in humans, the drug has a twofold higher $beta_1$-selectivity than atenolol. A 10-mg dose is equivalent to atenolol 100 mg. The drug has a half-life of 12-14 h and thus has a long duration of action beyond 24 h. Bisoprolol is 50% lipophilic, metabolized by the liver, and 50% hydrophilic, excreted by the kidney. The concentration of unchanged bisoprolol in rat brain is lower than that of metoprolol or propranolol but higher than that of atenolol after dosing. A study in patients with LV dysfunction, CIBIS II, showed a markedly significant reduction in overall mortality with bisoprolol therapy; the study was stopped because of significant favorable effects.

Bisoprolol treatment reduced all-cause mortality by 32% ($p = 0.00005$) and sudden death by 45% ($p = 0.00$ 1). There was a 30% reduction in hospitalization caused by worsening HF . The combination of bisoprolol and 6.25 mg HCTZ (Ziac) has been approved by the U.S. Food and Drug Administration (FDA) as first-line initial therapy for hypertension. The blood pressure-lowering effect is superior to that of atenolol. The 5-mg dose seems equipotent to atenolol 50 mg. Most important, this is a genuine 24-h acting, once-daily drug.

Bisoprolol has been shown to reduce ambulatory systolic and diastolic pressures by 43% and 49%, respectively, more than atenolol during the early morning hours during which catecholamine surge causes an increased risk of ischemia and sudden death. In response to pressure stimuli and increased catecholamines, there is less rise in blood pressure, and excessive increase in exercise blood pressure is controlled. Renal blood flow is enhanced. Coldness of the fingers and toes appears to be less than that observed with atenolol, metoprolol, and propranolol.

Perioperative Protection

Bisoprolol, in an RCT of 1351 high-risk patients undergoing vascular surgery, significantly reduced events ($p < 0.001$). Bisoprolol was started 1 wk preoperatively and continued for 30 d postoperatively; there were two deaths and no MIs versus nine deaths and nine MIs in the untreated group.

Drug name: *Carvedilol*

Trade names: Coreg, Eucardic

Supplied:	6.25, 12.5, 25 mg
Dosage:	Heart failure: 3.125 mg, then 6.25 mg twice daily; titrate over weeks to 25 mg twice daily
	Hypertension: 12.5 mg daily; titrate over weeks to 25-50 mg daily, max. 50 mg
	Angina: 12.5 mg twice daily; titrate over days to weeks to 25 mg twice daily

Carvedilol is one of the most effective beta-blockers available and is rated by the author as the best choice for use in acute MI and post-MI prophylaxis particularly if LV dysfunction is present. It is also recommended for hypertension, particularly in diabetics or prediabetics, because it has been shown to improve insulin sensitivity compared with metoprolol.

Dosage

The dosage must be individualized during up-titration. Titration also depends on the occurrence of hypotension and bradycardia. The dose of digoxin, diuretic, and ACE inhibitor or angiotensin II receptor blocker must be stabilized before commencing carvedilol. The initial dose is 3.125 mg twice daily for 2 wk and then, if tolerated, 6.25 mg twice daily. The dose can be doubled every 4 wk to the highest level tolerated by the patient. Observe the patient for dizziness, light-headedness, and hypotension for 1 h after the initiation of each new dose. The maximum dose is 25 mg twice daily in patients weighing less than 85 kg and 50 mg twice daily in those weighing over 85 kg. Carvedilol should be administered with food to slow the rate of absorption and to reduce the incidence of orthostatic effects.

If symptoms of vasodilation occur (e.g., dizziness and lightheadedness), reduce the dose of diuretic and then, if necessary, the ACE inhibitor. If symptoms do not resolve, reduce the carvedilol dosage. Symptoms of worsening HF (e.g., edema, weight gain, and shortness of breath) should be treated with increased doses of diuretic. The carvedilol dose should be reduced if symptoms persist.

Action

The ratio of $alpha_1$- to beta-blockade for carvedilol is 1:10, compared with 1:4 for labetalol, a more powerful $alpha_1$-blocker that causes a greater degree of vasodilation and orthostatic hypotension; antioxidant and antiproliferative properties have been reported. The drug improves ventricular function without upregulating downregulated myocardial beta-receptors in the failing heart.

Indications

The drug is approved by the FDA for the management of class I-III HF. The U.S. Carvedilol Heart Failure Study in patients with NYHA class II and III HF showed a 27% reduction in hospitalization and a 65% reduction in mortality risk.

1. The drug is contraindicated in patients with decompensated class IV HF.
2. As with other beta-blockers, the drug is contraindicated in patients with asthma, severe COPD, second-or third-degree AV block, and sick sinus syndrome. Carvedilol is not advisable in patients

with clinically manifest hepatic impairment. Mild hepatic injury has been noted. Stop the drug and do not restart if the patient has evidence of liver injury.

3. *Class IV patients,* when stabilized and free of fluid overload, can be started on 3.125 mg, a dose titrated slowly up over several weeks.
4. Ejection fraction and myocardial energetics take about 8 wk to improve as a biologic effect with beta-blocker therapy. The ventricle goes from a more spherical shape to a more normal elliptical shape, and LV mass decreases; these changes indicate that the pathologic remodeling process is being attenuated and in many cases reverts back to normal. Itis important to initiate therapy as early as possible while there is myocardial viability left and it is possible to recover a biologic effect on myocytes.
5. End-diastolic volume decreases; myocardial contractility increases, after about 3 mo of therapy; mechanical work improves, as reflected by stroke work; and myocardial O_2 consumption decreases.
6. Myocardial efficiency increases in patients who receive a beta-blocker as well as an ACE inhibitor. Plasma norepinephrine increases over time in patients treated with an ACE inhibitor. The combination of a beta-blocker and ACE inhibitor is complementary and cardio-protective in patients with MI, HF, diabetes, hypertension, and LVH.

Drug name:	*Esmolol*
Trade name:	Brevibloc
Dosage:	IV infusion within range 50-200 μg/kg/min

Dosage

The dose is 5-40 mg, usually 3-6 mg IV infusion over 1 min (30-300 μg to maximum of 500 μg/kg over 1 min) and then maintenance infusion 1-5 mg/min (maximum 50 μg/ kg/min). If control of the clinical situation is not achieved and more rapid titration is necessary, a further bolus may be given followed by an increase in the maintenance infusion rate. If mild hypotension is present, the maintenance dose should be reduced to 1-3 mg/min. Hypotensive effects of the drug usually disappear within minutes of cessation of the infusion. Esmolol IV 50-200 μg/kg/min has an effect equivalent to that of propranolol 3-6 mg IV.

Action

Esmolol is a cardioselective ultrashort-acting beta-blocker that is quickly converted by esterases of red blood cells to inactive metabolites. The drug has a short half-life of about 9 min that is stable in patients with mild HF, hepatic dysfunction, or renal failure. The drug action dissipates within 20-30 min of administration.

Indications

1. The drug is used for SVT in the perioperative period and for SVT not terminated by adenosine or in patients with contraindications to the use of adenosine or verapamil, especially during acute MI or other ischemic syndromes.
2. Esmolol is superior to verapamil in the management of recent-onset atrial fibrillation or flutter.

The drug is given only under continuous monitoring of cardiac rhythm and blood pressure under close supervision. Hypotension is the major adverse effect, and in one study this occurred more often with esmolol than with IV propranolol, but, with careful titration, esmolol is superior, given that adverse effects dissipate rapidly. Contraindications include:

1. Severe hypotension or cardiogenic shock.
2. Asthma or severe COPD.
3. Other contraindications to the use of beta-blocking agents.

Drug name:	*Labetalol*
Trade names:	Normodyne, Trandate
Supplied:	50, 100, 200, 400 mg
Dosage:	50-100 mg twice daily; titrate over weeks to 200-400 mg twice daily

A combined alpha- and beta-blocker, this drug is useful in the management of hypertension of all grades including hypertensive emergencies.

Disadvantages: Labetalol causes significant postural hypotension and must be given two or three times daily in large doses. Side effects include a lupus-like illness, a lichenoid rash, impotence, and, very rarely, hepatotoxicity manifesting as raised levels of transaminases, abnormal liver function test results, hepatitis, and fatal or nonfatal hepatic necrosis.

Drug name:	*Metoprolol*
Trade names:	Betaloc, Betaloc SR, Lopressor, Toprol-XL
Supplied:	50, 100 mg
	Betaloc SR: 200 mg
	Toprol-XL: 50, 100, 200 mg
Dosage:	50-200 mg twice daily
	Long-acting: 50-200 g once daily, max. 300 mg

Toprol XL, with its 24-h duration of action, is important for coverage of early morning catecholamine surge and early morning fatal MI. This drug is superior to other sustained- release metoprolol tartrate formulations, whose action may not continue throughout the 24-h period. Metoprolol is beta$_1$-cardio selective. In patients with bronchospastic disease, metoprolol, in doses lower than 150 mg daily, causes less bronchospasm than a nonselective beta-blocker at equivalent beta-blocking dose.

If bronchospasm is precipitated, it will respond to beta$_2$-stimulants, whereas there would be a poor response if a nonselective beta-blocker were used. Cardio selectivity confers advantages when beta-blockers are given to patients with labile diabetes. Metoprolol has been shown to reduce the incidence of VT during acute MI. The Goteborg metoprolol trial has documented the beneficial effect of metoprolol on survival during the early phase of MI (from d 1 for 90 d). Metoprolol caused a 36% reduction in mortality rate during the 90 d of therapy, and benefit was maintained for 1 yr. Metoprolol causes less sedation and drowsiness than atenolol. In addition, metoprolol is lipid soluble and appears to be more cardioprotective than water-soluble beta-blockers, which achieve lower concentration in the brain. The MERIT-HF study has established metoprolol as indicated in patients with class I-III HF. It is most important to start with small doses: metoprolol succinate (CR/XL) controlled-release 12.5 mg once daily, and titrate up slowly over 4-8 wk to a target dose of 150-200 mg.

Drug name:	*Nadolol*
Trade name:	Corgard
Supplied:	40, 80, 120, 160 mg
Dosage:	40-80 mg once daily, max. 160 mg

Dosage

The manufacturers suggest a maximum of 240 mg for angina and 320 mg for hypertension. These high doses are not recommended by the author. Because the drug has a very long half-life and is excreted entirely by the kidneys, accumulation commonly occurs in patients with mild renal dysfunction and in patients over 65 yr of age. In practice, a dose of 80 mg is equivalent to approximately 160 mg of propranolol. The dose must be reduced in renal failure and in the elderly. Also, the time interval between doses should be increased, that is, one tablet every 36 or 48 h may suffice in renal failure.

Use a nonrenally excreted beta-blocker in the presence of severe renal failure. Like propranolol, nadolol is a nonselective beta-blocker. It has a weak lipid solubility and therefore is almost completely excreted by the kidney. No large RCT has been done to prove its effectiveness in reducing CVD mortality or morbidity. Because of its long half-life it is given as a one-a-day tablet; this may be important when compliance is a problem. *Insomnia* and *vivid dreams* occur much less frequently with nadolol and atenolol compared with propranolol and other lipid-soluble beta-blockers, but they can occur with sotalol. The dose of atenolol, nadolol, and sotalol should be reduced in renal failure, and with severe renal failure, the time interval between doses should be increased.

Drug name:	*Propranolol*
Trade name:	Inderal
Supplied:	Inderal: 10, 20, 40, 80, 120 mg Inderal LA: 80, 120, 160 mg capsules
Dosage:	40–80 mg twice daily; titrate over weeks to 80-120 mg twice daily LA: 80-160 mg once daily, max. 240 mg

Indications

In the new millennium, it is no longer justifiable to prescribe propranolol because bisoprolol, carvedilol, and metoprolol have less severe side effects and have been proved to reduce morbidity and mortality in a wide range of cardiac patients including those with very sick hearts (HF class II-III and some class IV patients). Propranolol is contraindicated in patients with HF or EF $< 35\%$. This beta-blocker has been in use since 1964 and is still the most frequently prescribed beta-blocker worldwide. Unlike other beta-blockers, the drug is approved by the FDA for many indications: angina, acute MI, postinfarction prevention, arrhythmias, hypertension, IV use, anesthetic arrhythmias, hypertrophic cardiomyopathy, syncope, anxiety and essential tremor, migraine prophylaxis, and thyrotoxicosis. The long-acting preparation provides full 24-h coverage.

The BHAT study proved the beneficial effects of propranolol in post-MI patients for the prevention of reinfarction, sudden death, and fatal and nonfatal MIs. Cigarette smoking alters blood levels of propranolol and may mask CP effects. In nonsmokers, however, the drug has been proved effective, whereas only timolol has been proved effective in RCTs and in the post-MI patient. The drug is strongly lipid soluble and therefore has a high uptake in the brain; this may be the reason for fatigue, the rare occurrence of depression, and vivid dreams. Lipid solubility, brain concentration, and $beta_1$- and $beta_2$-blockade provide cardioprotection. In smokers, the salutary effects of propranolol are masked.

If depression, fatigue, or mild memory impairment occurs, switch to bisoprolol, metoprolol, or timolol.

Drug name:	*Sotalol*
Trade names:	Sotacor, Betapace, Beta-Cardone
Supplied:	80, 160, 240 mg
Dosage:	40–80 mg once daily; increase over days to weeks to 80-160 mg once or twice daily

Indications

In the United States, this drug is approved for oral use in patients with life-threatening ventricular arrhythmias (sustained VT). This agent is not recommended for the treatment of asymptomatic ventricular premature beats. In the United Kingdom and Canada, the drug is indicated for hypertension and for the management of arrhythmias, including paroxysmal atrial fibrillation, or for the maintenance of sinus rhythm in place of quinidine. The response to sotalol predicts the response to amiodarone during serial drug testing in patients with sustained VT. Sotalol is unique among the approved beta-blockers.

The drug has all the effects of a nonselective beta-blocker plus an added class III antiarrhythmic effect: the drug lengthens the duration of the cardiac action potential and prolongs the QTc interval of the surface electrocardiogram. The drug appears to be more effective than other beta-blockers in the control of numerous bothersome ventricular premature beats and sustained VT. However, some studies indicate no difference in efficacy. The drug has been shown to cause an 88% reduction in ventricular ectopic beat frequency at the optimal titrated dosage. Torsades de pointes have been precipitated as a rare complication, mainly in patients with hypokalemia.

Torsades occurred, however, despite therapeutic plasma sotalol concentration and normal serum potassium level in the absence of diuretics. *Caution* is necessary: do not administer sotalol with nonpotassium-sparing diuretics and drugs that cause QT prolongation. The drug represents a significant advance in the management of some ventricular tachyarrhythmias, including recurrent VT tachycardia or VF.

Contraindications

Congenital or acquired long QT syndromes are contraindications, as well as other contraindications to beta-blockade.

Drug name: *Timolol*

Trade names: Blocadren, Betim

Supplied: 5, 10 mg

Dosage: 5-10 mg twice daily, max. 20 mg twice daily

This noncardioselective drug has some advantages over propranolol. First-pass hepatic metabolism is 60%, and 40% of the drug is excreted unchanged in the urine. Variation in plasma level is only sevenfold. The drug is six times more potent than propranolol, so for a given dose, a better plasma level is achieved with less variation. It has moderate lipid solubility.

Timolol can be given twice a day with a fair certainty that plasma levels will be adequate. It has proved to be efficacious and safe in the reduction of raised intraocular pressure when used topically. Timolol is the first beta-blocker to have been shown beyond reasonable doubt to reduce cardiac mortality in the post-MI patient. A remarkable 67% reduction in sudden cardiac deaths was achieved by timolol in the post-MI study.

Newer Beta-Blocker

Nebivolol

Nebivolol is the most highly beta$_1$-selective agent to be tested. The drug stimulates the endothelial L-arginine/nitric oxide pathway and thus causes vasodilation. Results of trials in HF and in hypertension are awaited.

Indications: essential hypertension.

Dosage: 5 mg daily; for the elderly, initially 2.5 mg daily, increased if necessary to 5 mg daily.

Cautions: reduce dose in renal impairment, in the elderly, and with hepatic dysfunction.

ANNOTATION

1. More than 12 beta-blockers are available worldwide. As emphasized earlier, there are subtle and, fortunately, important differences among them.
2. Agents with ISA are not cardioprotective drugs and do not enter the contest.
3. Cardioselective agents cause fewer adverse effects, and, particularly in diabetic patients, they are usually the agents of choice but carvedilol, a nonselective agent, exhibited improved insulin sensitivity compared with metoprolol in GEMINI and is recommended by the author provided that the patient is not hypoglycemia prone and postural hypotension is not a problem.
4. Of the cardioselective agents, only bisoprolol and metoprolol (both lipophilic) have been shown to reduce CHD mortality and events significantly in the following well-conducted RCTs: CIBIS II, MERIT-HF, and SOLVD (beta-blocker renoprotection as good as ACE inhibitor). The noncardioselective agent *carvedilol* was studied in the successful CAPRICORN and

COPERNICUS trials. The common threads are $beta_1$ and lipophilicity, which augment brain concentration and may protect from sudden death. In the timolol study, there was a 67% reduction in sudden death. (Timolol is $beta_1$- and $beta_2$-selective and lipophilic; it appears to be a forgotten beta-blocker in North America.)

5. Of the agents–bisoprolol, carvedilol, and metoprolol–that should be recommended based on the aforementioned logical approach, bisoprolol carries some advantages for the hypertensive patient: genuine one-a-day administration, quelling of early morning catecholamine surge, and better control of early morning and exercise-induced excessive-rise in blood pressure than atenolol. It does not cause postural hypotension, as does carvedilol. *Bisoprolol* is the most highly cardioselective (of the three) and is therefore relatively safe to use in diabetes, COPD, and perioperatively, as is Toprol XL, a proven 24-h metoprolol formulation that has been well tested in the United States for hypertension.
6. Bisoprolol, carvedilol, and timolol followed by metoprolol succinate (sustained release) appear to be the winners overall for the management of hypertension.
7. Carvedilol showed marked benefit and safety in the CAPRICORN post-MI study and the COPERNICUS HF study and is the beta-blocker of choice for patients with HF and post MI with LV dysfunction. For hypertension, Toprol XL or bisoprolol, given once daily, has advantages over carvedilol, which usually should be administered twice daily.
8. The older beta-blocking drugs including atenolol, pindolol, oxprenolol, and nadolol should become obsolete.

7

Inhalation Technology

Pharmacologically active substances have been administered to humans for thousands of years. However, it is only in relatively recent times that the true nature of drug delivery to the human body has become an exact science. Today, drugs are delivered via many portals into the body using a variety of dosage forms which allow medicament administration, for example, orally or parenterally. One relatively recent pharmaceutical development is delivery via the lung. Although the breathing of "*vapors*" has been used since ancient times as a way of relieving respiratory problems, the effective acceptance and mass commercialization of inhaled respiratory medicines was not achieved until 1948 when Abbot Laboratories developed the Aerohaler for inhaled penicillin-G powder and then revolutionized the field in 1955 with the advent of the pressurized metered dose inhaler (pMDI).

Since this inception, the range of inhaler products and medicaments has grown and expanded to encompass alternative drug delivery systems, namely those based on the dry powder inhaler (DPI), and combination products. The lung offers a unique and challenging route for drug delivery for the treatment of respiratory diseases, such as asthma, chronic obstructive pulmonary disease (COPD), and cystic fibrosis. The organ offers a high surface area, circa 100 m^2, offering the possibility of high absorption and non-hepatic drug delivery, with obvious pharmacological advantages. Whilst inhalation pharmaceutical products may appear to operate on simplistic assumptions, the challenge for pharmaceutical technologists, formulators, and device engineers has been how to produce a product, which meets the demands expected by the pharmaceutical regulatory authorities and, importantly, patient acceptability.

As well as developing such a product there are challenges in pharmaceutical development in terms of material and formulation characterization and scale-up. The methods used to characterize inhalation products, as with other solid dosage forms such as tablets, are relatively simple and have been developed in recognition of the particular challenges for characterizing products for administration to the body. All pharmaceutical products must exhibit efficacy, safety, stability, and conform to regulatory drug product requirements. In addition, solid dosage forms tend to have their own set of characterization methods, for example, tablets are often tested for their hardness and disintegration time. The characterization methods for inhalation products have developed to include tests for performance based on, for example, fine particle fraction (FPF) and emitted dose.

The development of such tests has progressed, in part, due to the complexity of the inhalation drug delivery process: the tests have to mimic the inhalation process from emission from the device to deposition in various regions in the lung. In general, the practicable delivery of an effective respiratory dose of any medicament is achieved if the therapeutic component has an aerodynamic diameter of less than 6 μm. Although the deposition, adsorption, diffusion, clearance, and residence of particles in the lung is still not fully understood, it is generally accepted that this particle size range will produce a therapeutic outcome, with minimized upper bronchial impaction and expulsion due to limited sedimentation time in the alveoli. The evaluation of the deposition profiles of particulates in the lungs and airways has been extensively reported. However, it is generally accepted that the inspirational airflow which inhalation aerosols and dry powders are exposed to affects their lung deposition distribution. Understanding particulate deposition in the lung is a complex issue.

A physical basis for the deposition of aerosol particles in the human lung became available when Findeisen published the first lung deposition model. Important parameters for lung deposition are: the particle size distribution; the inhalation and exhalation flow rates, tidal volume, and the shape of the upper airways. The structure of the bronchial and alveolar regions in normal lungs is considered to be of minor influence; however, this is probably not the case for diseased lungs. It is generally agreed that particles smaller than 0.1 μm in diameter will deposit due to a diffusion processes, if the residence time is sufficient. Furthermore, the distance a particle travels by diffusion transport increases with decreasing particle size and increasing respiratory rate. Diffusional deposition

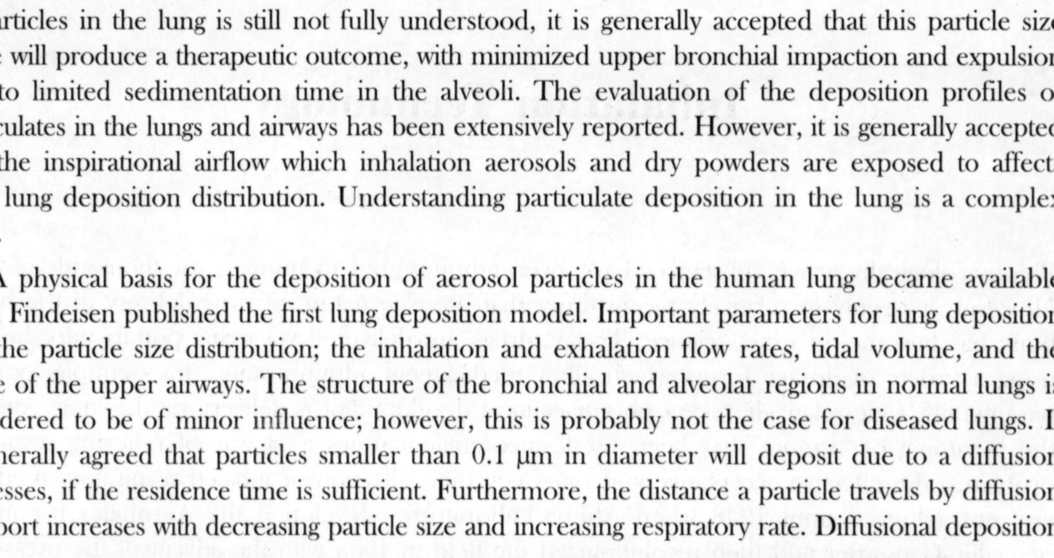

Fig. 7.1. Deposition as a function of various sizes of inhaled particles in the human respiratory tract as calculated by the ICRP Task Group on Lung Dynamics for nose breathing at a rate of 15 BPM and a tidal volume of 1450 cm^3.

therefore decreases with increasing particle size up to about 1 μm and becomes negligible for larger particles. Sedimentation processes become dominant for particles larger than 0.1 μm diameter, and deposition increases with particle size, particle density, and respiratory rate. Consequently, in the size range of 0.1-1 μm, particles are simultaneously deposited by gravitational and diffusional transport.

Inertial transport is the effective mechanism for larger particles, and deposition due to impaction increases with particle size, particle density, and airflow rate. Thus, in the size range above 1 μm, particles are deposited due to impaction and sedimentation. In general it can be concluded that deposition in the extra-thoracic and upper bronchial airways, through which the inhaled air passes at high velocity, is a result of impaction mechanisms while sedimentation processes dominate deposition in the lower bronchial airways and the gas exchange region where the residence time of the inhaled air is large.

As the size of physical objects is reduced, their bulk properties and behavior are greatly modified. For example, sugar is available as granulated sugar and icing sugar. However, icing sugar (~15 μm) and granulated sugar (~500 μm) have very different cohesive properties. This phenomenon is similar to particulates for inhalation. Part of the reason is that the ratio of surface area to mass ratio increases as particle size decreases, leading to an increased chemical and physical (increased surface energy) activity. This creates difficulties for the processing, handling and aerosolization of particulates in the dry state and non-aqueous liquid suspended states since the powders often form agglomerates, hindering the aerosolization of primary respirable particles.

The intrinsic cohesiveness of such fine powders makes the accurate delivery of particulate medicaments to the lung a real challenge. A variety of approaches are currently utilized to overcome this challenge and optimize formulations for achieving a therapeutically effective delivery to the lungs. This chapter will review the current state of the art of the three main inhalation delivery platforms (DPIs, pMDIs, and nebulizers), including novel exploratory delivery systems. Please note that this chapter will not extensively review the multitude of patents filed in this field.

DPI TECHNOLOGY

The Montreal protocol of 1989 banning the use of chlorofluorocarbons (CFCs) was a laudable achievement for the protection of the environment. However, while this ban was aimed at industries that used large volumes of these harmful materials, it also affected highly beneficial medicinal products, such as pMDIs, which had traditionally contained low levels of these CFCs. The consequence of this has been that existing pMDI products have had to be reformulated to contain hydrofluroalkane systems and new formulations must be developed to contain the replacement propellants. The challenges associated with such efforts have led to an increased popularity in alternative inhalation technologies, namely *dry powder inhalers*. As the name implies, DPIs contain and deliver the active medicament as a dry powder of suitable aerodynamic size for respiratory therapy.

Dry powder particles of suitable size range, generally considered between 1 and 6 μm, can be readily produced. However, such particles have high surface area to mass ratio, making them highly cohesive/adhesive in nature. Consequently, the drug must be formulated in such a way that the energy input during inhalation is sufficient to overcome the contiguous adhesive and cohesive particle forces and aerosolize the powder for respiratory deposition. Although, in principle, this approach may appear straightforward, and there exist many DPI products which achieve this, the physico-chemical nature and interactive mechanisms of the components in a DPI system are still relatively poorly understood.

Many commercial products have, by pharmaceutical standards, relatively poor efficiencies, with often less than 20% drug being delivered to the lung. This has generated a significant amount of research activity in the fields of pharmaceutics, powder technology, surface, aerosol and colloid science which has focused on understanding the interactions in DPI systems. In simple terms, DPI technology can be categorized into four areas; Formulation; active pharmaceutical ingredient (API) powder, Device; and Manufacturing. It is clear that understanding and engineering the physicochemical properties of the materials and processes in DPI technology will allow us to influence, and control, the aerodynamic efficiency and thus therapeutic efficacy. The DPI device will play a pivotal role in efficient API aerosolization, since its geometry and configuration will influence the flow and sheer forces acting on the formulation.

Many design approaches have been used to achieve and improve drug liberation, as can be seen from any cursory survey of the patent and product literature. Currently, marketed devices can be generally classified according to their method of energy input, formulation approach and dosing regime, and how these products achieve reproducible delivery of respirable API particulates. Common approaches employed to assure such reproducible delivery of the API (delivered dose) are by formulation of a micron sized drug with a larger inert carrier, agglomeration with an excipient of similar particle size or by simple agglomeration of the drug (where metering dose constraints allow). These formulation approaches may be termed as carrier and agglomeration based systems,

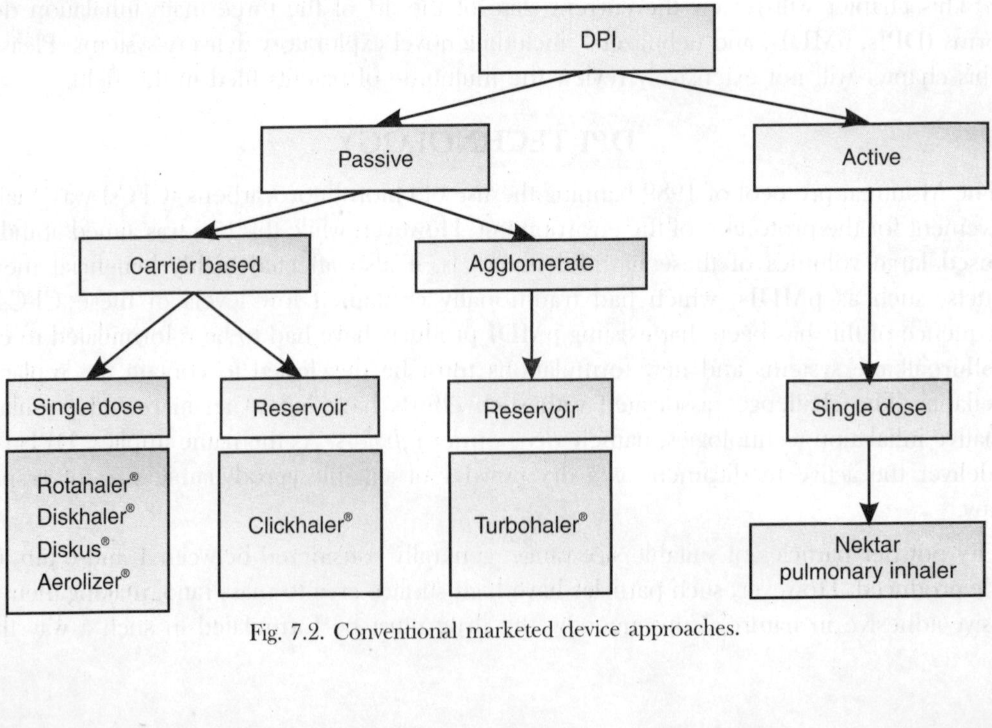

Fig. 7.2. Conventional marketed device approaches.

respectively. Both systems result in improved powder flow, ease of metering and consistent drug mass entrainment.

The delivery of the drug particulates to the lung from passive inhaler products is achieved after liberation of the drug from the formulation during the inhalation actuation. In addition, recent technological developments in spray drying and particle engineering have resulted in co-processed single dose DPI formulations combining, for example, a mixture of excipients in the commercial insulin product, Exubra, or by the addition of excipients which promote aerosolization efficiency through altering the particle physical morphology.

Carrier-based Systems of DPI

The majority of DPI products use the carrier-based formulation system. This approach has proved successful for a range of dosing regimes from high dose, that is, up to several hundred micron, for example, 400 μg for albuterol sulphate to low dose, for example, ≤12 μg for formoterol. In the 1970s, Hersey theorized that when small particles were mixed with larger particles (such as conventional carrier material) a so-called ordered mix was formed. Although these observations were deemed relevant to low dose tablet formulations, this theory was later expanded in the 1980s and 1990s by Staniforth, and has since become an underlying theory for carrier based DPI powder formulations. In general, this approach has been adopted by the pharmaceutical industry to produce many successful inhalation products, and, interestingly, as a way of preparing low dose blends for tableting applications.

These simple DPI formulations are typically prepared by mixing micronized drug with larger lactose carrier particles. The drug adheres to the larger particles in such a way that results in good blend uniformity and flow. Additionally, and importantly, when the formulation is delivered to the patient via a device, the drug particles are liberated to provide a reliable efficacious dose to the patient. This technique has been proven to be versatile and has been used to develop products which contain low dose drugs products, for example, Novartis 12 μg formoterol Foradil product, equivalent to a c.a. 0.05% w/w drug/carrier blend. As can be seen, the formulation consists of a mixture of larger carrier particles together with associated lactose fines and drug particulates. The relative surface distribution of drug particles to carrier material will be very low and yet the product provides good dose uniformity.

The same is true for higher dose formulations which contain several hundred micrograms of drug which contain higher ratios of drug to carrier. However, the important point with drug loading in DPIs is that the drug will be associated with the surface of the carrier, making the surface ratios much higher than mass ratios. Obviously, for all DPI products, carriers and drugs with consistent physiochemical characteristics are required to maintain product performance. This dependency on material characteristics has implications for drug aerosolization (liberation) performance if variations in API adhesion exist throughout the carrier system.

Agglomerated Drugs

Agglomerated systems are not as commonly used as the carrier-based systems but the technique can be successfully employed to produce DPI products. Agglomerates containing pure drug, or a mixture of drug and fine particle excipients, can be prepared by spheronization under controlled conditions. The key requirement of the agglomerates is that they have sufficient strength to survive the manufacturing processes, filling, and metering, but still allow efficient de-agglomeration and aerosolization during delivery to the patient. The most obvious, and successful, example of this formulation approach is Astra Zeneca's Turbuhaler where the agglomerated formulation contains only the drug (e.g. 100, 200, and 400 μg dose budesonide Pulmicort) or with the addition of micronized lactose excipient (6 and 12 μg dose formoterol Oxis). As with carrier based systems, these formulation approaches result in improved reproducibility in emitted dose and respiratory deposition efficiency.

However, both the efficiency of respiratory deposition will be dependent upon the energy imparted into the system upon inhalation (i.e. by the inspiration energy of the patient), where in general, agglomerated based systems require a higher pressure drop across the device to achieve this goal. For all DPIs, this is the real challenge for formulators and engineers: how to balance the mixing of particulate materials so they form a stable processable formulation but ensure that the interactions are weak enough to allow liberation of the drug particulates during inhalation. Although these principles of formulation seem simplistic, it is possible to put this technological and scientific challenge into perspective when considering the forces involved in a typical DPI system. For example, it is estimated that the force acting on a 1-μm spherical particle (adhered in a carrier system) in a 10 m/s laminar air flow, comparable to the axial velocity in an Aerolizer DPI, may be estimated as 2 nN while the adhesion force, using empirical calculation, is of the order of 100 nN.

Clearly, with such disparity in expected and predicted particle exposure and detachment forces, efficient aerosolization efficiency is a scientific and technological challenge. Subsequently, efforts to improve the performance of DPIs have focused on the engineering of excipient and API physico-chemical properties to overcome these forces and, modifications in device design, to improve energy translation and efficiencies.

Bases of DPI

To date, devices have been primarily passive, in that they rely on the patients inspiratory flow to generate the energy for drug entrainment and liberation. Such devices often contain engineered components which can facilitate drug liberation. Historically, these devices aerosolized a powder as a fixed single formulation dose or by aerosolization of a single formulation dose from a rotating capsule (e.g. Aerolizer Novartis), where drug particle liberation is presumably achieved by exposure of the formulation to the airstream, and in some cases, a baffle or grid. To improve the aerosolization efficiency of these early devices many attempts have been made to increase aerosolization efficiency, including more tortuous flow pathways and increased pressure drops (resulting in increased particle acceleration and turbulence), for example, the Turbuhaler. However, in terms of delivered dose, it is fair to say that there has been no great leap in efficiency improvements.

In addition to these advances, the incorporation of multiple blister or reservoir systems has resulted in improvements in the ease of use by the patient. More recently, active devices, utilizing pressurized gas, have been marketed that potentially increase aerosol efficiency and remove the requirement for patient energy input (e.g. as used in Nektar's Pulmonary inhaler). However, such systems have a unit price and ease of use consideration. Ultimately, whichever approach is adopted, advantages and disadvantages will exist and the efficiency of the system will be related to the formulation and design ingenuity.

Efficient Therapeutic Outcomes

As previously discussed, the efficient respiratory deposition of the API powder will be dependent on the physico-chemical nature of the particulate system and, in addition, the physical and environmental conditions to which they are exposed. At this point it is important to state that there are limited products presently available which contain engineered excipients or drugs. However, there has been much research activity into the modification of the powder and material components of DPIs.

Improved carriers

One of the most commonly studied areas in DPI formulation design is that of carrier modification. Difficulties (both technical and regulatory) in altering the physical properties of micron sized drug particulates has led to the more popular approach of modifying excipient carrier materials to achieve improvements in blend uniformity and delivered drug dose. It must be stated, however, that in general, this approach is still only led to limited empirical observations concerning a material descriptor and formulation performance, and it is unclear to what degree such observations have impacted pharmaceutical manufacturing processes. A series of approaches have indicated that the optimum performance characteristics of a DPI product require a formulation which exhibits adequate flow (for processing and dosing) and a certain level of excipient fines.

Blending technique

It is reasonable to expect that the macro-, micro-, and nanoscopic morphology of an excipient carrier will play a major role in efficient drug blending and aerosolization since micron sized particles will regularly encounter adhesive forces that are greater than those required for their liberation. Subsequently, the most common investigated approaches to improve drug removal from carrier particles during inhalation are to (i) reduce the contact area between drug particle and carrier, (ii) alter the particle geometry of the carrier, and (iii) reduce the surface energy of the contiguous surfaces. Of these three approaches, by far the greatest attention has been focused on modifying the carrier morphology in order to achieve changes in the contact geometry between drug and carrier.

In simple terms, by varying the rugosity of a carrier it becomes possible to vary the degree of adhesion and, thus, alter the efficiency of drug liberation. Although roughness is believed to be one of the dominating factors for the performance of such systems, it is generally very difficult to investigate the relationships between one material factor, in this case roughness, and performance without altering another potentially influencing factor, such as particle shape or quantity of intrinsic carrier fines. This makes quantifiable interpretation of any observations difficult. Furthermore, the potential therapeutic improvement of altering the carrier roughness has, in general, been investigated via comparison of off-the-shelf lactose excipients with materials produced by some variation in batch or process parameters. However, in general, the hypothesized influence of carrier morphology on aerosolization performance is generally in good agreement with recent literature. For example, recent work by Flament et al. has suggested a linear reduction in the FPF of micronized terbutaline sulphate with increased roughness, measured in this case by microscopic luminescence, of a 63-90 μm sieve fractioned lactose samples.

Earlier work, by Kawashima et al., also related the surface area and roughness parameters of a series of lactose carriers (approximated median diameter 60-65 μm with similar geometric standard deviations) to the aerosolization performance of micronized pranlukast hydrate. In general, an increased carrier roughness and surface area resulted in a reduction in drug aerosolization efficiency. Interestingly, a deviation from linearity was observed for lactose type F, and was attributed to a variation in amorphous content and/or increased drug-carrier contact area (e.g. as in the smooth carrier). Other recent studies have reported similar observations. For example, Zeng et al. reported that when the surface structure of a re-crystallized lactose carrier was modified (by etching with a 95% w/w ethanol solution), the induced surface cavities resulted in a decrease in FPF of blended albuterol sulphate.

It is interesting to note, however, that the addition of fine lactose, approximated as 5% (between 5 and 10μm), rectified this reduction. In other recent studies, the macroscopic etching of commercial grade lactose via mixing in an ethanolic solution or via temperature controlled surface dissolution has been shown to improve the aerosolization of various micronized APIs. As with the studies presented previously, it is envisaged that the improvement is, in part, due to a reduction in macroscopic roughness, limiting the contact geometry between drug and carrier surfaces. This potential for particle entrapment in large pits, crevices or craters is decreased as the macroscopic roughness is reduced, as can be seen by atomic force microscopy (AFM) topographical images of a commercial lactose carrier pre- and post-etching. In addition to direct surface modification, significant research has focused on the relationships between carrier geometry, for example, crystal habit, dimensions, and performance.

In recent studies, Zeng et al. have suggested that the crystal shape of lactose carriers can have significant impact on the aerosolization performance of attached drug particulates. In general, an increase in elongation (aspect) ratio was reported to increase the aerosolization performance (FPF) of albuterol sulphate, although a decrease in blend uniformity was observed. Again in these studies, it is suggested that a reduced carrier roughness resulted in increased fine particle drug. Furthermore, it is interesting to note that during these investigations, another important factor influencing aerosolization performance was noticed; the presence of fines. Apart from direct morphological and surface roughness parameters, one of the key factors that have been suggested to influence aerosolization performance of adhered drug particulates is the presence of similar sized excipient fines, for example, particles with a volume median diameter <5 μm.

The "positive" effect of excipient fines for drug aerosolization would also be expected since the surface of a commercial grade lactose carrier, will contain a distribution of "*energy*" sites that will promote particle adhesion to different degrees, and thus affect drug particle liberation in DPI carrier based systems. These sites of high adhesion may be attributed to a combination of "*particle entrapping*" crevices in the surface morphology of the carrier, crystalline phases with higher surface energy and the presence of amorphous material. Recent research, investigating the presence of these "*active sites*" in carrier systems, has suggested that an energy distribution exists for such sites, which, for low dose formulations, significantly influences drug aerosol performance. This proposed variation in drug adhesion can be observed when directly measuring the adhesion forces experienced by a single micronized drug particle, over the surface of a commercial lactose carrier. Clearly, the distribution of such sites could lead to increased particle adhesion and thus variability in content uniformity and drug aerosolization efficiency.

The addition of excipient fines to a simple carrier formulation, may improve the aerosol performance due to "*filling*" these "*active sites*" (by displacing existing drug particles), and/or by improving aerosol performance by the formation of fine-drug agglomerates. The presence of both these morphological features can be observed via high resolution *Scanning electron micrograph* (SEM) of a carrier particle surface. The mechanism for improved performance in these systems is dependent on both the previously described formulation factors, and the relative cohesive/adhesive balance between the components involved. Recently, direct measurement of inter-particulate forces in conventional DPI systems has shown that a clear balance may be observed between the adhesion and cohesion profiles of each component in the formulation, and these factors may be used to predict the performance of the formulation.

Such observations are also in good correlation with previous studies that predict ordered mixing in tableting ingredients. Regardless of the mechanism, it has been shown, as discussed previously, that the addition of fine excipient material to an inhalation formulation, results in improvement in performance outcomes (i.e. improved aerosolization efficiency). Recent work by several groups has suggested that the presence of carrier fines, significantly improve drug aerosolization performance by either filing these "*active sites*" or forming discrete agglomerates. For example, studies by Lucas et al. indicated that, regardless of the mixing order, the addition of fine lactose particulates to a formulation increased the aerosol performance. Similar observations were later described by Louey et al. Islam et al., and Adi et al. Of note, Islam et al. suggested that, while the carrier size and surface features had an influence on aerosol performance, the addition of fines into these formulations dominated performance.

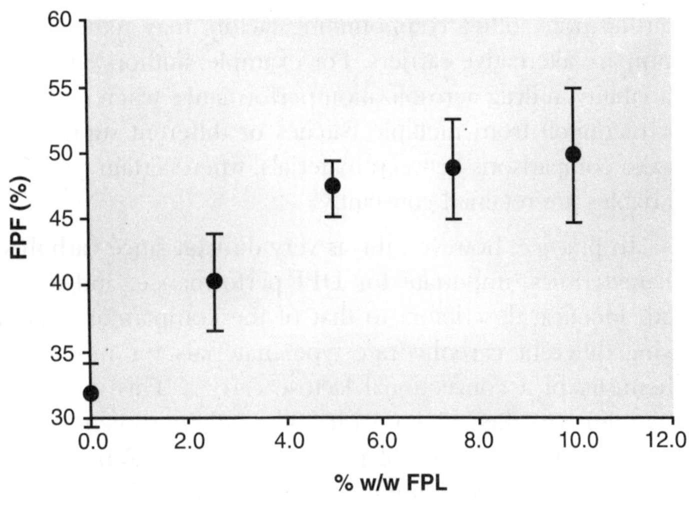

Fig. 7.3. The influence of fine-particle lactose (FPL) additive on the aerosolization performance of a protein-carrier formulation.

More recently, similar studies, investigating the influence of carrier milling, on aerosolization efficiency of blended APIs, again suggested that the introduction of fines dominated the performance compared to other factors, such as particle size and amorphous content. Although not fully understood, the presence of a certain level of fines clearly results in improved drug aerosolization performance and the influence of these small particulates on active site filling and/or spontaneous agglomeration is still under investigation. For a more in-depth review of the influence of fines on DPI performance, the reader is referred to a recent review by Jones and Price. Another approach to improve the aerosolization efficiency of blended drug particulates from carrier based systems is to alter the surface chemistry of the parent carrier. Work by Tee et al. and Steckel and Bolzen have investigated the influence of various sugars and sugar alcohols (from polyols, such as mannitol, to more complex disaccharides, such as lactose) on the aerosol performance of different model drugs.

The theory behind this approach is that any particular crystalline system has a finite surface energy. Thus, by changing the nature of the crystalline habit, morphology or surface chemistry, the relative particle adhesion of an API will be altered. For example, Tee et al. reported that, unsurprisingly, different sugar alcohols, such as mannitol, altered the aerosol performance of a micronized albuterol sulphate formulation. Interestingly, these studies also highlighted the positive effect of adding different amounts of sugar or sugar alcohol fines to these formulations. Other studies have investigated the influence of different crystalline carrier materials on drug aerosol performance and have reported that materials such as sugars and sugar alcohols can increase/ decrease aerosol performance. However, it is important to note, that such observations are difficult to compare, since different drugs, carrier size distributions and analytical techniques are used. Furthermore, other compounding factors may ultimately influence performance when trying to compare alternative carriers. For example, authors such as Steckel et al. have reported significant variability in drug aerosolization performance when evaluating a single carrier material (e.g. lactose or mannitol) from multiple batches or different suppliers. That is not to say, it is impossible to make comparisons between materials, when certain carrier descriptors are altered, as long as most variables are retained constant.

In practice, however, this is very difficult since carbohydrates have different moisture sorption characteristics, important for DPI performance, and it is challenging to produce sieve fractions with identical descriptors to that of the comparator material, namely lactose. One alternative to using different carbohydrate type materials for modifying performance is to alter the surface chemistry of a conventional lactose carrier. This may be achieved by varying the crystallization conditions, to alter the crystal habit or polymorphic form, or by the addition of ternary agents that act to control the force of interaction. For example, materials such as magnesium stearate and leucine are reported to improve the aerosolization efficiency of carrier based system by modification of the adhesive forces between the drug and the carrier.

This change in adhesive behavior can be observed in formulations of beclometasone dipropionate which suggested that the addition of magnesium stearate into the surface of a smoothed lactose surface increased the respirable dose of by a factor of 3.5 (29 and 102 μg doses for the smoothed and magnesium stearate treated lactose carriers, respectively). Similarly, investigation of the cumulative adhesion energies (measured by colloid probe microscopy, indicated a decrease in the 90th percentile adhesion energy by the same factor (3.5) (from a 112×10^{-9} nJ to 32×10^{-9} nJ for magnesium stearate treated systems). More recent work by Iida et al., has reported similar improvements, and, improved performance after storage of lactose- magnesium based formulations at elevated humidities.

In general, however, regardless of direct carrier adhesion or agglomeration formation, the ultimate aerosol performance of a DPI powder will be based on the inter-particulate forces acting between drug and carrier before, during and after processing and after aerosolization. As can be seen form the overview above, significant advances have been made in understanding these forces (e.g. recent work using colloid microscopy has been utilized to predict carrier adhesion and/or agglomerate formulation, and in general, the aerosol efficiencies of recent DPI systems are greatly improved compared to the early devices.

Aerodynamic Methods

As previously discussed, in order to achieve efficient respiratory delivery, the drug powder requires an aerodynamic diameter of <6 μm to avoid impaction and/or sedimentation in the upper airways. Historically, the method of producing particles in this size range is through high energy milling or micronization of a larger crystalline starting material. However, this method of production leads to irregular particle morphology and the potential for the formation of amorphous regions in the sample surface. Such regions can be highly unstable, and for small pharmaceutical molecules, this may result in spontaneous re-crystallization when exposed to environments that facilitate a lowering of the glass transition (such as elevated humidity) to ambient conditions. Since these micronized particulates have a high surface area, the instability in any surface amorphous regions may lead to variations in surface energetics and, where re-crystallization occurs, the potential for particle fusion, which may affect the particle size distribution, and subsequent FPF.

Consequently, there has been an impetus in the physical sciences to gain a greater understanding of surface induced amorphous material and to detect its presence at the low levels. In a drive to reduce surface instability, amorphous content, and high interfacial energies, in DPI drug, a significant amount of research into particle engineering has occurred. There are several approaches which have been successfully used to prepare respirable drug particles that have shown improved therapeutic outcomes, when compared to conventional micronized drugs. The most common approaches are to either precipitate material of the required size range (i.e. controlled crystallization), or to flash evaporate the drug from a solution with/without excipients, that facilitate a stable particulate system with chosen physical attributes.

Bulk precipitation therapy

As previously discussed, the conventional bulk precipitation of small molecule type drugs will result in a product which exhibits a particle size distribution which is unsuitable for respiratory delivery. In order to precipitate respiratory sized particles from solution in a single-step, a high degree of micro-mixing and molecular diffusion is required. Many techniques have been developed recently to overcome the growth kinetics, and generally either involve high energy input, for rapid micro-mixing, and/ or rapid diffusion of solvent and anti-solvent. One area that has received considerable recent attention is the use of supercritical fluids to precipitate particles of respirable size. Supercritical fluids are at a temperature and pressure that result in them behaving like both a gas and liquid (i.e. they can diffuse through materials easily, like a gas, whilst maintaining a high solvation capacity, as a liquid). Furthermore, by modifying the temperature and/or pressure within this supercritical region, large variations in properties such as density can be controlled. Since physical properties, such as density can be directly related to the solubility (or anti-solvent power), a range of technologies have been patented that use supercritical fluids to prepare particulates of a respirable size range.

One of the most successful approaches, in applying this technology to dry powder inhalation was the research group at Bradford University (UK) whose spin-off company, Bradford Particle Design (later acquired by Inhale (now Nektar)), utilized a process referred to as Solution Enhanced Dispersion by Supercritical Fluids (SEDS). The controlled precipitation of particles using supercritical fluids has been used to produce different drug polymorphs and powders with more suitable morphologies for efficient aerosolization. In addition recent studies, comparing the aerosolization efficiency of SEDS and micronized drug powders from lactose carrier blends, has suggested, that while the aerosol performance of similarly sized SEDS and micronized powders were similar, the physical stability of the SEDS material was improved. The difference in aerosolization performance between the micronized and SEDS sample was attributed to mechanically induced amorphous content, present in the micronized sample, re-crystallizing, and causing particle fusion.

This would be expected to occur when the environmental storage humidity was increased sufficiently to reduce the glass transition to ambient temperatures (albuterol sulphate glass transition humidity at 25°C is estimated to be approximately 50% RH) and could thus be observed in investigations conducted after storage at higher humidities (>60% RH). Other methods capable of producing discrete particle precipitation, rapid nucleation, and high rates of micro-mixing include sonication, high gravity precipitation, and impinging jet methodology. Recent work by Kaerger and Price has suggested that aerosol droplets produced via electro-hydrodynamic or an air pressure atomizer could be captured in a anti-solvent and crystallized via sonic energy.

This method of atomization and crystallization via sonication (SAXS) was shown to produce uniform particles with nanoscopic morphology within a suitable size range for respiratory delivery (1-5 μm). High gravity precipitation has also been recently investigated and relies on rapid mixing, in a high gravity rotating packed bed (HGRPB) rector, to produce micron and sub-micron particulates. As with all these high energy systems, such as SAX described above, the HGRPB relies on a high energy input (in this case intensified mass and heat transfer) to control the nucleation and crystallization of small particulates at a size suitable for respiratory therapy. Recently, this method has been successfully utilized to produce respiratory sized albuterol sulphate crystals, which depending on the operating conditions, can exhibit FPFs up to 55%.

Evaporating Methods

An alternative to precipitation and crystal formation of APIs with required size distributions for respiratory delivery is the rapid evaporation of the particles by spray drying. Spray drying has conventionally been used in the production of foodstuffs and pharmaceutical excipients, however, in recent years it has gained popularity as a means of preparing APIs in a suitable form for respiratory deposition. The most obvious commercial example of API spray drying is the recent Exubera formulation. By modifying the drug solution concentration and operating conditions it becomes possible to, with relative ease, alter the particle size or morphology of the final product. This has obvious implications, as reported by Chew et al., where the altering of the spray drying conditions resulted in the control of the degree of particle corrugation of an serum albumin (BSA), and thus, the aerosolization efficiency.

Additionally, the aerosol performance can be improved by the addition of excipients so as to vary the particle density, aerodynamic size, contact area, and physico-chemical stability of the formulation. For example, a mixture of excipients is employed during the spray drying process for the preparation of the insulin formulation for the inhaled product Exubera. The excipients act as bulking agents and aid stabilization of the formulation. The addition of other excipients, such as "blow-out" fluorocarbons, during the spray drying of formulations results in porous particles with reduced density and contact area. This approach was used in Nektars PulmoSphere technology. In a recent study, dry powder budesonide Pulmo Spheres were reported to exhibit improved in vivo respiratory deposition when compared to a conventional micronized formulation. These hollow, porous systems are designed to have reduced contact area and lower density.

The culmination of these two factors will result in improved powder aerosolization efficiency and reduced aerodynamic diameter. However, it is important to note, such systems will still have a relatively high surface area to mass ratio and therefore will presumably be highly cohesive/ adhesive in nature. An approach to overcoming this is to produce larger particles, which have volume diameters of the order 20 μm but with densities that result in aerodynamic diameters of around 5 μm.

Improved DPI Techniques

The performance of any DPI product will be dependent upon the relationships between the device design characteristics, the formulation and the airflow generated by the patient (and in the case of active systems, the device). The basic role of the device is two-fold; to facilitate patient manipulation and to provide a path for the formulation from the reservoir or storage unit to the patient. Additionally, in the case of reservoir based products, it should provide a stable environment for the formulation and an appropriate metering system. Therefore, the device component of a DPI product offers opportunities to modify the performance of the formulation by design engineering, for example, incorporating gauzes and meshes and modifying the nature of the airflow and aerosol generation of the formulation pre and post emission from the dose container system. The original DPI device, the Spinhaler, consists of a simple capsule which contained the medicament, a single drug, disodiumcromglycate.

The capsule is inserted into the device, pierced and a combination of the airflow generated by the patient and the Spinhaler device ensured the delivery of the drug to the patient. This "*simplistic*" concept of using a piercable formulation-containing a capsule in device has continued to be successfully used in DPI products, with particular application to the so-called "once-a-day" medications. For example, the Spinhaler concept was the basis for the development of GlaxoSmithKline's Rotohaler and, more recently, the HandiHaler (Boehringer-Ingelheim) and the Aerolizer (Novartis) and a multi-capsule product, the Aerohaler (Boehringer-Ingelheim). As well as successfully producing single dose capsule based DPI products the industry has also invested in developing multiple dose and multi-drug products. Perhaps the best known multiple dose products are Astra Zeneca's Turbuhaler and GlaxoSmithKline's Diskhaler and Diskus. Whilst they are termed multiple dose products, their achievement of dose multiplicity is very different.

The Turbuhaler is based on a device containing a reservoir of powder, which can be pure drug or a blend with lactose, which is metered immediately prior to delivery to the patient. The Diskhaler and Diskus multiple unit dose products employ discs and strips of individual blisters of medicament formulation respectively. The blister is pierced (Diskhaler) or removed (Diskus) immediately prior to delivery to the patient. In the case of the Diskhaler, discs of blistered medicament can be replaced in the device, whereas the Diskus cannot be replenished when the blister strip is empty. Since the introduction of these products, many pharmaceutical companies have continued to design more elaborate devices for delivery of generic drugs, drug combinations, and new chemical entities. In particular, a cursory review of the literature reveals that there have been developments in marketed multiple dose devices such as Chiesi's Pulvinal and Innovata's Clickhaler. Often such devices are based on the principles of the first generation products but have evolved to include patient friendly features such as dose counting, for example, in the Turbuhaler, the Clickhaler and the Twisthaler.

In addition, the concept of replenishable single dose capsule formulations has been extended to multiple dose DPIs with the marketing of the Novolizer, which has a replaceable multi-dose cartridge. Perhaps one of the most interesting applications of DPI technologies is the delivery of high molecular mass materials, such as proteins, to the body, offering systemic drug delivery. One such recent development is Inhale's pulmonary Exubera system. In addition to being an active device, the product allows the delivery of insulin to the lung via a respirable mixture of excipients and insulin. The obvious advantage is that this avoids the parenteral route for insulin treatment. Active devices also remove the need for patient energy input, and provide a means of reproducibly aerosolizing the powder, which maybe inhaled via a spacer unit or coordinated with the patient's inspiration.

For example, Vectura's Aspirair utilizes a compressed air reservoir (manually charged by the patient) to aerosolize the powder when the patient inhales. Activation in this system is achieved via a breath sensor, and the aerosol pathway is designed so the powder is deagglomerated in a vortex type assembly. Another approach, employed by Britannia Pharmaceuticals, utilized a pressurized air or a CO_2 canister to aerosol high dose (25-250 mg), formulations through a venturi, which could be inhaled via steady or tidal breathing. As an alternative to coordinated breathing or automated activation of powder aerosolization, the use of a holding chamber may be utilized. An example of this approach, is Nektar's Pulmonary inhaler, which generates a standing cloud of aerosolized powder, via pressurized gas (manually charged by the patient), which can be inhaled via controlled or slow tidal inhalation.

AEROSOL SYSTEMS

Liquid and propellant aerosol systems can be rudimentarily sub-classified into nebulizer and pMDI technologies. The main difference between these two systems is that nebulizers utilize an external energy source to produce aerosolized fine particulate droplets of the formulation whilst, pMDI systems incorporate a supercritical propellant into the formulation which provides the energy for aerosolization. Both of these approaches have advantages and disadvantages can contain either solution or suspension based formulation strategies and have undergone technological advances in recent years.

Modified pMDI Technology

Although the respiratory tract has been a therapeutic target for many centuries, the pMDI is generally considered the first generation of marketed inhalation products that adhere to the vigorous regulatory requirements. Marketed since their introduction by Riker Laboratories, Santa Barbara (now 3M Healthcare) in 1956 and many authors have described their basic form and functionality. Fundamentally, pMDIs rely on aerosol propellant technology, in which a high vapor pressure gas (typically a liquid propellant) is contained in a pressurized canister, in a super-cooled state. By metering a given volume of propellant containing formulation (typically 25-100 µl) within an integral metering valve, a reproducible dose of drug can be emitted and delivered.

The high, internal energy of the propellant results in rapid expansion of formulation through the metering stem and vaporization via an actuator orifice. Subsequently, micron sized droplets are produced containing the formulation that are suitable for respiratory delivery. In pMDI based systems, the drug can either be homogeneously suspended (colloidal system) or solubilized in a propellant (either with or without the addition of co-solvents). Until 1995, all marketed pMDI products contained CFCs as the delivery propellant. However, concerns over the possible detrimental ozone depleting effects of CFCs, reported in 1974, resulted in 150 nations signing the Montreal protocol in 1987, which committed the signatories to cease CFC production by 1996.

In order to continue to use and develop pMDIs, pharmaceutical companies have subsequently committed significant resources to the development of CFC-free pMDI systems. The candidates that emerged from this research as potentially suitable CFC replacements were short-chain hydrofluoroalkanes (HFAs): specifically, 1,1,1,2,3,3,3 Heptafluoropropane (HFA-227) and 1,1,1,2 Tetrafluoroethane (HFA-134a). Although these materials were accepted as reasonable replacements for the previously used CFCs, the HFA propellants possessed different physical and chemical proprieties and thus made a simple formulation propellant exchange impossible, spawning a whole new series of challenges for the formulation of pMDIs. Even though these challenges still exist, significant time and effort has been invested in overcoming these formulation hurdles, and many HFA formulations are currently available as marketed products or are currently under development.

Advancement in pMDI Systems

Due to the poor solvent properties of HFAs, suspension based systems are a popular formulation route for current pMDIs. In general, research in this area has focused on two main themes over the past 5 years: particle engineering and the application physical stabilizers. As with DPI systems, the drug particles used in suspension based systems should ideally have a diameter of less than 6 µm to achieve a therapeutic affect upon administration. As previously discussed, these particulate systems have a high surface area to mass ratio and therefore tend to be highly cohesive/adhesive in nature. Although, effects such as humidity plays a minor role in pMDI systems, the presence of solvation, van-der Waals, and electrical double layer forces, as well as density effects, may result in suspension instability and uncontrolled agglomeration and/or caking.

Until recently, suspension pMDIs were mostly comprised of micronized drug particles. As previously discussed, micronization is a high energy process that leads to a broad particle size distribution and allows poor control over particle morphology and density. Although this approach is still used in current pMDI suspensions, research has focused on improving internal stability by particle engineering, especially in light of increasingly stringent regulatory standards proposed for pMDIs. Since many examples of particle engineering technologies exist only a limited few that are specifically related to pMDI suspension technology are discussed here. Media milling is a concept that has found some application in biopharmaceutical processing and may be applicable to pMDI suspensions.

Media milling incorporates the suspending medium (i.e. HFA) into the processing step, such that the final production step is conducted in situ and the final product can be extracted at the mill output. The adaptation of this technique to in situ pressurized milling has had some success. For example, work conducted by Lizio et al. suggested that this methodology could be successfully used to produce a pMDI suspended peptide formulation containing particles with a volumetric mean diameter of 3.1 µm, without generating degradation products or contaminants from the milling process. Similarly, DuPont workers, reported increased formulation stability of in situ media milled budesonide formulations when compared to conventionally micronized particulate suspensions. Another approach for improving stability via modified particulate milling is to process the drug with pharmaceutical stabilizers by co-grinding. For example, Williams et al. investigated the effect of co-grinding a model drug, triamcinolone acetonide (TAA), with a polymeric surfactant (Pluronic F77).

The co-processed material exhibited reduced propellant solubility, increased formulation stability and improved aerosol dispersion (resulting in increased FPF) when compared to a TAA control. Apart from co-processing methodologies, specific particle engineering techniques may be employed to improve suspension stability. In general, any of the particle engineering methodologies discussed in the DPI section of this chapter may be utilized. However, of note is the use of porous particles. As with DPI formulations, the control of the density, particle size distribution, and porosity of the powder can affect the relative surface area to mass ratio thus modifying the adhesion characteristics. However, in addition, for pMDI systems the rate of sedimentation (or creaming), will be directly related to the densities of the media and the particulates.

Thus, by modifying the density of the API powder it may become possible to "*density match*" the formulation and thus create a stable system. This approach was employed in Nektar's PulmoSphere technology. Furthermore, by incorporating a porous surface topography and spherical morphology, contact area effects are minimized. The monitoring of particle surface energy is another possibility that has been explored in order to improve the suspension stability in pMDIs. For example, the atomic force microscope colloidal probe technique may be used to measure the forces between individual micron sized particulates in model propellants, similar to those used in pMDIs, namely, pressurized HFAs. Furthermore, these techniques can be expanded to investigate the relationship between drug-drug interactions, surface free energy, and in vitro performance. An alternative option to particle engineering is the use of HFA soluble surfactants and/or stabilizing agents.

The molecules in these materials reduce particulate interactions by accumulating at the interface of the colloidal particles and act to either reduce interfacial tension, produce steric hindrance between particulates, or form colloid bridges to increase flocculation. Particle cohesion in this model system is dependent on both the concentration and the molecular weight of the polymer used. It is not surprising that there has been a significant focus on HFA stabilizing agents for pMDI applications since they may provide a more immediate, and less costly, alternative to particle engineering. Indeed, historically, CFC formulations were stabilized using surfactants such as oleic acid. Unfortunately, these molecules were incompatible with the replacement HFAs and alternatives have been sought. Such excipients include long chain polymers, such as PEG and polyvinylpirrolidone; fluorinated carboxylic acids or ester surfactants and hydrophilic surfactants. Furthermore, the solubility of conventional CFC pMDI surfactants (e.g. oleic acid, lecithin) may be utilized by using specific co-solvents.

Suspension stability based methods

The most obvious approach for overcoming suspension stability issues in pMDIs is to solubilize the drug in the propellant, with or without co-solvents. This approach has obvious advantages. Since the drug is effectively present as a molecular dispersion any issues with flocculation, sedimentation, Oswald ripening, particle-device/component adhesion, and metering consistency are overcome. However, it is important to note, new challenges are faced with solution pMDI systems. For example, in most cases, the drug will not be directly soluble in the non-aqueous propellant and will require a considerable amount co-solvent for solublization. Furthermore, the formulation may be chemically less stable (and require stabilizers to reduce degradation) or may be more prone to temperature cycle instabilities and precipitation. Apart from the improvement in physical stability and metering consistency, a key advantage of using solution pMDI technology is the potential to accurately control the particle size distribution of the resulting aerosol. For suspension based systems, the aerosol particle size will ultimately be governed by the drug particulate size and adhesive/cohesive interactions.

In comparison, the spray pattern and final aerosol particle size distribution of a solution based system will be governed by the vapor pressure of the propellant, the proportion and volatility of co-solvent and the dimensions of the actuator orifice. By controlling these parameters in a system containing volatile co-solvents, the final particle size distribution will depend on the drug concentration in the aerosol droplets. The most successful example of pMDI solution based systems is 3Ms beclomethosone dipropionate formulation; Qvar which utilizes ethanol as a co-solvent. As discussed above, in comparison to suspension based pMDIs, the FPF of the aerosol cloud is not be dependent on micronized API size (and interactions) but on the size of the evaporating propellant/co-solvent droplets. Subsequently, the aerosol mass median aerodynamic diameter of Qvar beclometasone is reported as 1.1 µm (less than most conventional suspension pMDI formulations), resulting in improved respiratory deposition (between 50% and 60% as measured in human studies). As with the Qvar system, the use of volatile co-solvents in pMDI solution formulations may yield particles with small aerodynamic diameters (of the order 0.8-1.2 µm).

Although this size range is clearly suitable for respiratory delivery, it will generally result in a product targeted largely to the alveoli, which for generic API formulations, may not be comparable to existing deposition patterns and clinical efficacy. One method of controlling the aerosol particle size of solution based pMDI formulations is by addition of a non-volatile component into the formulation. The presence of a non-volatile additive in a soluiblized HFA-volatile-co-solvent system

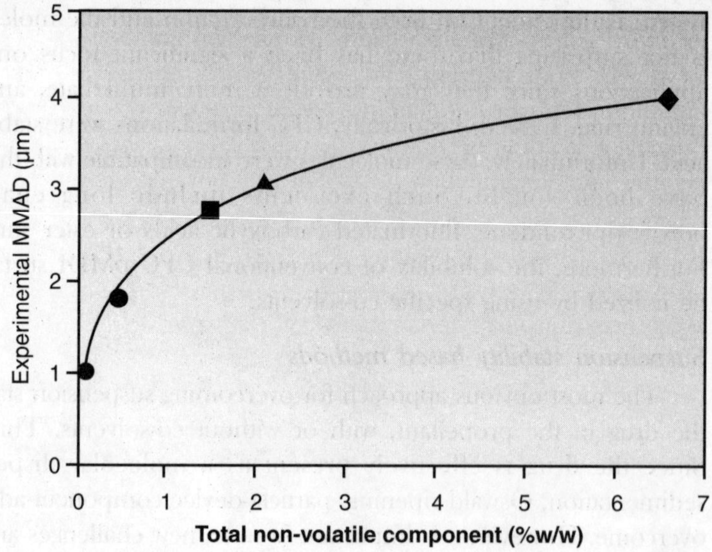

Fig. 7.4. Effect of total content of non-volatile component on MMAD of BDP pMDI solution formulation in HFA.

will result in an increase in the final droplet size. Since evaporation is dependent on the initial solution composition, the integration of a non-volatile component, such as glycerol, into the primary pMDI solution will result in the final aerosol containing both drug and non-volatile additive. Subsequently, after evaporation of the volatile components the resultant particulate size will be dependent on the concentration of the non-volatile component which will dominate the particle size.

This approach has been successfully employed in the Modulite technology developed by Chiesi, which, depending on the non-volatile concentration of glycerol or polyethylene glycol, has resulted in the ability to modify the aerodynamic diameter of the resultant aerosol. In addition to simple co-solvents, and the addition of non-volatile agents which modify aerosol particle size, significant advances have been made using alternative solubilizing agents. For example, hydrophilic and hydrophobic excipients that are HFA-soluble have been recently developed. Examples of these novel excipients include oligolactic acids (OLAs). These acids, first introduced as suspension aids, are short-chain polylactides that, in the absence of co-solvents, degrade to endogenous lactic acid. The OLA's, with an average of 6-15 repeat units, interact with the API to form API/excipient complexes that are highly soluble in HFA propellants (>5% w/w). These OLAs have no detectable toxicity, as measured by radio-labelled distribution, and inhalation safety studies.

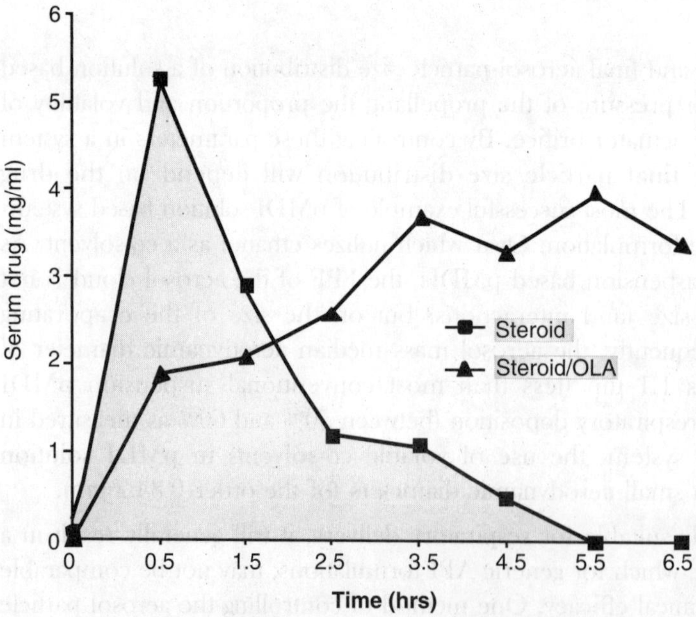

Fig. 7.5. Sustained release of Butixicot (steroid) using OLA as a sustained-release matrix following pMDI aerosolization.

A further advantage of this new excipient class is the potential for OLA-drug matrix formation to provide the potential for sustained drug release in a very simple, economical manner, potentially increasing the efficacy of many compounds. One drawback of OLA excipients is that they are water insoluble, making the pre-formulation and solubilization of polar drugs difficult. Consequently, another group of hydrophilic excipients based on functionalized PEGs are available for use with biopharmaceuticals or small molecules in HFA pMDIs. These compounds, generally used at low concentrations relative to the drug, are also readily soluble in both HFA 134a and 227 (alone, or with minimal amounts of ethanol). Furthermore, these functionalized PEGs also have the potential to form ion-pairs with drug salts. For example, 3M workers examined the ability to prepare HFA-soluble ion-pairs and reported significant improvements in solubility in HFA/ethanol systems.

Other solubility improving compounds include mono-amides and mono-esters, two ubiquitous substances in the body that degrade to endogenous or known biocompatible compounds. These new mono-functionalized excipient-drug complexes, when combined with ethanol solvents, have recently been reported to increase solubility of a drug to 1.5 wt%, which corresponds to an ex-valve dose of over a milligram. In addition to solution based systems, other "*molecular dispersions*" may be formed using reverse micelles and microemulsion technology. These systems have the same advantages in terms of aerosol performance, as pure solution based systems, but may have the potential for incorporation of molecules that will not be soluble using conventional co-solvents.

Although these formulation strategies are not currently utilized in pMDI products, a limited number of studies have investigated these systems in HFA propellants. For example, Butz et al. reported that water soluble compounds could be emulsified in a water-in-fluorocarbon emulsions (using a perfluorooctyl bromide dispersed phase and perfluoroalkylate-dimorpholinophosphate stabilizing surfactant). The subsequent emulsion was readily dispersible in all proportions in both HFA 134a and 227. Other molecular dispersants include cyclodextrins. Cyclodextrins have been used extensively to form inclusion complexes with many substances since the complex exhibits higher aqueous solubility and improved chemical stability. This approach can be utilized in pMDI systems, as reported by Williams and Liu, who investigated a novel technique of incorporating aspirin in a hydroxypropyl-β-cyclodextrin inclusion complex in HFA 134a.

Key components

Apart from the pMDI formulation approaches discussed above, product efficiency, patient compliance, and usability will also be governed by the functionality of the device itself. In simple terms, pMDI devices contain 5 key components, namely, the pressurized canister, the actuator device/canister housing, the valve components, crimp seal components, and actuator.

Dose counting

In comparison to multi-dose DPI systems, which integrated dose counting mechanisms into the earliest devices, pMDIs have, until recently, had no convenient way for patients to track the number of doses remaining in the canister at any given time. Indeed, in early pMDI systems, patients could keep track of pMDI usage by recording each dose taken manually on a record sheet, and subtracting the total from the labeled number of doses. Alternatively, patients could test the fill volume by a rudimentary float-test. Such unreliable methods often lead to patients throwing away a product which still contained an acceptable number of doses or using a product beyond the recommended number of doses, the latter being potentially dangerous as patients could be inhaling sub-therapeutic doses.

A recent survey of 342 adult asthmatics conducted in 2004 by Dr. Bradley E. Chipps, of Capital Allergy and Respiratory Disease Centre in Sacramento (California), reported that 62% of patients had no idea that they were supposed to keep track of the dose status of their pMDI. Amongst those who knew, only 24% were aware of what was left in their pMDI, 25% of subjects found their MDI empty when they needed it, and 8% of those people ended up calling 911 emergency for help. In 2003, the FDA issued guidance for the industry concerning the incorporation of dose counting technologies in pMDI devices. This announcement prompted the industry to accelerate development of several dose-counter technologies, resulting in a surge in activity and the filing of more than 30 patents relating to counting systems by several drug delivery device companies. These included Bespak, KOS, Trudell Medical, and Valois Pharmaceutical Divisions. These companies offered proprietary technologies such as the requiring 50% less force to release a dose. This is especially important for the elderly and young children. Commercially, in 2004, GlaxoSmithKline led the way in applying these new FDA dose counting guidelines by launching the Seretide Evohaler in the United Kingdom.

Patient coordination

Many patients find it difficult to coordinate the actuation of their pMDI with inhalation maneuvers. Overcoming this potential hurdle was one of the key attractive features that have been promoted as an advantage for passive DPI systems. Furthermore, inspiratory flow rate can influence the dose emitted from an inhaler, the amount inhaled, the oropharyngeal deposition, and the regional lung deposition of inhaled medications. In the late 1960s, development began on the first breath-actuated inhaler. Having seen the difficulty that some patients experienced when coordinating pressing and breathing, an opportunity was identified for a device that would actuate as the patient breathed in. This would eliminate the need for coordinated actuation or the use of add-on devices such as aerosol-holding chambers. The first Autohaler device was launched as the Duohaler in 1970. The Autohaler was pocket-sized and easy for patients to use, however some complained that the loud "click" that sounded when the mechanism fired was disconcerting.

The device did not gain instant popularity and was shelved until work began on an improved device in the 1980s. Subsequently, pirbuterol and albuterol HPA (IVAX) are now available in the Autohaler. Similar breath actuated pMDI device designs include the Baker Norton Easyhaler, the more recent Xcelovent and CCLs Integrated Breath Actuated Inhaler with dose counter. In general, these devices have a mechanical flow trigger that, after priming, actuates the release of the drug when the patient's inhalation airflow reaches a required flow rate (e.g. 30 l/min). As expected, this approach has been shown to improve drug delivery in adults and children with poor coordination. Just as coordination influences patient compliance, so does the aerosol plume velocity and temperature. Since pMDI formulations are based on the volatile evaporation of a unit dose, the resultant aerosol may travel at high velocity and be "cold".

This relationship can be visualized when comparing the plume velocities of commercially available pMDIs, where plume forces have been reported as varying between 29 and 117 mN and temperatures between –32°C and +8°C (when measured 5 cm from the end of a actuator mouthpiece). These patient "feel" issues may result in a reluctance and/or difficulty for effective patient inhalation, since the Freon effect, in combination with particle impaction, is felt at the back of the throat; possibly causing a gagging effect. Developments in the design of actuators, notably the use of actuators with smaller orifice diameters which produce a much slower, "warmer" spray, may improve patient co-ordination and compliance.

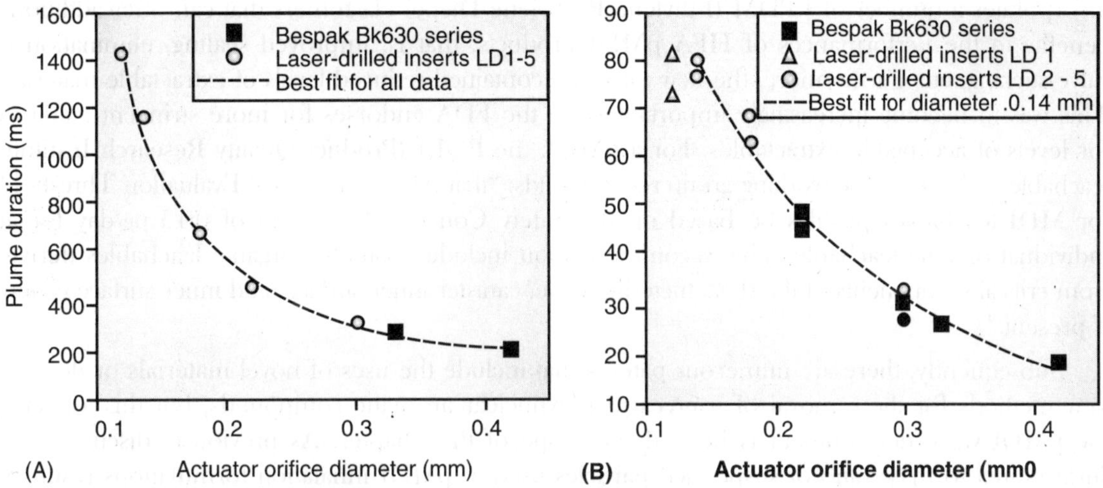

Fig. 7.6. Relationship between actuator orific diameter and (A) plume and (b) FPF for a solution-based pMDI.

By using actuators with small orifice diameters, it is possible to produce a relatively slow and "warm" spray from HFA pMDIs, compared with traditional CFC products. This approach makes it easier for a patient to coordinate the act of firing the pMDI with an inhalation maneuver. Furthermore, for solution based systems, if the unit dose remains constant, and the orifice diameter is reduced, the reduction in actuator orifice diameter will result in increased FPF. For solution based systems, the relationships between, for example, co-solvent, non-volatile component and actuator orifice were developed by Lewis et al. into a set of empirical equations that could describe the performance of a pMDI based on formulation variables.

Improvements to increase therapeutic outcomes

Other advances in pMDI device technology have been a result of the modification of canister components, including valve, crimp assembly and internal canister materials. For example, conventional metering valves are designed in such a way, that a metered volume of liquid is retained, by surface tension, in the chamber after an actuation. The nature of such valve designs results in priming issues, and where suspensions are used, homogeneity issues. To overcome this, valves are being developed with either more "*open valves*" or virtual metering tanks, which only form a closed system on actuation. These improvements increase dose reproducibility, remove priming effects, and potentially increase therapeutic outcomes. Furthermore, these developments in valve technology, particularly the "Easi-fill" valve, can be incorporated into new pMDI designs with relative ease. Since pMDI formulations are pressurized systems, the crimp assembly and internal seals are important components since they should, for obvious reasons, avoid propellant leakage, and moisture ingress. As with the formulation issues that arose with the conversion of CFCs to HFAs, the simple switch using conventional canister components was not straight forward, since the solvent properties of HFA propellants (and co-solvents) affected the degree of elastomer swelling (or shrinkage) and increased propellant leakage when compared to the CFC based formulations.

In order to overcome such issues, alternative elastomer materials with improved HFA compatibility were required. The Spraymiser Valve (3M) is a new generation of valve that incorporates a number of EPDM (Ethylene Propylene Diene) elastomers that can offer additional benefits to the performances of HFA pMDI products, that is, improved sealing, elimination of valve sticking, etc. Furthermore, the new materials contained reduced levels of extractable materials. This was to become increasingly important since the FDA endorses for more stringent guidance for levels of acceptable extractables. For a pMDI, the PQRI (Product Quality Research Institute) leachable and extractable working group recommends: "that AETs (Analytical Evaluation Threshold) for MDI leachables profiles be based on the Safety Concern Threshold of 0.15 μg/day for an individual organic leachable. This recommendation includes potential organic leachables derived from critical components of the dose metering valve, canister inner surface, and inner surface coating if present."

Subsequently, there are numerous patents that include the uses of novel materials or describe new methods for the removal of sources of polynuclear aromatic compounds, but this aspect of the pMDI valve development is beyond the scope of this chapter. As previously discussed, the surface area to mass ratio of suspended particles used in pMDI inhalation formulations results in a high adhesion potential. Consequently, exposed material surfaces inside the pressurized canister may facilitate adhesion and thus reduce emitted doses throughout the lifespan of a device. Although this may not be critical for conventional dose pMDIs, where small variations in drug loss may not influence the required pharmacopoeial tolerances, it may be critical for lower dose or more adhesive APIs. An example of the relative differences in the median adhesion forces between single micronized albuterol sulphate particles and three materials used in pMDI canisters shows that there is a difference of two orders of magnitude in the drug adhesion to the components. In view of these potential issues, a series of coatings have been developed, namely fluorinated polymers, which are now utilized in the Ventolin HFA product.

These coatings are generally made of relatively inert organic materials, such as perfluoro-alkoxyalkane, epoxy-phenol resin, and fluorinated-ethylene-propylene polyether sulfone, which have low surface energies and are thus are less likely to cause particle adhesion. The major developments in pMDI formulation and technology have been reviewed and an overview provided as to how the latest advances in aerosol technology have been used in order to improve upon existing inhaler performance. There are a multitude of formulation and device factors to consider when developing a pMDI. For these reasons a multiplicity of approaches have been applied for the determination of the effect of such variables on performances. As a result, although much is known about pMDIs at the empirical level, a systematic approach between in vitro measurements and in vivo clinical outcomes has been clearly missing.

Technology for Asthma Therapy

Nebulizers have been used for asthma therapy for many years and are used for the delivery of fine droplets of drugs to the lungs. For this reason they have been optimized for aerosol delivery in the upper range of inhalation delivery, that is, 1-6 μm. Nebulizers can produce a constant stream of aerosol particles, from both solutions and/or suspensions which can be inhaled via tidal breathing. Furthermore, the aerosol droplets produced may be of a smaller range than pMDIs, making them capable of easily penetrating the small airways. Nebulizers enable treatment of patients that require higher doses of drugs (i.e. antibiotics), that can, until now, only be formulated to be delivered via a nebulizer (i.e. DNAse for the treatment of cystic fibrosis), or have difficulties with conventional inhalers, such as the very young or elderly patients.

However, Nebulizer systems have many draw backs. For example, historically, nebulizers have been cumbersome, often being used at home or in a clinical setting. Furthermore, they deliver medication over a long timescale and have a complexity of device assembly with low efficiency and high variability in drug delivery. The most common type of nebulizer is the air-jet nebulizer. Jet nebulizer devices utilize a pressurized gas to create the aerosol particles. The gas stream is passed over the solution, creating a liquid film. This film breaks apart due to breakage of surface tension forces, forming aerosol particles, with the aerosol droplet size being dependent on the compressed air pressure applied. Subsequently, respiratory sized droplets can easily be produced once the large droplets (>10 um) have been filtered out by impaction on the device surfaces or baffles. The major disadvantage of jet-nebulizers is their constant aerosol output during the inhalation, exhalation, and breath-hold, which is not matched to the patient's respiratory effort.

Other disadvantages include: a residual volume (0.5-1.0 ml), a long therapy time (10-15 min for completion), a variable mass output throughout nebulization and an increase in osmolarity of the aerosol/liquid with nebulization time. Of all the jet nebulizers on the market, the most popular systems are probably those produced by Pari GmbH, for example, the Pari LC Plus and the Pari LC Star products. These systems belong to the so called jet enhanced nebulizers. Conventional jet nebulizers aerosolize continuously, resulting in higher losses of dose and shorter nebulization times. Since the aim is to be able to deliver the drug only during inhalation, novel breath- operated nebulizers have been developed. These novel nebulizers have been shown to increase the in vivo mass of the inhaled drug compared with conventional jet nebulizers.

Breath-synchronized nebulizers have been reported to effectively reduce drug wastage during the expiratory phase, both in vivo and in vitro. Examples of breath operated jet nebulizers include the AeroEclipset and the Smartstream. These devices are designed to deliver drug during the whole inspiration, whereas new adaptive electronic dosimetric jet nebulizers, such as the HaloLite are operated using a different principle, for example, delivering drug during the first ~50% of each inspiration, or the delivery of aerosol only during inhalation, as in the conventional jet nebulizer, the AKITA. Furthermore, improvements in constant output nebulizers have also been made with the advent of the breath-enhanced nebulizers, for example, the PARI LC Star and the Ventastream nebulizer. In these new devices the aerosol is produced at a higher rate during inhalation than during exhalation, using valves to control air flow and mixing.

In general, the level of waste of the nebulized dose, in these systems, is reduced by at least 50% in comparison with the original assisted open vent nebulizers. Since droplet size distribution and output rate are also influenced by the physical properties of the drug solution (suspension) and air flow rate from the compressor, it can, be concluded that there are high variations in the performance of different types of such nebulizers. Adaptive aerosol delivery technology is an approach developed to address the highly variable drug delivery characteristic of conventional delivery systems. These novel types of nebulizers adapt to the individual breathing pattern of the patient (determining the shapes of the inspiratory and expiratory flow pattern) and deliver the aerosol only during inhalation. Such systems pulses aerosol delivery only during inhalation and each pulse is matched to the previously determined inspiration time.

The breathing pattern is continuously monitored and the system adapts to the changes in pattern. Furthermore, such systems are programmed to deliver a preset metered dose. An example of this new technology can be found in the HaloLite AAD system. In a recent study, utilizing a modified venturi Venstream nebulizer which incorporated an AAD system, nebulized budesonide was administered to 125 Spanish children with moderate to severe asthma for 24 weeks in a double-blind, randomized, parallel group study. A total of 75% of the children received 100% of the programmed dose. Although there was no statistically significant difference between the treatment regimens, there was a clear improvement in the overall health score. Similarly, in another study involving 47 children aged less than 3 years old, a higher number of patients achieved successful treatment with the HaloLite AAD system (81%) compared with the Ventstream nebulizer (66%). Overall the AAD technology appears to be a promising approach for the challenge of delivering reproducible doses of aerosolized drug to the lungs. An alternative popular type of nebulizer is the ultrasonic nebulizer.

Ultrasonic nebulizers generate high frequency ultrasonic waves, produced by a rapidly vibrating piezoelectric crystal. A given solution presented to the piezoelectric crystal surface is broken into fine droplets where the aerosol droplet size is inversely proportional to the power of the acoustic frequency. In a similar way to jet nebulizers, baffles within the nebulizer remove large droplets. Ultrasonic nebulizers have many advantageous over compressed air systems, including the small size of device, the minimal patient coordination required, and rapid drug delivery. However, it is important to note, that these systems also have disadvantages, including the high device costs, susceptibility to mechanical breakdowns and potential for contamination. Depending on the method (direct or indirect) used to produce the aerosol droplets the ultrasonic class of nebulizers can be subclassified into passive and active vibrating devices. Passive vibrating mesh ultrasonic devices (e.g. the Omron MicroAir NE-U22) are small battery-operated nebulizers which utilize a piezoelectric crystal in an ultrasonic horn to force drug solutions through a mesh of hundreds (or thousands) of micron-sized holes to creating an aerosol.

The example given here, in comparison to many conventional devices, results in primary droplets that are sufficiently small so that no baffles are required to filter the spray, thus allowing the use of a smaller volume fill. Active vibrating mesh ultrasonic nebulizers ensure the mesh in contact with the reservoir fluid is vibrated directly by a piezoelectric crystal to generate the aerosol cloud. These devices use a perforate membrane and a micro-pumping action to draw jets of fluid though the holes in the membrane, dispersing the jets into a drug cloud. The size of the aerosol droplets is controlled by the shape/size of the holes and the surface chemistry and composition of the drug solution. Examples of such types of nebulizers are: the AeroNebPro and AeroNebGo, the e-Flow and the TouchSpray. In comparison, a novel class of nebulizer device which has been investigated by Battelle Pharma, Ohio, is the Electrodynamic Aerosol Devices (EAD). This novel device uses a powerful electric field to create near monodisperse, low velocity charged aerosols from a liquid formulation.

As the solvent in the droplets evaporate in entrained air, the surface charges on each droplet repel each other and cause the droplets to separate, producing an aerosol cloud with a small particle size. The limitation of this technique is that electric charge is known to affect aerosol deposition, both in vitro and in vivo. To the authors' knowledge no devices with the EHD technology are on the market as yet, however, early performance results have shown promise. In general, there have been few recent advances in nebulizer technology. In 2003 Boehringer Ingelheim introduced a new nebulizer, the Respimat Soft Mist Inhaler which is a propellant-free, multidose system. This nebulizer system uses an innovative approach for inhalation therapy combining a mixed form of nebulizer and pMDI.

The medication is stored in a sealed plastic container inside of the cartridge. When the dose-release button is pressed, the energy released from the spring forces the solution through a "*uniblock*", releasing a slow-moving aerosol mist. Observations concerning the use of this kind of system with suspensions has not been reported. Furthermore, recently, Aradigm launched AERx Pulmonary Delivery System in 2000. This nebulizer generates an aerosol by mechanically pressurizing a disposable blister pack, forcing the liquid through an array of micromachined holes. This device has been used with a vast array of drug molecules with various successes including: morphine, rhDNase, insulin, and fentanyl. Finally, Chrysalis Technologies has developed a novel proprietary capillary aerosol generation system in which the aerosol is formed by pumping the drug formulation through a small, electrically heated capillary. This device enables thermally stable drugs and vehicles to be evaporated and subsequently condensed in entrained air in a controlled fashion, producing a soft aerosol.

Although this device has the advantage of being able to exhibit very high delivery efficiencies (fraction of milliliter), one envisaged disadvantage is that it is unlikely that all drugs can be aerosolized in this way without some degree of drug degradation occurring. At the present, there is a great deal of debate between health professionals in how to decide which nebulizer to use for a particular application or situation. Pharmaceutical companies still supply drugs to be used for nebulization independently of which device is going to be used for its aerosol delivery. This approach may seriously affect the intended therapeutic benefit. At the moment it is considered to be sufficient to merely characterize nebulizer performance using a simple saline solution at or near physiological concentration. It seems reasonable that in view of this testing methodology and the examples given here, that there is a need for a more stringent evaluation standardization test that will improve the future health care of patients.

CONCLUDING REMARKS

In recent times the number of DPI products on the market has increased significantly, with many more technologies still in development. Additionally, there have been numerous patents filed and a significant body of research literature published. In general, devices and/or technologies are available, or in development, which deliver generic drugs, new chemical entities and combination products, such as the Advair Diskus, which contains two therapeutic agents. It seems likely that the industry will continue to develop new products based on these approaches. In addition, efforts will continue for the delivery of larger molecules such as proteins and peptides to the lung. However, as the potential applications of DPI expands, for example, by the delivery of complex molecules to the lung, further studies into all aspects of DPIs must be undertaken in order to gain a deeper understanding of this drug delivery system. The evaluation of the published literature and patents confirms that there have been many advances in nebulizer technology over the past 10 years.

Recognition of the importance of breathing pattern in aerosol delivery and deposition has led to the design of devices that allow targeting of deposition to specific areas of the airways and novel systems incorporating patient feedback. Such systems have provided control of factors that affect drug deposition and dose and have resulted in the development of new products that improve therapeutic outcomes. In the early 1960s, inhaled therapy was regarded as a specialized therapeutic route, limited to the treatment of severe lung disease. This opinion has been "rewritten" in the past 50 years, with the perceptions regarding inhalation therapy changing to, not only the treatment of respiratory lung diseases, but also systemic delivery. Consequently, pulmonary delivery is becoming an increasingly popular delivery route for both small molecules and the delivery of peptide and proteins.

With increased knowledge concerning the safety and consequent acceptance of delivery via the pulmonary route, there has recently been a significant level of research activity in this area undertaken by pharmaceutical companies and academic institutions. This approach has already revolutionized inhaled drug therapy, with many different therapeutic approaches via inhalation already possible and others that may be developed for clinical use in the near future. Advances in formulation/device technology may aid the development of novel inhalational delivery systems, but a great deal of research is still required to make this delivery route truly profitable and effective in future.

8

Symptomatic Arrhythmias

The rational basis of antiarrhythmic therapy ideally requires a knowledge of

1. The mode of action of the drug with which control of the arrhythmia is to be attempted.
2. The mechanism of the arrhythmia:
   (a) The site or sites where such disturbances are present.
   (b) Disturbances in impulse generation (enhanced automaticity or ectopic tachy-arrhythmia).
   (c) Disturbances of impulse conduction (reentrant arrhythmias). Most of the evidence suggests that reentry is the mechanism for sustained ventricular tachycardia (VT).
3. The clinical situation:
   (a) Acute, persistent or paroxysmal.
   (b) Associated with a low blood pressure < 90 mmHg, congestive heart failure (CHF), or distressing symptoms.
   (c) Associated with cardiac pathology, accessory pathway or secondary to hypoxemia, acute blood loss, electrolyte or acid-base imbalance, extracardiac conditions as diverse as thyrotoxicosis, chronic obstructive pulmonary disease, or acute conditions such as pyrexial illness, pneumothorax, or even rupture of the esophagus.

Prevention of episodes of arrhythmia and treatment of the acute attack are considered separately. Except in emergency, the first step in therapy is to remove or correct a precipitating cause when this is feasible, thus reducing the need for antiarrhythmic medication. Precipitating factors include heart failure (HF), ischemia, digoxin toxicity or administration of beta-stimulants or theophylline, sick sinus syndrome, atrioventricular (AV) block, thyrotoxicosis, hypokalemia, hypomagnesemia, hypoxemia, acute blood loss, acid-base disturbances, and infection.

Not all arrhythmias require drug treatment, and the need for therapy is considered carefully for each individual bearing in mind the limited effectiveness of drugs and the occurrence of adverse effects, which is particularly important during long-term management–and the occasional unexpected proarrhythmic effect of some drugs. In addition, especially in the management of ventricular arrhythmias that are difficult to control, it is important to achieve target plasma concentrations known to be associated with suppression of the arrhythmia. This requires careful titration of dosage dictated by salutary effects and sometimes by concentration monitoring. *Note.* Plasma concentration for some drugs does not reflect the full activity of the drug or metabolites.

Table 8.1 Electrophysiologic classification of antiarrhythmic drugs

Class	*Drugs*	*Duration of action potential*	
I	Membrane-stabilizing, inhibit fast sodium channel Restriction of sodium current	A. Quinidine Disopyramide Procainamide	Slightly prolonged (slows phase 0 of the action potential)
		B. Lidocaine Mexiletine Aprindine Phenytoin Tocainide	Shortened (minimal slowing of phase 0)
		C. Flecainide Lorcainide Encainide Propafenone	Unaffected or slight effect (slow phase 0)
II	Inhibit sympathetic stimulation	Beta-blockers	
III	Delayed repolarization	Amiodarone Bretylium Bethanidine	Prolonged as major action
		Clofilium Sotalol	Prolongation of the effective refractory period
IV	Calcium antagonists Inhibit slow calcium channel Restriction of calcium current	 Verapamil Diltiazem	

TAXONOMIC FEATURES

Antiarrhythmic drugs may be considered from three distinct points of view:

1. According to their *site of action.*
2. According to their *electrophysiologic action* on isolated cardiac fibers. Vaughan Williams indicated that class IB drugs are rapidly attached to sodium channels during the action potential. Thus, few channels are available for activation at the commencement of diastole and the effective refractory period (ERP) is prolonged. During diastole, however, the drugs are rapidly detached. At the end of diastole, most channels are drug free. Thus, there is no slowing of conduction velocity in the ventricle or His-Purkinje system. Class IC drugs detach very slowly from their

binding to the channels during diastole. This action eliminates some sodium channels, producing slower conduction; ERP is not prolonged. Class IA drugs are intermediate between classes IB and IC. Note that this classification is based on drug action rather than the drugs themselves. Amiodarone has class I, II, and III actions. Experience during the past 20 yr indicates that drug actions are much more complex than those given in this classification, but the classification is retained because it has been used worldwide for over 30 yr and it simplifies our understanding of drug action.

3. The Sicilian Gambit classification is based on identification of the mechanism, determination of the vulnerable parameter of the arrhythmia most susceptible to modification, definition of the target likely to affect the vulnerable parameter, and selection of an antiarrhythmic that will modify the target. Any classification scheme tends to be arbitrary and should not be accepted by all.

SYMPTOMS OF ARRHYTHMIAS

The following advice on diagnosis is confined to relevant clinical clues and is of necessity brief. The diagnosis is usually established by careful examination of multiple leads of the electrocardiogram and information derived from carotid massage as appropriate in doubtful cases.

Arrhythmias with NARROW QRS Complex

Regular

1. *Ventricular rate 100-140/min.* Consider *sinoatrial tachycardia.* P-wave morphology is identical to that during normal sinus rhythm. The arrhythmia accounts for about 5% of cases of supraventricular tachycardia (SVT).

   or

   Paroxysmal atrial tachycardia (PAT) with block. The P wave is often buried in the preceding T wave.

2. *Ventricular rate 140–240/min.* Consider AV *nodal reentrant tachycardia* (>70% of regular SVTs). Retrograde P waves are usually buried within the QRS or at the end of the QRS complex, with a short R-P interval causing pseudo-S waves in II, III, AVF and pseudo r^1 that mimics Rsr^1 in V_1

   or

   Wolff-Parkinson-White (WPW) syndrome with AV reentry (about 15% of SVTs)

   or

   atrial flutter

   or

   ectopic atrial tachycardia.

Irregular

1. Atrial fibrillation (AF).
2. Atrial flutter (when AV conduction is variable).

3. Multifocal atrial tachycardia (chaotic atrial tachycardia). Varying P-wave morphology, P-P intervals vary, atrial rate 200-130/min.

Arrhythmias with WIDE QRS Complex

Regular rhythm

VT or SVT with functional aberrant conduction. SVT with preexisting bundle branch block, and preexcited tachycardia (SVT with anterograde conduction over an accessory pathway). Regular wide QRS tachycardia is considered VT if the following clues apply:

1. Predominantly negative QRS complexes in the precordial leads V_4 to V_6: diagnostic for VT.
2. Presence of a QR complex in one or more of the precordial leads V_2 to V_6: certainly VT.
3. Absence of an RS complex in all precordial leads: negative QRS complexes in all precordial leads V_1 to V_6: negative concordance V_1 to V_6 = VT.
4. AV relationship different from 1:1: more QRS complexes than P waves; consider VT. Consider SVT with aberrant conduction only in the absence of any of (a) or (b) and with the following present.

QRS morphology similar to intraventricular conduction defect (IVCD) apparent on electrocardiography (ECG) while not in tachycardia:

1. In basic sinus rhythm

   or

   Visible on atrial ectopic beats = *SVT with aberrant conduction*. Fixed relationship with P waves, or the presence of a q in V_6, suggests but does not prove a supraventricular origin.

2. Other clues in favor of VT include lead V_6: QS or rS; r:S ratio less than 1. Lead V_1, V_2: if the tachycardia has an LBBB shape and

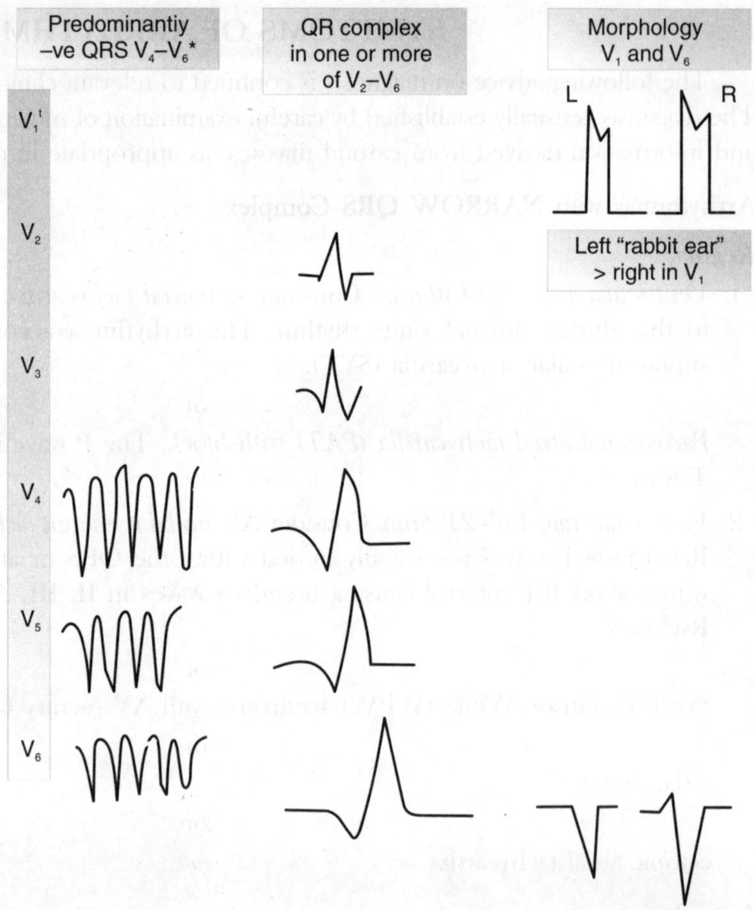

Fig. 8.1. Electrocardiographic hallmarks of ventricular tachycardia.

an r wave that is smaller than the r when in sinus rhythm, or notched or slurred downslope of the S wave; left "rabbit-ear" taller than the right.

Alternative diagnoses to VT or SVT with aberrance:

(a) Atrial flutter and WPW conduction.

(b) WPW tachycardia (rare type with anterograde conduction, through accessory pathway, preexcited tachycardia).

Irregular rhythm

Atrial fibrillation (AF) and WPW syndrome with a rapid conduction, rate 200-300/min. *Note*: The rapidity of the ventricular response (*regular* or *irregular, narrow* or *wide*) should alert the physician to the diagnosis of WPW. AV block or dissociation excludes the presence of a bypass tract. AF and IVCD; a previous ECG is needed for comparison.

CLINICAL APPROACHES

AV Nodal Reentrant Tachycardia (AVNRT)

SVT usually arises from reentry mechanisms involving the AV node (AVNRT) and occasionally an accessory pathway, AV reciprocating tachycardia (AVRT), the sinus node, or the atrium. Consequently, drug therapy is directed toward slowing or blocking conduction at some point within the reentry circuit. Whenever the AV node or sinus node is involved directly in the reentry circuit, as is usually the case, SVT is frequently terminated by maneuvers or drugs that increase vagal activity or drugs that slow the velocity of propagation of impulses in the region of the sinoatrial (SA) or AV nodes. AVNRT in patients aged < 35 yr usually occurs in an otherwise normal heart, with a good prognosis. It may occur in patients with organic heart disease, e.g., ischemic or rheumatic heart disease, and can be life threatening. The episode is characterized by an abrupt onset and termination. The heart rate varies from 140 to 240/beats/min and the rhythm is regular.

Termination of the acute attack of AVNRT

Vagal maneuvers. Many patients learn to terminate the arrhythmia by gagging or by the Valsalva maneuver (expiration against a closed glottis) or Müller maneuver (sudden inspiration against the closed glottis) or facial immersion in cold water. The Valsalva maneuver is effective in approx 50% of patients with AVNRT. *Warning.* Eyeball pressure has also been used to cause reflex vagal stimulation but is not recommended as retinal detachment may occur.

Carotid sinus massage either causes a reversion to sinus rhythm or has no effect at all. This all-or-none effect is in contrast to the slowing that results when atrial flutter or fibrillation is present. Carotid sinus massage is not recommended in the elderly or in patients with known carotid artery disease. In patients over age 35 yr, if a history of carotid disease is suggested by transient ischemic attacks or carotid bruits on auscultation, do not massage the carotid. The patient must be supine with the head slightly hyperextended and turned a little toward the opposite side. Locate the right carotid sinus at the angle of the jaw. Using the first and second fingers, apply firm pressure in a circular or massage fashion for 2-6 s. Carotid massage is discontinued immediately upon termination of the arrhythmia because prolonged asystole may otherwise supervene in rare patients.

If unsuccessful, massage the left carotid sinus. *Warning.* Never massage for more than 10 s and do not massage the right and left carotid simultaneously. In some patients, restoration of sinus rhythm is clearly a matter of great urgency because of hemodynamic deterioration or angina, and direct current (DC) cardioversion is performed either before any drug is given or when the need becomes apparent following unsuccessful drug therapy.

Drug management in the acute situation

Adenosine

Adenosine (Adenocard) 0.05-0.25 mg/kg given IV is an effective and relatively safe agent for the termination of acute reentrant SVT. Adenosine terminates AVNRT and AVRT in up to 90% of cases. The 6-mg and 12-mg doses cause a 60-80% and 90-95% termination, respectively. The drug has both therapeutic and diagnostic value. The drug has replaced verapamil when very rapid conversion of the arrhythmia is required, as in cases with HF or hemodynamic compromise, thus avoiding DC countershock in many individuals.

It can be given to patients with known SVT and aberration; little harm ensues if the diagnosis is in fact VT, whereas verapamil use is detrimental. Nonetheless the drug is contraindicated in patients with a wide QRS complex tachycardia unless the diagnosis of SVT with aberrancy is certain. The drug has no appreciable hypotensive or negative inotropic effect and slows AV nodal conduction.

Dosage. Usually IV bolus 6 mg (0.05-0.25 mg/kg) over 2 s, rapidly flushed into a *peripheral vein,* should cause reversion to sinus rhythm in 1 min; the action of the drug is only about 1/2 min and, if needed, a 12-mg second bolus injection is given approximately 2 min after the first injection. A further 12-mg dose may be repeated in 3-5 min, as the arrhythmia occurs in 10-33% of cases. The very short half-life, less than 2 s, allows rapid dose titration to be achieved. Adenosine has been safely given centrally by means of a catheter positioned in, or near, the right atrium. The initial dose should be 3 mg followed every minute by 6, 9, and 12 mg until termination of the tachycardia. Chest pain occurs more frequently with central injections than with peripheral administration (17% versus 10%).

Interactions. Dipyridamole potentiates the effects of adenosine. Thus, the dose of adenosine must be reduced in patients taking dipyridamole including Aggrenox; significant hypotension may be caused by the combination. Theophylline is an antagonist.

Adverse effects. Transient headache, facial flushing, and dyspnea, but bronchospasm may last more than a half-hour in *asthmatics.* The drug has minor proarrhythmic effects and may cause atrial and ventricular premature beats; rarely SVT or atrial flutter may degenerate to AF. Atrial flutter is not an indication for adenosine; 1:1 conduction and extreme tachycardia may ensue.

Symptoms. Asthma, severe chronic obstructive lung disease (COPD), wide QRS complex if the diagnosis of SVT with aberrancy is uncertain and in heart transplant recipients.

Verapamil

Verapamil IV is effective in more than 90% of episodes of AVNRT (narrow complex); here there is no question of VT, so the drug is safe provided the contraindications listed below are excluded. The drug has proved useful in patients with normal hearts and without the following complications:

1. Hypotension.
2. Known WPW syndrome or suspicion of WPW in view of very fast ventricular response > 220/min.
3. Acute myocardial infarction (MI), CHF, or cardiomegaly.
4. Left ventricular dysfunction, ejection fraction (EF) < 40%.
5. Sinus or AV node disease.
6. Wide QRS tachycardia (>0.10 s).
7. Suspected digitalis toxicity.
8. Beta-blockade present.

In these situations, adenosine is the drug of choice.

Symptoms. Verapamil is contraindicated in patients with wide QRS tachycardia, WPW syndrome, severe hypotension, HF, known sick sinus syndrome, digitalis toxicity, and concurrent use of beta-blockers or disopyramide.

Dosage. IV 0.075-0.15 mg/kg, i.e., 5-10 mg, is given slowly over 1-2 min. The depressant effect on the AV node may persist for up to 6 h, and a second dose may produce complications. If the arrhythmia persists following Valsalva maneuver or right carotid massage or recurs without hemodynamic deterioration in patients with a normal heart, a second dose not exceeding 5 mg may be considered after 30 min. If restoration of sinus rhythm is clearly urgent, adenosine is tried, and then, if needed, DC conversion. IV infusion 1 mg/min to a total of 10 mg or 5-10 mg over 1 h; 100 mg in 24 h.

Prolongation of AV conduction induced by verapamil can be reversed by atropine. Calcium gluconate or chloride is useful in the management of hypotension, circulatory collapse, or asystole owing to sinus arrest or AV block. Atropine may be of value in this situation.

Propranolol

Propranolol 1 mg IV given slowly and repeated every 5 min to a maximum of 5 mg; the usual dose required is 2-4 mg. *Metoprolol* is given in a dose of 5 mg at a rate of 1-2 mg/min repeated after 5 min if necessary to a total dose of 10-15 mg.

Digoxin

Because the effect of digoxin takes more than 2 h to appear, it is not often advised when rapid restoration of sinus rhythm is required.

Diltiazem IV

Diltiazem IV is as effective as verapamil and causes fewer adverse effects; both agents cost approximately $15 per treatment compared with $100 for two 6-mg doses of adenosine. Diltiazem has a role in patients who have well-defined SVT.

Dosage. Initial bolus 0.25 mg/kg over 2 min; if needed, rebolus 0.35 mg/kg.

Phenylephrine

Phenylephrine (Neo-Synephrine) is an alpha-sympathetic agonist. The drug is now rarely used because of better alternatives, especially adenosine. The drug has a role only in young patients with a normal heart, when adenosine has failed, when the blood pressure is <90 mmHg, and when cardioversion is considered undesirable. The resulting increase in blood pressure stimulates the baroreceptor reflexes and increases vagal activity, often resulting in termination of the arrhythmia.

Symptoms. Patients with MI, severe cardiac pathology, and narrow-angle glaucoma.

Dosage. Phenylephrine is administered as repeated IV bolus injections; 0.1 mg is diluted with 5 mL 5% dextrose and water (D/W) and given over 2 min. Blood pressure is measured at 30-s intervals. Arterial blood pressure must not be allowed to exceed 140 mmHg. Sufficient time (1-2 min) should elapse after each bolus to allow the blood pressure to return to its baseline value before subsequent doses are administered.

Dose range. 0.1-0.5 mg. Higher doses have been used but are not recommended. Administration by IV drip infusion may result in variation in drug rates that may cause an unacceptable increase in blood pressure.

Chronic maintenance of AVNRT

Recurrent prolonged or frequent episodes may require chronic drug therapy. The following may be tried:

1. *Digoxin.* This drug is especially useful in patients with left ventricular dysfunction (LV) and in those with resting systolic blood pressure < 110 mmHg in whom beta-blockers, verapamil, or diltiazem may cause symptomatic hypotension or bradycardia. Also, the drug is inexpensive and available as a one-a-day tablet. It is not used in patients with WPW syndrome.
2. A *beta-blocker.* Choose a once-daily preparation: 50 mg, metoprolol succinate sustained release (Toprol XL) 50 mg. If AVNRT recurs on this regimen, verapamil 80-120 mg orally usually aborts the attack within 1 h and avoids bothersome emergency room visits.
3. Digoxin plus a one-a-day beta-blocker may be necessary in patients resistant to (1) and (2).
4. *Verapamil* 80-120 mg three times daily, *Isoptin* SR 120–180 mg daily, and *diltiazem* (Cardizem CD) 180-240 mg are effective in <50% of patients. This is an expensive regimen, and compliance may be a problem.
5. *Flecainide* (200-400 mg daily) has been shown to decrease freedom from recurrent tachycardia in up to 80% of patients, compared with 15% in individuals administered placebo. Pooled studies indicate that flecainide is effective in approximately 77% of patients with AVNRT and in 66% of those with AVRT.
6. *Amiodarone,* low dose, may be used in refractory cases before contemplating ablative therapy of an accessory pathway.
7. *Pill in the pocket*: No prophylactic treatment is given.

   At the onset of an episode, the patient takes:

   (a) Verapamil 80-120 mg or a beta-blocker e.g., metoprolol tartrate (rapid acting) or bisoprolol 5 mg orally.

(b) Flecanide

or

(c) The combination of diltiazem and propranolol terminated 80% of SVT episodes within 2 h of administration.

These regimens are safe and efficacious in patients with normal hearts. Recurrent AVNRT resistant to pharmacologic therapy should prompt consideration of catheter ablation.

Multifocal Atrial Tachycardia (Chaotic Atrial Tachycardia)

This arrhythmia is caused by frequent atrial ectopic depolarizations. The arrhythmia is characterized by variable P-wave morphology and P-P and PR interval. The atrial rate is usually 100-130/min, and the ventricular rhythm is irregular. The diagnosis is made by demonstrating three or more different P-wave morphologies in one lead. The arrhythmia is usually precipitated by acute infections, exacerbation of COPD, electrolyte and acid-base imbalance, theophylline, $beta_1$-stimulants, and, rarely, digitalis toxicity.

Digitalis is usually not effective, and treatment of the underlying cause is most important. If the ventricular response is excessively rapid, slowing may be achieved with verapamil given orally. Magnesium sulfate is an effective alternative if verapamil is contraindicated. Adenosine is ineffective.

Paroxysmal Atrial Tachycardia (PAT) with Block

Episodes are usually associated with severe cardiac or pulmonary disease. PAT is a common manifestation of digitalis toxicity. The atrial rate is commonly 180-220/min. AV conduction is usually 2:1. The rhythm is usually regular. The ventricular rate of 90-120/min may not cause concern, and the P waves are often buried in the preceding T wave, so the diagnosis is easily missed. If the ventricular rate is 90-120/min and the serum potassium (K^+) level is normal, digoxin and diuretics should be discontinued, and often no specific treatment is required.

If the serum K^+ concentration is <3.5 mEq (mmol)/L and a high degree of AV block is absent, IV potassium chloride 40 mEq (mmol) in 500 mL 5% D/W is given over 4 h through a central line. If the serum K^+ level is <2.5 mEq (mmol)/L, KCl is best given in normal saline to improve the serum potassium level quickly. Other therapies are outlined under treatment of digitalis toxicity.

Atrial Premature Contractions

Atrial premature contractions (APCs) often occur without apparent cause. Recognized causes include stimulants, drugs, anxiety, hypoxmia, HF, ischemic heart disease, and other cardiac pathology. APCs in themselves require no drug therapy. Treatment of the underlying cause is usually sufficient. In patients with no serious underlying cardiac disease, reassurance is of utmost importance. Stimulants such as caffeine, theophylline, nicotine, nicotinic acid, and other cardiac stimulants as well as alcohol should be avoided.

When heart or pulmonary disease is present, APCs may predict runs of SVT, AF, or atrial flutter, and the resulting increase in heart rate may be distressing to the patient. Digoxin may be useful and, rarely, disopyramide may be necessary. If mitral valve prolapse is associated, sedation or a beta-blocker may be useful.

Atrial Flutter

Underlying heart disease is usually present; however, hypoxemia owing to a pneumothorax, atelectasis, and other noncardiac causes may precipitate the arrhythmia. Atrial flutter tends to be unstable, either degenerating into AF or reverting to sinus rhythm. The atrial rate is usually 240-340/min. The ventricular rate is often 150/min with an atrial rate of 300, i.e., 2:1 conduction. Therefore a ventricular rate of 150/min with a regular rhythm should alert the clinician to a diagnosis of atrial flutter. The sawtooth pattern in lead II should confirm the diagnosis.

Carotid sinus massage may increase the degree of AV block, slow the ventricular response, and reveal the sawtooth P waves as opposed to P waves separated by isoelectric segments in PAT with block. Rarely a 1:1 conduction with a rapid ventricular response is seen, especially in patients with preexcitation syndromes or in patients receiving a class 1 antiarrhythmic agent.

Therapy

Atrial flutter is easily converted to sinus rhythm by synchronized DC shock at low energies of 25-50 joules. Electrical cardioversion is often indicated and should be performed if the patient is hemodynamically compromised or if the ventricular response is >200/min or the patient is known or suspected to have WPW syndrome. If the patient is hemodynamically stable with a ventricular response < 200/min, propranolol may be used to slow the ventricular response. The benefit of propranolol or metoprolol is that patients who can undergo electrical cardioversion may do so easily, whereas following digoxin DC shocks have been reported to be hazardous. If underlying heart disease is present, digoxin has a role in the acute and chronic management.

Digoxin converts atrial flutter to AF, and the ventricular response is nearly always slowed to an acceptable level provided sufficient digoxin is used. Removal of underlying causes may be followed by spontaneous reversion to sinus rhythm. Verapamil or diltiazem is effective in slowing the ventricular response and may occasionally cause conversion to sinus rhythm. Digoxin, verapamil, and beta-blockers are contraindicated in patients with WPW syndrome presenting with atrial flutter.

Quinidine, procainamide, or disopyramide must not be used alone for the conversion of atrial flutter to sinus rhythm because these drugs, especially quinidine, increase conduction in the AV node and may result in a 1:1 conduction with a ventricular response exceeding 220/min. If quinidine is administered, it must be preceded by adequate digitalization to produce a sufficient degree of AV block. Propafenone and flecainide have been shown to convert atrial flutter to sinus rhythm in 20% and 33% of patients, respectively.

Atrial Fibrillation

AF is the most common sustained arrhythmia observed in clinical practice. In most patients, drug action to control the ventricular response provides adequate therapy. In some patients it may be necessary to achieve reversion to sinus rhythm with drugs or DC conversion. A ventricular response < 80/min in an untreated patient should raise the suspicion of disease of the AV node or concomitant sick sinus syndrome, particularly in the elderly. AF with a slow ventricular rate that becomes regular, indicative of the presence of a junctional pacemaker, should raise the suspicion of digitalis toxicity.

Features

In acute AF, with a fast ventricular rate, especially if there is hemodynamic compromise or with rates exceeding 220/min, WPW syndrome may be the cause, and drugs that block the AV node are contraindicated. Drug therapy to slow the ventricular response in patients with AF, in whom conduction using the accessory pathway can be excluded, includes the following:

1. *Diltiazem* IV is the drug of choice for urgent rate control in patients with AF; a constant IV infusion brings the ventricular response under control reliably. Sinus rhythm is achieved in only about 15%, and hypotension occurs in up to 33% of patients. Diltiazem IV followed by *procainamide* IV bolus may cause reversion to sinus rhythm.
2. *Esmolol* slows the rate adequately over 20 min, and sinus rhythm may ensue. The drug causes hypotension in up to 40% of patients. Esmolol and digoxin are effective, and hypotension is much less common than when esmolol alone is used. Digoxin appears to protect from hypotension.
3. *Ibutilide* (Corvert). Ibutilide fumarate injection is indicated for the conversion of AF or flutter to sinus rhythm.

   Dose. A 1-mg dose is given over 10 min with cardiac monitoring for patients > 60 kg (0.01 mg/kg < 60 kg); a 1-mg dose is repeated in 20 min.

   Caution. The drug should not be used in patients with a low serum potassium level or prolonged QT interval because torsades de pointes (TDP) may be precipitated. There is a 5% incidence of torsades in patients with ischemic heart disease. If AF is of less than 2 days' duration, anticoagulants are not indicated; in patients with AF persistent for more than 2 d, anticoagulants are recommended for 4 wk, and then conversion is attempted.

   Caution. Stop the infusion as soon as AF converts to sinus rhythm or if new or worsening ventricular arrhythmias develop. The patient with continuous cardiac rhythm monitoring should be observed for at least 4 h following infusion or until the QT interval has returned to baseline. If the arrhythmias do not terminate within 10 min of the end of the initial infusion, a second 10-min infusion of 1 mg may be administered 10 min after completion of the first infusion. QT interval should not be >440 ms, and the serum potassium level should be above 4 mEq (mmol)/L. Ibutilide is a methane sulfonamide derivative with structural similarities to sotalol; its primary mechanism of action is a class III effect.

   Interactions: Do not administer to patients taking sotalol or amiodarone.
4. *Dofetillde* (Tikosyn) is a methane sulfonamide drug similar to ibutilide but is administered orally and inhibits only the rapid component of the delayed rectifier potassium current, whereas ibutilide also has an effect on the early fast inward Na^+ current.

   Indications. Oral therapy for conversion of and maintenance of sinus rhythm after conversion in patients with persistent highly symptomatic AF or atrial flutter. The drug is not useful in paroxysmal AF.

   Adverse effects and precautions are similar to those for ibutilide, with a 3.3% incidence of torsades.

Contraindications include:

(a) QT interval > 440 ms.

(b) Thyrotoxicosis.

(c) CHF, renal failure.

(d) The presence of drugs that increase the QT interval.

(e) Calcium antagonists, cimetidine.

5. *Azimilide* is a newer agent undergoing clinical trials because it appears to be more effective and has fewer side effects than ibutilide and dofetilide; also, it can be used IV for conversion or orally for long-term prevention, with a distinctly lesser incidence of torsades (1% versus 3%). The difficult to manage paroxysmal AF may benefit. It is tolerated in renal dysfunction.

Paroxysmal atrial fibrillation

The patient should be anticoagulated to prevent embolization. IV disopyramide (not approved in the United States) or procainamide given with digoxin may restore sinus rhythm in <33%. If the rate <150 is well tolerated, it is advisable to observe the rhythm for 8-12 h because spontaneous rhythm is common, particularly if the patient has taken an extra dose of sotalol (40-80 mg) prior to coming to the emergency room.

DC cardioversion should be immediately available if drug therapy fails, when the ventricular rate is rapid, or when hypotension and HF ensue. *For the prevention of recurrent episodes*: sotalol low dose 40-60 mg is beneficial in 40-50% of patients. Although low-dose amiodarone is more effective than sotalol (approx 75%) and pulmonary side effects run approx 3%, other side effects may emerge. Clinical trial results are awaited for azimilide.

Chronic atrial fibrillation

In most patients with chronic AF of more than 1 year's duration, slowing of the ventricular response will suffice.

1. A beta-blocker is a good choice even in patients with class I-III HF or asymptomatic LV dysfunction.
2. Verapamil or a beta-blocker slows the ventricular rate during exercise, whereas digoxin often fails to do so. Verapamil SR has a role in selected cases but may cause constipation or other adverse effects and is considerably more expensive than digoxin. *Beta-blockers are preferred.*
3. The combination of digoxin and bisoprolol 5 mg or metoprolol long acting (Toprol XL) 50 mg is often beneficial.
4. *Sotalol is not advisable for chronic AF* because the drug may cause conversion to sinus rhythm in a few patients, which may lead to an increase in thromboembolism; also, sotalol carries a risk of TDP and is no more effective than other beta-blockers in controlling the ventricular rate. The three drugs are contraindicated in patients with WPW syndrome and AF or atrial flutter.
5. *Digoxin* is inexpensive and is suitable in many elderly people who do not engage in much physical activity. If the patient is asymptomatic, the rate at rest is <80, and on a 4-min walk the

rate is <110, digoxin should suffice, particularly if a beta-blocker has caused tiredness; verapamil should be avoided in the elderly because it causes bothersome constipation and may cause HF in some. The ventricular rate may be slow in the elderly because of concomitant AV node disease. Sick sinus syndromes should be anticipated, and pacing may be considered.

Synchronized DC cardioversion

The advisability of attempting DC conversion of AF is always considered carefully. In selected acute cases, reversion to sinus rhythm may be warranted. One should first consider whether it would be worthwhile to restore sinus rhythm with DC conversion. The question is necessary because reversion to sinus rhythm may cause embolization or AF may recur because the underlying heart disease is unchanged. The incidence of embolization following cardioversion is about 2%. Except in special circumstances, cardioversion is usually contraindicated in patients with:

1. AF of duration >1 yr.
2. Valvular heart disease in patients due to undergo surgery within the next few weeks.
3. Digitalis-induced arrhythmias.
4. Sick sinus syndrome.
5. Advanced AV block.
6. Left atrium > 5 cm, as sinus rhythm is usually not maintained.

If AF is of <48 hours' duration, anticoagulants are believed not to be necessary. If AF is of >3 days' duration, administer warfarin for 3 wk before and for 8-12 wk after elective cardioversion. Digoxin is maintained for the period before conversion and is interrupted 24-48 h before conversion. Sotalol or quinidine is commenced immediately after conversion when there is believed to be a high probability of recurrence of AF.

Sotalol is more effective than quinidine in the prevention of recurrent AF and for the maintenance of sinus rhythm. Amiodarone may be commenced before conversion if this drug is selected. Cardioversion is considered only when conditions contraindicating its use are absent or the patient's life is threatened by the rapidity of the ventricular response or loss of atrial transport function, such as

1. Heart rate >150/min believed to be causing HF, chest pain, or cardiogenic shock.
2. Patients with hypertrophic cardiomyopathy or severe aortic stenosis (in whom atrial transport function is of great importance).
3. AF in patients with WPW syndrome.

Rate control versus rhythm control

No clear advantage exists between rhythm control and rate control.In five RCTs no significant differences were recorded in primary end points. Adverse outcomes with a rhythm-control approach were seen in the two largest studies. None of the benefits expected from rhythm control were documented, unfortunately. The strongest risk factor for stroke was a lack of anticoagulation, because of the tendency to stop anticoagulants in patients who reverted to sinus rhythm, resulting in increased stroke rate.

Role of anticoagulants

Approximately 70% of AF occurs at age 65-85. The risk of stroke in patients with AF who have one or more risk factors for stroke is approx 5%, and anticoagulants are recommended to keep the International Normalized Ratio (INR) at 2 to 3. Bleeding occurrence is associated in virtually all cases with an INR > 3.

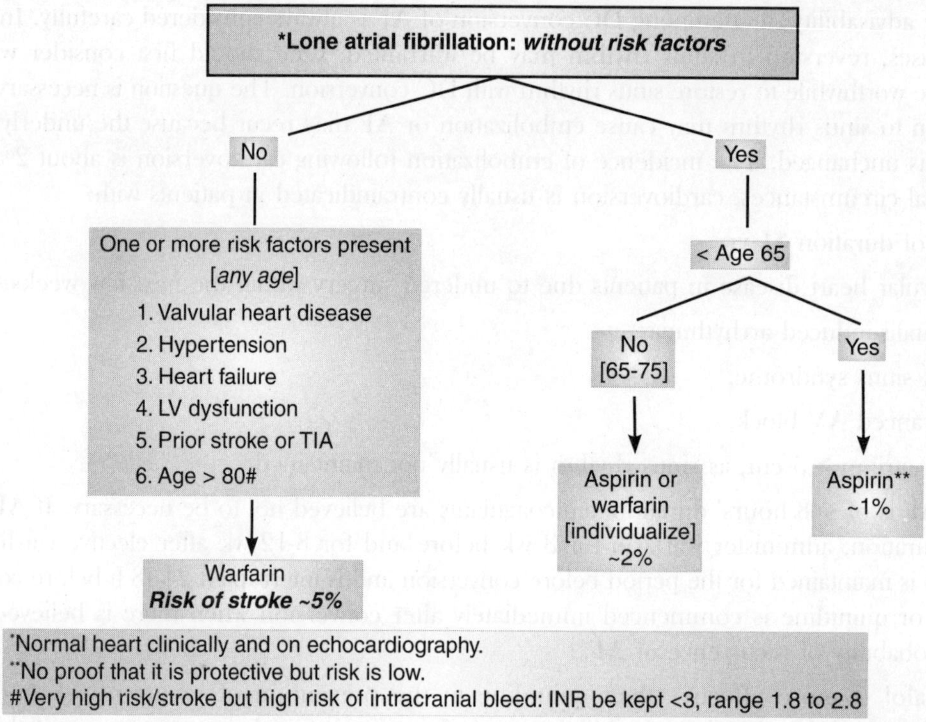

Fig. 8.2. Guidelines for the prevention of embolic stroke in patients with chronic or paroxysmal atrial fibrillation.

Risk factors include: Significant valvular heart disease, particularly mitral stenosis and moderate mitral regurgitation, LV dysfunction, HF, hypertension, prior stroke or transient ischemic attack, and age > 75 yr.

In patients < 65 yr with lone AF (no cardiac abnormality risk factors), the risk of stroke is low (<1%), and coated acetylsalicylic acid (ASA) 325 mg is recommended. In patients age 65-75 with lone AF, the risk is approx 2%, and ASA versus warfarin should be determined on an individual basis because trials have not yielded clear answers. All patients with risk factors should receive anticoagulants, but patients older than 75 have an increased risk of intracranial hemorrhage and the INR should be closely monitored to keep it between 1.8 and 2.8.

Arrhythmias in Wolff-Parkinson-White Syndrome

Patients with WPW syndrome may present with atrial flutter or AF as well as with AVRT. During sinus rhythm the short PR interval and delta wave are characteristic. A circus movement through the AV node with retrograde conduction along the accessory bundle, reaching the AV node again via the atria, produces the typical AVRT pattern. This variety usually responds to verapamil. However, verapamil should not be given to prevent paroxysmal tachycardia until proven safe by electrophysiologic (EP) testing because the drug has been reported to accelerate the ventricular response following the development of AF or atrial flutter. In some patients with AF or flutter with WPW conduction, the impulses are conducted at high frequency through the accessory pathway.

A rapid ventricular response of 240-300/min can occur with a risk of precipitating ventricular fibrillation. The rapidity of the ventricular response should alert the clinician to the diagnosis of WPW syndrome. In this subset of patients with WPW syndrome, drugs that block impulses through the AV node (digoxin, verapamil, diltiazem, and beta-blockers) do not slow the response and are contraindicated. Furthermore, digoxin and verapamil may dangerously accelerate the ventricular rate. Verapamil or digoxin may precipitate VF. DC conversion is indicated, and drugs that block conduction in the bypass tract–flecainide, disopyramide, procainamide, aprindine, propafenone, and especially amiodarone–slow the response. IV amiodarone can rarely increase the ventricular rate or cause hypotension and caution is required.

Lidocaine could occasionally increase the ventricular response in patients with WPW syndrome presenting with AF or flutter. EP testing is advisable in symptomatic patients and in those with AF, flutter, or rates exceeding 220/ min. Patients shown to be at risk of sudden death should have the bypass tract obliterated cryothermally or by other ablation techniques that produce a cure. Fortunately, serious life-threatening arrhythmias are rare. Asymptomatic patients in whom the delta wave disappears rarely have problems.

VENTRICULAR ARRHYTHMIAS

Ventricular arrhythmias are best considered under benign, potentially fatal, and fatal arrhythmias.

Premature Ventricular Contractions: Benign Arrhythmias

The Cardiac Arrhythmia Suppression Trial (CAST) indicated that treatment of post-MI PVCs and potentially fatal arrhythmias with flecainide or encainide caused an increase in mortality. Beta-blockers are the only safe antiarrhythmic agents that have been shown to improve survival in patients with postinfarction ventricular arrhythmias regardless of their ability to suppress PVCs. Amiodarone has been shown to have a modest beneficial effect.

Consideration

1. The significance of PVCs as risk markers for sudden death varies enormously according to the clinical context.
   (a) There is no evidence as yet that suppression of PVCs reduces the incidence of sudden death.
   (b) The numerous antiarrhythmic agents all possess the capacity to produce serious side effects or arrhythmias, sometimes worse than the arrhythmias that they are supposed to suppress.
   (c) The added cost and inconvenience to the patient can be a burden and may not be justifiable.

2. The correlation between the suppression of PVCs or lack of it with suppression of recurrent VT is limited. This holds for PVCs in pairs (salvos of two) or triplets (salvos of three—nonsustained VT).

PVCs in the presence of the otherwise normal heart require no therapy.

Treatment of PVCs associated with cardiac pathology

1. There is suggested evidence that patients 10-16 d post-MI with frequent PVCs > 10/h have an increased 1-yr mortality rate. Treatment of this subset, which represents about 25% of post-MI patients, may reduce mortality, but antiarrhythmic agents other than beta-blockers have not been proved to be effective or safe.

VENTRICULAR TRACHYCARDIA

VT is a regular wide QRS complex tachycardia. In general, consider regular wide QRS complex tachycardia as ventricular unless there is strong evidence to the contrary; if there is doubt about the diagnosis (VT versus SVT with aberrant conduction), treat as VT; the accurate diagnosis of VT is sometimes difficult. In the treatment of VT, cardioversion is frequently employed, especially in patients with acute MI or when the rate is over 200/min or there is hemodynamic disturbance or coronary insufficiency.

Drugs are used cautiously to terminate the attack because of the combined effect of the arrhythmia and the drug on blood pressure. Cardioversion is immediately available during drug treatment. It is quite legitimate to use DC conversion as first-line elective therapy:

1. Lidocaine (lignocaine). If there is a failure to respond, try: (a) Procainamide, (b) Disopyramide, (c) Flecainide (perhaps second choice in the United Kingdom).
2. If hemodynamic deterioration occurs at any stage during the use of the antiarrhythmics, synchronized DC shock is immediately utilized.
3. If VT occurs during the first 48 h of acute MI, the IV antiarrhythmic is continued for 24 h after conversion.
4. If VT occurs after 48 h post-MI in the absence of recurrence of infarction or ischemic pain, it is likely that a permanent irritable focus possibly associated with an aneurysm is present. The arrhythmia is abolished with IV lidocaine or other agents, and an oral antiarrhythmic is commenced within a few hours and maintained for a minimum of 3-6 mo.
5. It is advisable to initiate therapy with beta-blockers if there is evidence of painful or silent ischemia or recent infarction. A beta-blocking agent is worthy of trial in sustained or non-sustained VT. *Beta-blockers* are antiischemic and can reduce symptoms and prevent VT regardless of their effect on ventricular ectopy. In addition, they have been demonstrated to reduce post-MI mortality and mortalityinpatients with CHF, aharbinger of fatal arrhythmias.
6. Bigger's division of ventricular arrhythmias into benign and malignant forms has merit.

Fatal Ventricular Arrhythmias

Treatment is obviously necessary in these three categories:

VF outside hospital

The survivors of outside-hospital VF represent a therapeutic dilemma. There are no clear answers to therapy; they are best managed by finding an agent or combination of agents that are effective on EP testing or a trial of amiodarone. Patients who survived an episode of out-of-hospital VF in the absence of acute infarction were enrolled in the Conventional versus Amiodarone Drug Evaluation (CASCADE) study. The trial comprised 113 patients treated with amiodarone and 115 given conventional antiarrhythmics. At 4-yr follow-up, the amiodarone-treated group showed

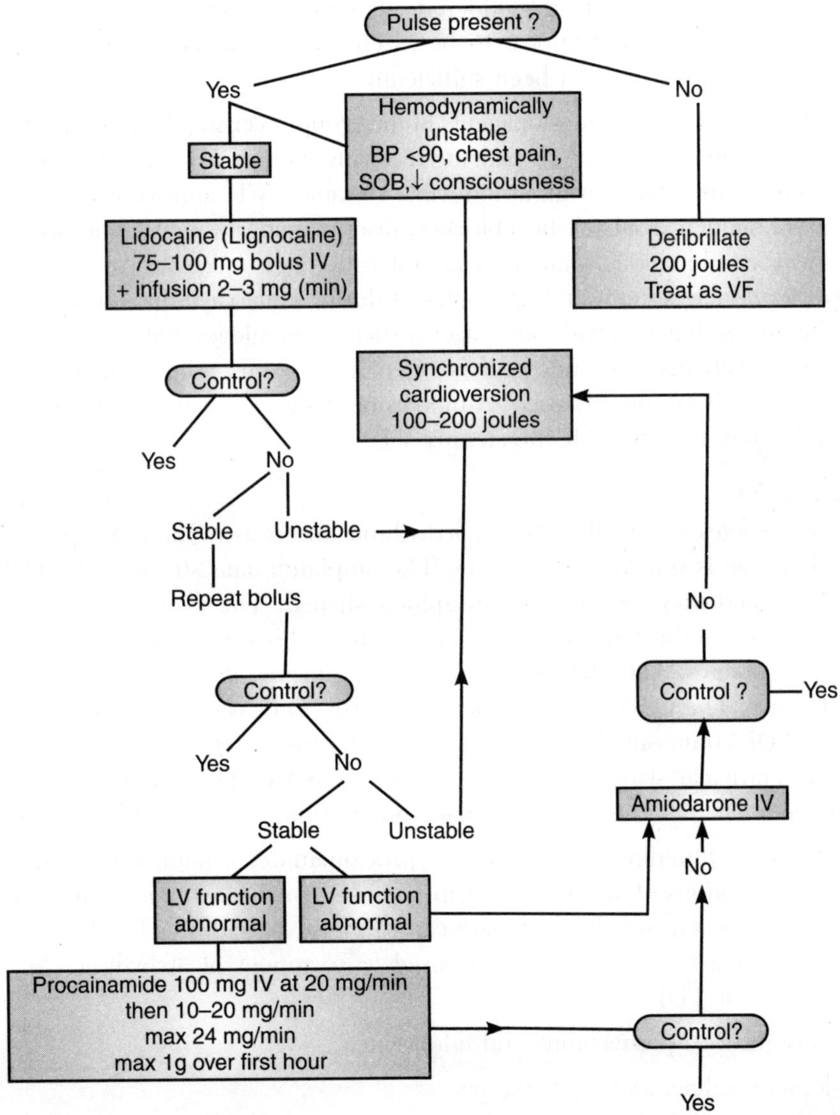

Fig. 8.3. Management of sustained ventricular trachycardia. BP, blood pressure; SOB, shortness of breath; VF, ventricular fibrillation.

improved survival and received fewer shocks from an implanted defibrillator; syncope followed by a shock from a defibrillator was less common.

Recurrent sustained VT

The choice of an oral antiarythmic depends on the physician's clinical experience and formulated policy with respect to the use of particular drugs and EP testing. EP testing may result in selection of a drug or combination that is more efficacious than empirical therapy and that does not include amiodarone or sotalol. Clinical studies support the empirical use of amiodarone and sotalol to prevent recurrent VT and sudden death, especially in patients with CHD and previous infarction. Amiodarone does not appear to be as effective in patients with HF or hypertrophic cardiomyopathy, and sotalol has not been sufficiently studied.

In a nonrandomized study, 3 of 44 patients in the group receiving therapy guided by EP studies died suddenly, compared with 8 of 18 patients receiving therapy not guided by EP studies. Also, patients who were asymptomatic without inducible sustained VT, and who were untreated, had a low probability of sudden death. A beta-blocker, disopyramide or mexiletine, may be tried first, followed by drug combinations, and then amiodarone. It is necessary to monitor the serum potassium (K^+) level, magnesium, and Q-T interval during antiarrhythmic therapy. A beta-blocker is often valuable in combination with other agents such as mexiletine but is not recommended for use in association with disopyramide and is often unnecessary when amiodarone is selected. Metoprolol or sotalol is commonly used. Amiodarone combined with mexiletine may also prove successful in refractory recurrent life-threatening VT.

Torsades de pointes

Torsades de pointes is a life-threatening arrhythmia and is associated with prolongation of the QT interval. The rate is usually 200-250/min. The amplitude and shape of the QRS complexes progressively vary, and they are dramatically spindle-shaped. The peaks of the R-wave direction change from one side to the other of the isoelectric line. This twisting appearance resulted in the name torsades de pointes. The normally conducted complexes show a prolonged QT interval or a prominent U wave. The QT interval is usually >500 ms. QTc is not an important measurement. The episode of TDP commonly lasts 5-30 s and may end with a return to normal rhythm, extreme bradycardia, or ventricular standstill or may degenerate to VF. The patient usually complains of syncope. Palpitations may be present. The many precipitating causes of TDP include:

1. *Antiarrhythmics.* Quinidine, disopyramide, procainamide, sotalol, rarely amiodarone, and especially combinations of amiodarone with quinidine or disopyramide, but in the absence of hypokalemia, amiodarone does not appear to cause torsades. The drug has been used successfully to treat torsades and can be used as treatment of arrhythmias in patients who have had recurrent TDP.
2. Coronary vasodilators: prenylamine and lidoflazine.
3. Astemizole, terfenadine, and pentamidine.
4. Psychotropic: tricyclic antidepressants and phenothiazines.
5. Electrolyte disturbances: hypokalemia and hypomagnesemia.

6. Bradycardia owing to complete heart block, sinoatrial block, or sinus bradycardia.
7. Congenital QT prolongation syndromes (with or without deafness).

Rare causes include subarachnoid hemorrhage, ischemic heart disease, mitral valve prolapse, and liquid protein diet.

Therapy. Because TDP is a brady-dependent arrhythmia, accelerating the heart rate is the simplest and quickest method to shorten the QT interval and usually results in control of the attacks. Temporary atrial or ventricular pacing (AV sequential pacing is preferable to ventricular pacing) should be instituted as soon as possible. Pacing rates within the range of 70-90/min are usually effective. Occasionally, higher rates up to 120/min are required. While preparing for pacing or when pacing is not available, a trial of isoproterenol IV infusion 2-8 μg/min may be given, and control is sometimes achieved within a few minutes.

Isoproterenol is contraindicated in acute MI, angina, or severe hypertension. If pacing is not readily available, other possible therapies include atropine, IV propranolol, or bretylium. *Magnesium sulfate bolus injection,* 2 g (10 mL of a 20% solution) over 1 min, causes reversion to sinus rhythm in some patients, particularly in patients with an acquired form of TDP. In the congenital prolonged QT syndrome, beta-blockers are of proven value, and a dose of propranolol 40–60 mg three times daily is often effective. Phenytoin has a role if betablockers are contraindicated. For resistant cases, permanent pacing plus beta-blockers or left stellate ganglionectomy is of value.

Treatment of the acquired prolonged QT syndrome is correction of the underlying cause and avoidance of the class I antiarrhythmics, especially quinidine, disopyramide, procainamide, and sotalol. A short-coupled variant of TDP has been described by Leenhardt and colleagues. This variant responds to verapamil IV and not to beta-blockers or amiodarone. Because of the high incidence of sudden death in this variant of torsades, an ICD is strongly recommended.

ANTIARRHYTHMIC DRUGS

Class IA

Drug name:	Quinidine
Supplied:	Quinidine sulfate: 200, 300 mg
	Quinidine bisulfate: 250 mg

The use of quinidine has decreased greatly since 1975, especially since the advent of other antiarrhythmic agents. The drug has been shown to increase cardiac mortality, and its use will decline further.

Dosage

Quinidine sulfate: 200 mg test dose to detect hypersensitivity reactions. If there are no adverse effects, give 200-400 mg every 3 h for three or four doses, then 6-hourly. Introduce controlled-release preparations only after suppression of the arrhythmia. Quinidine bisulfate: 250 mg; usual maintenance 500 mg twice daily. Sustained-release tablets: Quinaglute Dura-Tabs 324 mg; 1-2 tablets, two or three times daily.

Physiological role

Quinidine inhibits the fast sodium channel, slows phase 0 of the action potential, and depresses spontaneous phase 4 diastolic depolarization. Quinidine also inhibits the outward K^+ current and thus prolongs the duration of the action potential. Type I agents in general are potent local anesthetics on nerves and produce a depressant effect on myocardial membrane. The added antivagal action can cause acceleration of AV nodal conduction.

Clinical approach

1. Absorption from the gut is about 70%.
2. Peak plasma concentrations are achieved in 1-3 h.
3. The half-life is 7-9 h, but slow-release preparations are available. In liver or renal disease the half-life is increased; in HF the half-life is relatively unchanged.
4. Protein binding is 80-90%.
5. Metabolized mainly in the liver by hydroxylation; a small amount is excreted by the kidneys.
6. Plasma levels for antiarrhythmic effects are 2-5 μg/mL (3-5.5 μmol/L).

Precautions

Sinus arrest, SA block, AV dissociation, excessive QRS and QT prolongation with the precipitation of TDP or other reentry arrhythmias can occur. The risk of torsades may be decreased by giving quinidine only to patients with a normal serum K^+ level and a QT interval < 400 ms. Other adverse effects include nausea, vomiting and diarrhea, and thrombocytopenia. The idiosyncratic reaction and rate precipitation of ventricular fibrillation, especially in those undergoing medical cardioversion for AF, are well known. The drug decreases *ventricular fibrillation threshold*, and it does not seem reasonable to give priority to a drug that may increase the risk of VF. Quinidine may precipitate VT and cardiac arrest. The drug is contraindicated in WPW syndrome associated with atrial flutter or AF. Quinidine increases the serum digoxin level, and the digoxin dose should be decreased by 50%. The effect of coumarin anticoagulants is enhanced.

Drug name: Disopyramide

Trade names: Norpace, Rythmodan, Dirythmin SA

Supplied: Ampules: 100 mg in 5 mL

100, 150 mg and controlled release (CR-SR)

Dosage: Oral: 300 mg initially, and then 100-150 mg every 6 h; max. 200 mg every 6 h

Sustained action: 150-300 mg twice daily IV: 2 mg/kg over 15 min and then 1-2 mg/kg by infusion over 45 min (max. 300 mg in first hour; 800 mg daily); maintenance 0.4 mg/kg/h

Sustained release

Norpace CR 150 mg. Rythmodan Retard 250 mg; Dirythmin 100-150 mg; Dirythmin SA 150 mg.

Physiological role

Actions are similar to those of quinidine, but the drug has considerably more anticholinergic activity. Thus, urinary retention, constipation, and worsening of glaucoma are bothersome adverse effects.

Hemodynamic

The drug has a very significant negative inotropic effect, which is much greater than that of any other available antiarrhythmics. CHF may be precipitated, and this is a major drawback to the use of the drug in patients with poor ventricular function.

Clinical approach

1. Absorption from the gut is adequate; bioavailability is about 80%.
2. Peak plasma concentration is achieved in 1-2 h. The therapeutic range is 2-5 mg/L (2-5 μg/mL, 6-12 (μmol/L).
3. Half-life 6-8 h.
4. 50% of the drug is excreted unchanged by the kidney; about half is metabolized, and the metabolites may have some activity. One metabolite is powerfully anticholinergic.

Symptoms

The drug is indicated for the management of VT in the acute setting. It is used after trials of other agents such as lidocaine and procainamide have resulted in failure. The drug along with other class I agents does not decrease the incidence of VF in the acute phase of MI and does not reduce the incidence of cardiac death in the post-MI patient. Disopyramide acts selectively on the accessory bundle and may result in a clinically useful decrease in ventricular response or may terminate the episode of atrial flutter or fibrillation in WPW syndrome. The drug may be combined with digoxin in the management of supraventricular arrhythmias. The drug has a limited role in the management of recurrent ventricular arrhythmias and in general is not administered in combination with other antiarrhythmic agents, apart from digoxin. Disopyramide appears to have a role in the management of hypertrophic cardiomyopathy because of its negative inotropic effect.

Precautions

1. Do not use in the presence of severe renal failure or HF or in patients with poor LV contractility.
2. Glaucoma, myasthenia gravis, hypotension, and significant hypertrophy of the prostate causing urinary retention are contraindications.

HF, sinus node depression, dry mouth, and blurred vision are not uncommon sequelae. TDP or VF may also occur. Do not use in combination with diltiazem or verapamil.

Drug name:	*Procainamide*
Trade names:	Pronestyl, Procanbid
Supplied:	Oral capsules: 250, 375, 500 mg
	Sustained release (Procanbid): 500, 1000 mg tablets
	IV ampules: 100 mg/mL, 10 mL vial

The drug is indicated for the acute management of VT when lidocaine fails.

Dosage

IV 100-mg bolus at a rate of 20 mg/min, and then 10-20 mg/min, maximum 24 mg/min, to a maximum of 1 g over the first hour. Maintenance 1-4 mg/min. IV doses exceeding 24 mg/min commonly cause hypotension. Oral 500-mg loading dose then 375-500 mg every 3 h for 24-48 h, and then sustained-release tablets: 500-1000 mg every 12 h. Oral use should be limited to a maximum of 6 mo.

Clinical approach

1. Absorption is about 85%.
2. Half-life is about 3-5 h.
3. Rapid renal elimination occurs, and plasma metabolism is by hydroxylation.
4. The therapeutic plasma level is 4-8 mg/L (4-8 μg/mL or 17-34 μmol/L). Toxic levels range from 8 to 10 mg/L with severe toxicity about 16 mg/L (16 μg/mL or 67 μmol/L).

Symptoms

1. The drug may cardiovert acute-onset AF. Digoxin treatment is recommended as a pretreatment because procainamide can accelerate AV conduction. An increase in ventricular rate to >226/min can be life-threatening. The maximum rate of administration is 24 mg/min.
2. The acute management of VT unresponsive to the second bolus of lidocaine and AF with anterogradely conducting accessory pathway.

Precautions

1. The drug has a mild to moderate negative inotropic effect and can precipitate HF.
2. VT or VF may occur, or recur, at least in part as a result of the drug's proarrhythmic effects.
3. Lupus syndrome occurs in about one-third of patients after 6 mo of therapy; agranulocytosis may occur after prolonged treatment and appears to be more common with the slow- release formulation.
4. Avoid combination with angiotensin-converting enzyme (ACE) inhibitors because of enhanced immune effects. Cimetidine increases procainamide levels by inhibiting renal clearance.

Contraindications include AV block, hypotension, HF, and severe renal failure.

Cautions include asthma and myasthenia gravis. Reduce the dose or increase the dose interval in renal impairment.

Class IB

Drug name:	*Lidocaine*; *Lignocaine*
Trade names:	Xylocaine, Xylocard
Supplied:	Prefilled syringes: 5 mL, 50 mg/5 mL (1%); 10 mL, 100 mg/10 mL (1%).
	For dilution: 1 g/25 mL
	In the UK: lignocaine 0.2% in dextrose, 2 mg/mL, 500 mL or 1 L container; Xylocard 20 mg/mL, 5-mL syringe = 100 mg

Lidocaine is the standard agent for suppression of ventricular arrhythmias associated with acute MI or cardiac surgery.

Dosage

IV: initial bolus 1.0-1.5 mg/kg (75-100 mg). After 5-10 min, administer a second bolus of 0.75-1.0 mg/kg. Halve dose in the presence of severe hepatic disease or reduced hepatic blood flow as in cardiogenic shock, severe HF, during concurrent cimetidine administration, and in patients over age 65 yr. Blood levels are increased by concomitant administration of hepatic metabolized beta-blockers and halothane. The initial bolus is given simultaneously with the commencement of the IV infusion of lidocaine so that a lag between the bolus and the infusion does not occur. Commence the infusion at 2 mg/min. If arrhythmias recur, administer a third bolus of 50 mg and increase the infusion rate to 3 mg/min. Carefully reevaluate the clinical situation and rationale before increasing the rate to the maximum of 4 mg/min. *Maximum* dose in 1 h = 300 mg, IM 400 mg. *Preparation of IV infusion*: 1 g lidocaine for IV use is added to 500 mL or 1 L of 5% D/W.

Physiological role

The major action is depression of spontaneous phase 4 diastolic depolarization. There is little effect on action potential duration or conduction in normal circumstances, but conduction in diseased myocardium or following premature stimulation may be depressed. The drug is more effective in the presence of a relatively high serum K^+ level, so hypokalemia should be corrected to obtain the maximum effect of lidocaine or other class I antiarrhythmics.

Clinical approach

A single bolus is effective for only 5-10 min. Clearance is related to the hepatic blood flow and to hepatic function. Clearance is prolonged in the elderly in cardiac failure and hepatic disease. Therapeutic blood levels are 1.4-6 mg/L (1.4-6 μg/mL or 6-26 μmol/L). Central nervous system side effects, including seizures, may occur at concentrations >5 mg/L.

Symptoms

The main indication for lidocaine is in the acute management of VT. In the early *post-MI patient,* proven prophylactic therapy requires the use of high doses of lidocaine. Prophylactic lidocaine remains controversial, and is not cost-effective; most centers in the United States and Europe do not routinely use the drug IM lidocaine is efficacious but must be given to about 250 patients with suspected MI to save one from VF.

Ventricular arrhythmias owing to *digitalis intoxication*, phenolthiazines, or tricyclic antidepressants are managed effectively by lidocaine. Lidocaine crosses the placenta after IV or epidural administration. The drug appears to be safe during pregnancy if used in the smallest effective dose for the suppression of VT.

Precautions

Lidocaine has little hemodynamic effect. High concentrations may cause bradycardia and hypotension. Rare sinus arrest occurs in patients with sinus node dysfunction. Lidocaine has precipitated complete heart block in patients with impaired atrioventricular conduction. Nausea, vomiting, dizziness, twitching, and grand mal seizures may occur with repeated bolus injections in patients with impaired hepatic function.

Contraindications. The presence of grade 2 or 3 AV block or nodal or idioventricular rhythm. Lidocaine may suppress necessary escape foci in these rhythms and produce cardiac arrest.

Drug name:	*Mexiletine*
Trade name:	Mexitil
Supplied:	Ampoules: 10 mL, 25 mg/mL
	Capsules: 150, 200, 250 mg
Dosage:	IV (not in the USA): 100–250 mg at 25 mg/min followed by infusion of 250 mg as a 0.1% solution over 1 h, 125 mg/h for 2 h, then 500 mg/min
	Oral: initial dose 200-400 mg, followed after 2 h by 200-250 mg every 8 h (every 12 h in hepatic or severe renal failure, HF, or acute MI).
	Maximum dose 900 mg daily

Mexiletine is a primary amine that is fairly effective after IV and oral administration in the management of ventricular arrhythmias. Some studies indicate a lack of effectiveness and a high incidence of minor adverse effects, but low proarrhythmic effects.

Physiological role

Mexiletine has class IB activity. EP effects are similar to those of lidocaine. There is no significant effect on LV contractility. The drug has a relatively mild negative inotropic effect, and this is an advantage in patients with HF.

Clinical approach

1. Oral absorption is adequate.
2. Peak plasma concentrations are achieved in 2-4 h.
3. Half-life is 9-12 h and may be prolonged to 19-26 h in patients with HF, acute MI, and liver dysfunction.
4. About 15% of the drug is excreted unchanged in the urine, so the dose interval is increased in severe renal impairment to avoid toxicity.
5. Effective plasma concentrations: 0.75-2.0 mg/L (0.75-2 μg/mL or 3.5-9.3 μmol/L).

Precautions

Hypotension, bradycardia and transient AV block, and precipitation of arrhythmias owing to proarrhythmic effect may occur. Mexiletine is lipophilic and enters the brain, so neurologic side effects are common, especially confusional state, tremor, diplopia, paresthesia, ataxia, and nystagmus. Side effects occur in up to 70% of patients. Indigestion is common, but liver damage rarely occurs. Phenytoin and rifampin reduce plasma levels.

Contraindications. Severe LV failure, cardiogenic shock, hypotension, bradycardia, sinus node disease, AV block conduction defects, hepatic or severe renal failure, and epilepsy.

Drug name: *Phenytoin; Diphenylhydantoin*

Trade name: Epanutin

Supplied: 5-mL ampule, 50 mg/mL

Dosage

IV: 3.5-5 mg/kg at a rate not exceeding 50 mg/min via caval catheter. Repeat after 10 min if required. Do not exceed a rate of 50 mg/min. *Maximum* of 500 mg in 1 h. *Maximum* dose 1 g in 24 h. Dilute in 0.9% NaCl. Flush the vein immediately with normal saline to avoid thrombosis. *Oral* 500 mg daily for 2 d, and then 300-400 mg daily. However, chronic administration is not recommended because of the drug's poor anti-arrhythmic effect. The drug is not approved by the Food and Drug Administration for use as an antiarrhythmic.

Clinical approach

Phenytoin accelerates conduction by increasing the initial rapid phase of the action potential in Purkinje fibers. The drug also depresses phase 4 diastolic depolarization. The half-life of the drug is 24 h, and therapeutic plasma levels are 10-20 µg/mL (40-80 µmol/L). Reduce dosage in the presence of hepatic or renal disease.

Symptoms

The drug is no longer indicated in digitalis-induced arrhythmias. The drug is occasionally useful in the prolonged QT syndromes when beta-blockers are contraindicated, or it is occasionally combined with a beta-blocker. Indications are restricted because of the potential hazards of IV administration.

Precautions

Well-known central nervous system effects, hypotension, HF, and, rarely, respiratory depression may occur. Note that the drug is supplied in 250-mg vials with diluent containing 40% propylene glycol with a pH of 11. The drug precipitates if mixed with an alkaline solution such as dextrose.

Contraindications. Complete heart block and HF. Do not administer concomitantly with dopamine or lidocaine.

Class IC

Drug name: *Flecainide*

Trade name: Tambocor

Supplied: 400, 600 mg (UK 100, 200 mg)

Dosage: 100 mg twice daily, increasing if needed to 200 mg twice daily, reducing the dose after 3-5 d if possible

Elderly: 50 mg twice daily, max. 100 mg twice daily; reduce dose in renal or hepatic dysfunction

Action

The drug has class IC activity.

Clinical approach

1. Flecainide is well absorbed and is metabolized.
2. The half-life is about 20 h; half-life is prolonged with renal failure. Therapeutic levels are 400-800 mg/mL.

Precautions

The drug is effective in suppressing ventricular premature beats and nonsustained VT and blocks the accessory pathways in WPW syndrome. The drug is poorly effective against sustained VT, especially in patients with ventricular dysfunction. In this subset, the drug increases the frequency of sustained VT that is poorly tolerated and difficult or impossible to terminate even with prolonged resuscitation. Flecainide is effective in the prevention of recurrent AVNRT and AVRT. An overall efficacy for long-term control of PSVT (81% for AVNRT and 74% for AVRT) has been observed. Another study indicated that flecainide was effective in 74% of cases of AVNRT and in 66% of AVRT. The incidence of such serious drug-induced arrhythmia is about 20%. Among 588 patients, 33 developed new or worsened VT or VF, and there was difficulty resuscitating some patients. Other side effects include precipitation of or worsening HF, dizziness, and visual problems.

Contraindications. Patients with CHF or LV dysfunction, benign or potentially fatal ventricular arrhythmias.

Drug name:	*Propafenone*
Trade names:	Rythmol, Arythmol (UK)
Supplied:	150, 225, 300 mg
Dosage:	Oral: For patients > 70 kg, 150 mg three times daily after meals; lower dose in elderly or patients < 70 kg; increase if needed in 3-7 d to 300 mg two or three times daily
	IV (not in the USA): 2 mg/kg followed by an infusion of 2 mg/min
	Administer under hospital supervision with ECG monitoring and blood pressure assessment

The increase in mortality rate owing to flecainide observed in the CAST study has caused the drug to be removed from use for benign or potentially fatal ventricular arrhythmias. Flecainide is occasionally used for the prevention of atrial flutter or fibrillation or acute conversion, but *caution* is necessary because 1:1 conduction may occur as with quinidine if treatment with digoxin is not enforced.

Physiological role

This class IC drug blocks the fast sodium channel; membrane-stabilizing activity; prolongation of the PR interval owing to blocking at the AV node; prolongation of the PR and QRS intervals owing to prolonged ventricular conduction but with no effect on QT. The drug has additional mild beta-blocking effects.

Clinical approach

1. The oral preparation is nearly completely absorbed but is subjected to extensive first-pass metabolism.

2. Bioavailability ranges from 5% to 12% for the 150- and 300-mg tablets used in the United States.
3. Peak plasma concentration is observed within 2-3 h of administration.
4. The plasma half-life is 2-12 h, and about 95% of the drug is protein bound. Therapeutic plasma concentrations are highly variable: 0.2-3.0 µg/mL. Genetic variations in propafenone metabolism may explain the variable plasma half-life and poor circulation between plasma levels and therapeutic response.

Precautions

The drug is useful in the management of ventricular arrhythmias. Suppression of ventricular arrhythmias has been documented in several studies, but late proarrhythmic effects are unknown, as with other class IC agents. Propafenone is effective for some supraventricular arrhythmias, including WPW syndrome. The drug has a low proarrhythmic effect. The incidence of proarrhythmias appears lower than observed with encainide or flecainide, but fatal VT has been reported. Prolongation of the PR and QRS interval and conduction defects occur, as with other class IC agents. Other side effects include dizziness, lightheadedness, disturbance of taste, nausea, constipation, and abdominal discomfort; rare hepatitis and lupus-like reactions have been reported. Side effects occurred in about 33% of patients. Single reports of granulocytopenia and agranulocytosis have occurred. The precipitation of TDP is rare, and the drug was capable of controlling two cases of torsades.

Physiological activities

Propafenone may potentiate the effect of warfarin. Digoxin levels are increased from 35% at 450 mg propafenone daily to 85% at 900 mg daily. The negative inotropic effect of beta-blockers or calcium antagonists is additive. The drug must be used under strict supervision in combination with these agents. The drug has been combined with quinidine, procainamide, and amiodarone at reduced doses.

Contraindications

1. The drug has a negative inotropic effect, precipitates HF, and is contraindicated in patients with HF or severe impairment of ventricular function.
2. Disease of the sinus node, AV or bundle branch block, severe bradycardia, marked hypotension, cardiogenic shock, severe chronic obstructive lung disease.
3. Benign and potentially fatal ventricular arrhythmias similar for encainide and flecainide as influenced by the CAST study.

Class II: Beta-Blockers

Physiological role

From the EP standpoint, beta-blockers produce depression of phase 4 diastolic depolarization. Sympathetically mediated acceleration of impulses through the AV node are blocked, and beta-blockers are therefore effective in abolishing arrhythmias induced or exacerbated by increased catecholamine levels. There is a variable but often fairly potent effect on ventricular arrhythmias that may be abolished if they result from increased sympathetic activity such as may occur in myocardial ischemia and other situations. Beta-blockers cause a significant increase in VF threshold. Intravenous propranolol, metoprolol, and timolol have been shown to reduce the incidence of VF during acute infarction.

Beta-blockers and amiodarone are the only antiarrhythmics proved to reduce cardiac mortality. They suppress VT without complete suppression of VPCs. The frequency of VT was reduced by at least 75% in 8 of 17 patients receiving 50 mg atenolol daily. The drug was more effective in suppressing VT than in suppressing ventricular ectopy.

Drug name: *Sotalol*

Trade names: Betapace, Sotacor, Beta-Cardone

Supplied: 40, 80, 160 mg

Dosage: 40-80 mg once or twice daily; max. 240 mg, rarely 320 mg daily

Sotalol is a unique beta-blocker with additional class III activity. The drug is a more effective antiarrhythmic agent than other beta-blockers. After DC cardioversion of AF, sotalol is as effective as quinidine in maintaining sinus rhythm and is better tolerated. In 98 patients with post-DC cardioversion of AF, sotalol 160-320 mg daily provided maintenance of sinus rhythm in 52%, and in 85 quinidine-treated patients sinus rhythm was maintained in 48%. Twenty-six percent of the quinidine group and 11% of the sotalol group withdrew because of adverse effects.

Sotalol is useful for the management of paroxysmal AF. The drug is not recommended for slowing the ventricular rate in patients with chronic AF because other beta-blockers have similar beneficial effects, but they do not carry the risk of TDP. Sotalol is effective in suppressing ventricular couplets and frequent ventricular ectopy. In a randomized study, a 99% reduction in ventricular couplets was achieved with sotalol and 49% with propranolol. In small nonrandomized studies with up to 18 mo of follow-up, sotalol therapy has caused reduction in the recurrence of sustained VT and VF. In a study of 16 patients with recurrent sustained VT refractory to an average of 4.8 antiarrhythmic agents, sotalol was effective at high dosage, 320-960 mg daily, in suppressing inducible sustained VT.

Arrhythmias refractory to other antiarrhythmic agents including amiodarone have shown a beneficial response to sotalol administration. Patients evaluated with non-invasive or invasive means exhibit efficacy with sotalol similar to that of those treated empirically. Thus, both sotalol and amiodarone can be used as empirical therapy in patients with life-threatening ventricular arrhythmias. A head-to-head comparison of sotalol and amiodarone was carried out by the Amiodarone versus Sotalol Study Group. This multicenter trial studied 59 patients with documented VT who had failed a class I agent. Patients were randomly assigned to amiodarone and sotalol. At 1-yr follow-up, there was no significant difference between the groups. Treatment failures (death, recurrent ventricular arrhythmias, and side effects) were similar (13 of 29 with sotalol and 14 of 30 with amiodarone).

The author and the majority of cardiologists recommend sotalol rather than amiodarone as the first-choice class III agent for the management of life-threatening ventricular arrhythmias including after cardiac arrest. In the Electro-Physiologic Study versus Electrocardiographic

Monitoring (ESVEM) study, sotalol was superior to six sodium channel-blocking antiarrhythmic agents. The actuarial probability of a recurrence of ventricular arrhythmia and the risk of death from any cause, from a cardiac cause, and from arrhythmia were significantly lower in patients treated with sotalol.

Sotalol should not be used with potassium-wasting diuretics or other agents that cause hypokalemia or prolong the QT interval because TDP may be precipitated. Torsades caused by sotalol do occur, albeit rarely, in the absence of hypokalemia. A case of torsades occurring in a patient with a serum potassium level of 4.0 mmol/L and therapeutic sotalol concentration, failing to respond to conventional therapy including electrical cardioversion 15 times during a 12-h period, responded within an hour of dialysis. The incidence of TDP is approx 2% and is similar to that observed with quinidine. Of a total of 1288 patients treated with sotalol in controlled trials, 27 experienced TDP (2%).

Class III

Drug name: *Amiodarone*

Trade names: Cordarone, Cordarone X

Supplied: Tablets: 200 mg; ampules: 150 mg

Amiodarone is a benzofuran derivative that has two atoms of iodine and structural similarities to thyroxine (T_4) and triiodothyronine (T_3). The drug was originally developed in Belgium as a coronary vasodilator for the treatment of angina.

Dosage

IV (indicated only for recurring VF, unstable VT): 5 mg/kg over 2 h via a central line, maximum 1.2 g/24 h. Repeat if needed for 2 d, for life-threatening arrhythmias. Severe hypotension may occur with the IV preparation; this may be owing to its solvent. Oral: a low-dose regimen is used: 200 mg three times daily for 1 wk reduced to 200 mg twice daily for a further week or two, and then maintenance of 200 mg daily or the minimum dose required to control the arrhythmia after 4 wk. A 5 d/wk regimen may suffice in some. The total daily dose of drug should be lower in patients with HF. After 4 wk, aim for a serum level of 2 μg/mL amiodarone and 2 μg/mL of the active metabolite desethyl amiodarone. Initial loading with IV amiodarone 5 mg/kg followed by oral dosing can shorten the time to optimal arrhythmic control.

Physiological role

Electrophysiologic and antiarrhythmic effects: oral administration of amiodarone lengthens the action potential duration and thus the ERP in all cardiac tissues including accessory pathways. The drug has been labeled a class III antiarrhythmic agent because of the aforementioned major action. Its effects, however, overlap those of other classes of antiarrhythmics, e.g., class IA. Amiodarone produces potent sodium channel blockade during phases 2 and 3 of the cardiac action potential. Thus, the drug depresses myocardial and His-Purkinje conduction. The salutary antiarrhythmic effects are accompanied by a consistent lengthening of the QT interval. The latter effect, if conspicuous, may arouse suspicion of imminent side effects such as precipitation of VF or TDP. EP studies do not consistently predict the drug's clinical effect on ventricular arrhythmias.

Horowitz and colleagues and others noted a predictable effect. *Hemodynamic* consequences are usually negligible. Vasodilation produces a mild decrease in systemic vascular resistance. There is a very mild negative inotropic effect, and in clinical practice the drug does not usually precipitate HF, except in rare circumstances associated with cardiac slowing.

Clinical approach

1. Absorption from the gut is about 50%; bioavailability ranges from 22% to 86%. The drug undergoes extensive hepatic metabolism. Peak plasma levels are reached 6-8 h after an oral dose. The therapeutic effect is not well defined by plasma levels. However, after 4 wk of therapy, levels between 1.0 and 2.5 μg/mL appear to be associated with therapeutic effect and lowered toxicity.
2. The volume of distribution is high (approx 5000 L). The drug is almost completely bound to protein and binds extensively with virtually all body tissues. The concentration in the myocardium is 10-15 times that in plasma, and the concentration in the liver, lung, and adipose tissue is much higher. Because of the large volume of distribution, the drug is preferably given as a loading dose when used orally, if prompt control of arrhythmia is required.
3. Rosenbaum and colleagues have shown that, when the drug is administered orally, action depends on tissue stores and not on plasma levels. Others have demonstrated some relation between plasma levels and effect, and this measurement may be useful in guiding dosage.
4. Half-life is 30-110 d. Therapeutic effect generally occurs within 10-15 d of oral therapy but may take 3-6 wk with the use of low loading doses. At high loading doses a response may be seen within 48 h. Activity persists for more than 50 d after stopping the drug.

Symptoms

The drug is used for the management of refractory life-threatening arrhythmias. In patients with fatal arrhythmias, recurrent cardiac arrest, and HF, amiodarone therapy is justifiable. In the United Kingdom, but not in the United States, the drug is indicated for the management of patients with WPW syndrome who have life-threatening arrhythmias and is given IV in these situations. The drug given IV via a central line is useful in the management of AF and VT and is also beneficial with refractory VT caused by digitalis toxicity. IV administration has been shown to be useful in critically ill patients with VT-VF, and the drug appears to be effective in over 66% of refractory cases of recurrent VT-VF. A metaanalysis indicates that prophylactic amiodarone reduces the rate of sudden death in high-risk patients with recent MI or CHF with an overall reduction of 13%. Clinical studies of interest include the following:

1. In the Basel Antiarrhythmic Study of Infarct Survival (BASIS) during the 1-yr follow-up, there were only five deaths in the amiodarone-treated patients, 12 deaths in patients treated with class IA drugs, and 15 deaths in the control group ($p < 0.05$).
2. The Canadian and European Myocardial Amiodarone Trials (CAMIAT and EMIAT) did not show significant mortality reduction with the administration of amiodarone. In the Canadian Implantable Defibrillation Study (CIDS), the ICD surprisingly failed to demonstrate a significant survival benefit over amiodarone; at 3 yr, 60% of ICD patients were receiving a beta-blocker compared with 25% for amiodarone.

3. In the CASCADE study, amiodarone-treated patients who had survived an episode of VF in the absence of acute MI showed improved survival.
4. The Cardiac Arrest Study Hamburg (CASH) in patients following cardiac arrest has shown the ICD to be superior to amiodarone. After 2 yr of follow-up, there was a 37% reduction in overall mortality rate among patients assigned to ICD compared with drug therapy; metoprolol, however, was as good as amiodarone in decreasing the mortality rate for up to 2 yr following cardiac arrest.

Precautions

1. Corneal microdeposits of yellow-brown granules occur in 50-100% during long-term therapy. They do not affect vision, are reversible on discontinuing the drug, and constitute clinical evidence of the drug's impregnation. Periodic eye examinations are recommended during long-term use. There is some evidence that the deposits can be diminished by withdrawal of therapy for 1-2 wk or by ophthalmic drops containing sodium-iodine heparinate; optic neuritis and blindness have been reported.
2. Rashes, especially photosensitivity; permanent grayish blue discoloration of the skin occurs in over 10% of patients after approx 18 mo of high-dose therapy but is rare with current low-dose regimens. On discontinuing amiodarone, the discoloration regresses. Patients should be advised to use a sunscreen and to shield their skin from light.
3. Mild bradycardia, but rare precipitation of severe brady-arrhythmia in patients with sick sinus syndrome, may occur. Asystole and sinus arrest have occurred in patients with therapeutic serum concentrations of amiodarone.
4. Precipitation of TDP is rare and occurs mainly in patients with hypokalemia and with drugs that prolong the QT interval. Amiodarone has been used successfully to treat TDP but cannot be recommended as first-line or second-line treatment.
5. Pulmonary infiltrates and pulmonary fibrosis occur in 5-20% of patients receiving chronic doses, 400-800 mg/d. Pulmonary toxicity occurred in 6% of 145 patients given an average dose of 845 mg for 2 wk, and then an average of 410 mg daily for 18 mo, but the incidence is <1% with the low dosage advised earlier in this chapter. Periodic pulmonary function tests, in particular diffusion capacity (DCO), reveal early cases. Discontinuation of amiodarone and steroid therapy causes regression of pulmonary complications in some patients.
6. Hypothyroidism or hyperthyroidism occurs in about 3-7% of patients. The T_4 concentration is mildly raised, the T_3 level is very mildly decreased, and there is a marked rise in reverse T_3 levels in patients without clinical thyroid disturbance.
7. Proximal muscle weakness and neurologic problems, especially ataxia, are not uncommon.
8. Gastrointestinal symptoms and increase in transaminase levels are seen (10-20%). Hepatitis is rare. After 5 yr of amiodarone therapy, approx 50% of adult patients and treated children are observed to be continuing successfully on the drug.
9. *Drug interaction* with anticoagulants has been well documented. Amiodarone potentiates the actions of oral anticoagulants, and the combination may lead to serious bleeding. The

maintenance dose of warfarin should be halved when amiodarone and warfarin are used together. There is evidence that amiodarone increases plasma *quinidine* levels. The combination of these two agents, both of which tend to prolong the QT interval, is not recommended. A significant interaction between amiodarone and digoxin occurs. Serum digoxin concentration usually doubles, but ventricular dysrhythmias are suppressed by amiodarone. Halve the dose of digoxin when the two drugs are given together. The combination of amiodarone and verapamil or diltiazem should be avoided because severe depression of SA and AV nodes may occur. Interactions have been noted with flecainide, phenytoin, and theophylline.

Drugs that prolong the QT interval should not be combined with amiodarone: sotalol, class IA antiarrhythmics, tricyclic antidepressants, and phenothiazines. *Caution.* Potassium-losing diuretics must not be given concomitantly, as torsades may be precipitated. A 5-yr follow-up of 242 amiodarone-treated patients revealed the following: 59% adverse effects and withdrawal in 26%; 25% discontinued the drug. Few patients tolerated the drug on a long-term basis. In addition, blindness appears to be a rare but alarming adverse effect. Before commencing amiodarone, the physician must be aware of the amiodarone withdrawal problem. If the drug is discontinued because of a serious side effect, the effect may persist for months. Also, other antiarrhythmics then used may interact with amiodarone.

Bradyarrhythmias

Bradyarrhythmias that cause hemodynamic disturbances and are symptomatic or life- threatening and unresponsive to optimal doses of atropine require pacing.

Update: a recent study indicates that statin therapy may decrease the risk of atrial fibrillation; this would be a remarkable advance, if proved to prevent some cases of the most common and serious arrhythmias observed in clinical practice.

9

Glycemic Diseases

Diabetes mellitus is a devastating disease that is a leading cause of blindness, end stage renal disease, neuropathies, and peripheral vascular disease that lead to amputations. It is also a coronary heart disease equivalent and is the sixth leading cause of death in the United States. Diabetes is a national epidemic that has been diagnosed in approximately 16 million Americans, and it is estimated that an additional 6 million Americans have undiagnosed diabetes. The incidence of diabetes is on the rise, and is expected to continue, especially type 2 diabetes mellitus (T2DM). Diabetes can be simply classified into type 1 (insulin dependent) and type 2 (non-insulin dependent) diabetes mellitus.

Type 1 diabetes (T1DM) is characterized by autoimmune destruction of insulin-producing β-cells in the pancreas causing hyperglycemia. These patients require exogenous insulin replacement to maintain glucose homeostasis. Type 2 diabetes is characterized by insulin resistance and progressive β-cell failure. Patients with T2DM can be managed with oral therapy initially (insulin sensitizers - insulin secretagogues); however, a large number of patients will require exogenous insulin therapy along with insulin sensitizers due to progressive β-cell failure.

Insulin is an excellent choice, as its glucose-lowering potential is limited only by the potential side effect of hypoglycemia. Insulin therapy may have additional benefits, as the early use of insulin may slow the loss of beta cell insulin secretory function in type 2 diabetic patients versus sulfonylureas. Glycemic control is imperative to reduce the risk of complications associated with T1DM and T2DM. The American Diabetes Association recommends a hemoglobin A1c (A1C) level at a minimum below 7%, and as close to 6% as possible if it can be achieved without significant side effects.

Glycemic control primarily reduces micro-vascular disease (retinopathy, neuropathy, and nephropathy) shows a trend to reduce macrovascular events. However, glycemic control is very difficult to achieve as evidenced in clinical trials and in clinical practice. Insulin therapy is required to meet these goals in many patients, but multiple daily injections of insulin are perceived adversely by many patients and clinicians alike.

PULMONARY DELIVERY OF INSULIN

Insulin is a polypeptide that is denatured in the gastrointestinal tract when taken orally; therefore, parenteral administration is necessary. Patients are required to receive multiple doses of insulin either subcutaneously or intravenously. Since the introduction of insulin therapy, alternative delivery routes and systems have been investigated. Pulmonary delivery is attractive because insulin is permeable at the level of the alveoli. In January 2006, an inhaled dry powder formulation of fast-acting insulin was approved in the United States and Europe for the treatment of type 1 and 2 diabetes in adults. There are several other insulin inhalational systems under various stages of development. Differences between these systems include formulation, particle size, and delivery device. This chapter will review the ideal delivery, differing delivery systems, pharmacodynamic, pharmacokinetics, efficacy, and safety of inhaled insulin. The concept of inhaled insulin was first introduced in 1925 where it was studied in five diabetic patients.

Variability in bioavailability and system portability made pulmonary insulin delivery at the time unfeasible. Interest in pulmonary delivery resurfaced in the 1970s and 1980s in the clinical setting. With the advancements in drug delivery technology, the pulmonary delivery of insulin became more viable. Pulmonary delivery of insulin is ideal due to the large surface area of the lung available for absorption (up to 100 m^2); having high permeability at the alveolar surface, and a vast circulation. However, the lung prohibits deep penetration through several protective mechanisms. Most airborne particles are deposited in the upper airways where the mucocilliary elevator expels them from the lungs into the digestive tract. In order for a substance to reach the alveoli it must possess certain characteristics such as appropriate size, particle density, and morphology. Particles sizes of 1-3 μm are ideal for penetration to the alveoli and minimally deposit in the oropharynx. Particles smaller than 1 μm are mostly exhaled, and sizes larger than 10 μm have little chance of making it past the oropharynx. Furthermore, substances that do not have the appropriate density or morphology will be removed by macrophages in the alveoli.

Small nonporous particles are engulfed by macrophages quickly, whereas porous particles are removed at a slower rate. Mode of delivery to the agent's target can also hinder therapy via the pulmonary route. Many marketed delivery systems propel particles with too much velocity depositing the substance in the oropharynx instead of in the lower airways (i.e. pressurized metered dose inhalers). New inhaler technologies have advanced considerably and are aimed to provide high levels of dosing while minimizing losses within the device and environment. Moreover, advances in particle technology have improved reproducibility and enhanced delivery to the deep lung.

CLINICAL APPROACH

Several insulin delivery systems are in development or on the market. Each pulmonary insulin delivery system consists of three components: an insulin formulation, a unique insulin delivery "*packaging,*" and a unique delivery device. Though they all may be characterized as inhaled insulin systems, these unique characteristics require us to individually scrutinize each product. The inhalers range from handheld breath actuated devices to meter dose inhalers similar to those used for asthma agents. Insulin delivery packaging ranges from dry-powder blister packs or capsules, to liquid blisters or liquid in asthma type canisters. Diverse particle technologies have also been investigated for the delivery of insulin. The Exubera and AERx systems were first technologies investigated in the 1990s.

Use of Exubera

Exubera is the first Food and Drug Administration approved inhaled insulin product. It is approved for the treatment of adult patients with T1DM and T2DM. Exubera is currently available in a short acting formulation to control post prandial glucose excursions. In patients with T1DM, Exubera should be used in combination with long acting insulin, which provides basal insulin replacement. In patients with T2DM, Exubera can be used either as monotherapy or in combination with long acting insulin or insulin sensitizers. The delivery system administers a dry-powder through a device that provides a spacer into which particles are launched via a compressed air supply then slowly inhaled by the patient. The spacer helps decouple the high speed of the jet needed to disaggregate particles from particle impaction in the back of the oropharnx.

The aerosolized powder is contained in blister packets which are administered through the inhaler. Each blister pack contains either 1 or 3 mg of insulin. The 1-mg blister pack is equivalent to approximately 3 units of SC insulin and the 3-mg blister pack is equivalent to approximately 8 units. The Exubera system is designed to emit a specific dose of insulin from the mouthpiece. Up to 45% of the 1-mg blister contents and 25% of the 3-mg blister contents may be retained within the blister well. Patients must take great care in dosing insulin with the Exubera system since three 1-mg blister packs do not equal one 3-mg blister pack. Three 1-mg blister packs have a 30-40% higher C_{max} and area under the curve (AUC) compared to one 3-mg blister pack. In patients with diabetes, the serum insulin concentrations peak faster with Exubera (49 min, range 30-90 min) than SC regular insulin (105 min, range 60-240 min).

In healthy volunteers, the onset of glucose lowering activity occurs in 31 min with the maximum glucose lowering effect occurring at approximately 110 min. The duration of action is 6 h, which is longer than that of SC insulin lispro and comparable to SC regular insulin. In another study in 17 healthy male volunteers, Exubera's time action profile was compared to SC insulin lispro and SC regular insulin using the euglycemic insulin clamp method. When compared to SC regular insulin, Exubera had a faster onset of action (32 vs. 48 min, respectively; P = 0.001). The same could be said when Exubera was compared to SC insulin lispro (32 vs. 41 min, respectively; P <0.05). The time to maximal metabolic effect was comparable between Exubera and SC insulin lispro (143 vs. 137 min; P = NS) but was shorter compared to SC regular insulin (193 min; P <0.01). The duration of metabolic activity for Exubera was similar to

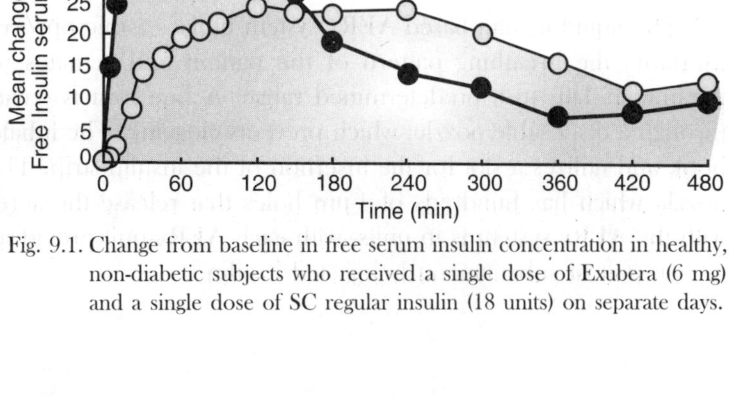

Fig. 9.1. Change from baseline in free serum insulin concentration in healthy, non-diabetic subjects who received a single dose of Exubera (6 mg) and a single dose of SC regular insulin (18 units) on separate days.

SC insulin regular, but significantly longer than SC insulin lispro. Exubera's insulin pharmacokinetic and glucokinetic reproducibility have been investigated in two studies. The reproducibility of Exubera was similar to SC regular insulin and the intrasubject differences between the two same-dose route of administration were small.

Influence of absorption

Underlying lung disease

Exubera has not been studied in certain patient populations such as diabetic patients with asthma or chronic obstructive pulmonary disease (COPD). Since it has not been studied, the use of Exubera in these patients is not recommended. When studied in non- diabetic patients with COPD, the plasma insulin levels were twice those in subjects without COPD.

Smoking

Smoking increases insulin levels by more than two times compared to non-smokers. Therefore, it is recommended that patients who smoke or have stopped smoking in the previous 6 months not receive Exubera. Within 1 week of discontinuation of smoking the bioavailability of Exubera decreased by approximately 50% when compared to smokers, but was still higher than non-smokers. These rapid changes in absorption make dose prediction, and the potential risk of hypoglycemia, higher in smokers.

Asthma

Asthma can greatly affect the ability to move air and thus deliver particles to the deep lung. Studies have consistently reported reductions (approximately 20%) in absorption of inhaled insulin in subjects with mild to moderate asthma when compared to subjects without asthma. Additionally, bronchodilator therapy, if administered prior to inhaled insulin therapy, can increase absorption by 25-50%. These large absorption differences are the basis for the current recommendation against inhaled insulin use in subjects with asthma.

Intercurrent respiratory tract illness

Clinically, concurrent, uncomplicated upper respiratory illnesses (bronchitis, upper respiratory infection, and rhinitis) did not affect rates of glycemic control or hypoglycemia. About 1 in the 20 subjects temporarily discontinued Exubera therapy during the illness and alternative therapy would have to be implemented in these subjects.

AERx System

The liquid insulin based AERx system utilizes a microprocessor controlled inhaler device that monitors the breathing pattern of the patient and actuates when a combination of breathing parameters falls in a predetermined range. A fine mist is created by driving the insulin solution through a disposable nozzle, which prevents clogging. The inhaler is about the size of a paperback book and utilizes a slot for the insertion of the insulin strip. The insulin strip uses a one-time use nozzle which has hundreds of 1-µm holes that release the aerosolized insulin uniformly. Dosing with the AERx system is in units, with each AERx unit providing the same glucose lowering effect as one unit of subcutaneously injected insulin.

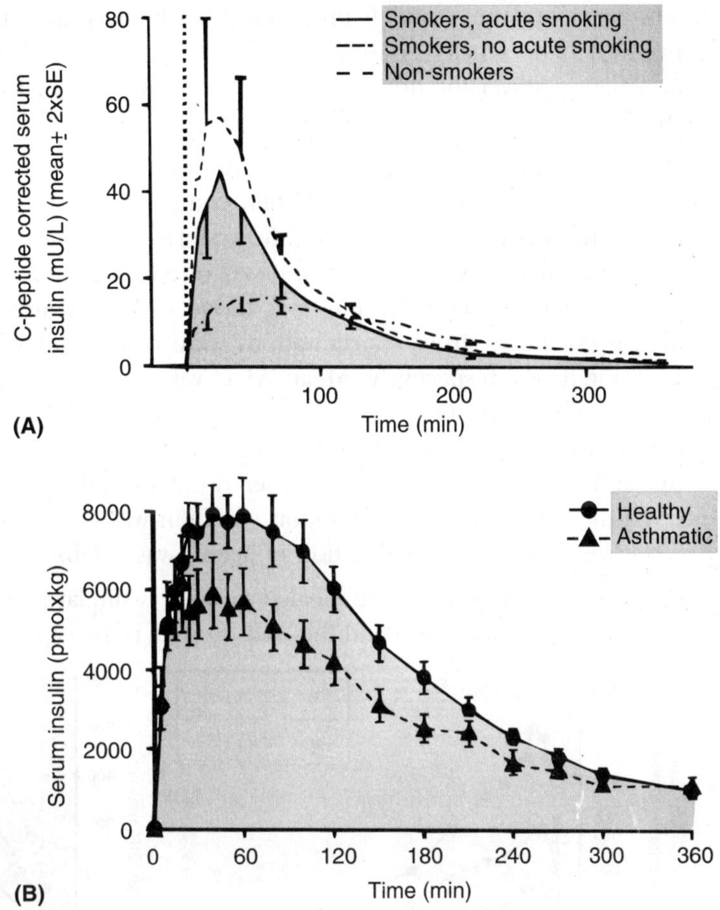

Fig. 9.2. Insulin pharmacokinetics in smokers and asthmatics using the AERx system.

The patient can select to give between 2 and 10 units of insulin with each strip and can be adjusted by one unit increments. The onset of action with the AERx system is approximately 10 min with peak insulin concentrations occurring approximately 1 h after inhalation. Dose effects the duration of action with a dose of 0.3 units/kg lasting 5h and a dose of 1.8 units/kg lasting 8 h. Pharmacokinetic studies found that inhaled delivery using the AERx system resulted in rapid absorption of insulin and a corresponding faster decline in plasma glucose compared with subcutaneous (SC) administration. Intrapa-tient variability is similar in T1DM patients given equivalent doses of either SC regular insulin or inhaled insulin. The speed of absorption and resultant hypoglycemic activity is dependent on the volume and depth of breath, hence the inhaler design. Studies in special populations with the AERx system have mirrored those with Exubera. This includes smoking, underlying lung disease, asthma, and intercurrent respiratory tract infection.

Pharmacokinetic parameters are altered in smokers who use the AERx system. Insulin AUC was 60% higher, C_{max} was three times higher, and absorption was more rapid in smokers compared to non-smokers. Studies in asthmatic patients found that inhaled insulin absorption was decreased compared to those healthy non-asthmatic subjects. With the reduced absorption there were smaller reductions in blood glucose, possibly implying that patients with asthma may require higher doses of inhaled insulin to achieve glycemic control, or bronchodilator therapy prior to each dose of inhaled insulin. In patients with respiratory tract infections it was found that insulin pharmacokinetics were not significantly different during or immediately after acute infection.

AIR System

The AIR system uses a semi-disposable inhaler to deliver a dry-powder insulin particle that is characterized by large size but low density keeping the aerodynamic diameter·small. Currently under development are the 2 and 6 unit capsules which contain 0.9 and 2.6 mg/capsule of insulin, respectively. The inhaler is approximately the size of a cell phone and utilizes breath actuated technology to deliver the dose. The insulin capsule is placed in the inhaler where it is punctured by the device.

The inhalation of air through the device provides enough energy to make the capsule spin, forming an aerosol for inhalation. When compared to SC insulin lispro in healthy subjects, the AIR system seems to be comparable in terms of pharmacokinetic and pharmacodynamic parameters. Subjects who were given doses of 2.6, 5.2, and 7.8 mg, the onset of action were 13, 13, and 15 min, respectively. This was found to be a significantly faster onset compared to SC insulin lispro. Peak serum concentrations were reached in 45, 30, and 45 min for the 2.6, 5.2, and 7.8 mg doses, respectively. Mean AUC values were similar between the inhaled AIR system and SC insulin lispro. C_{max} values were similar for the 7.8 mg dose compared to 18 units of SC insulin lispro, however, the 2.6 and 5.2 mg had lower C_{max} values compared to the 6 and 12 units of SC insulin lispro, respectively. Higher doses extends the duration of action. In subjects who received the 7.8 mg dose, the duration of action was 8 h compared to subjects who received the 2.6 and 5.2 mg dose whose duration of action was 6-7 h.

Intrasubject variability of the inhaled product is similar to SC insulin lispro. A major difference between the Exubera product and the AIR product is the capsule dose equivalency. Unlike Exubera,

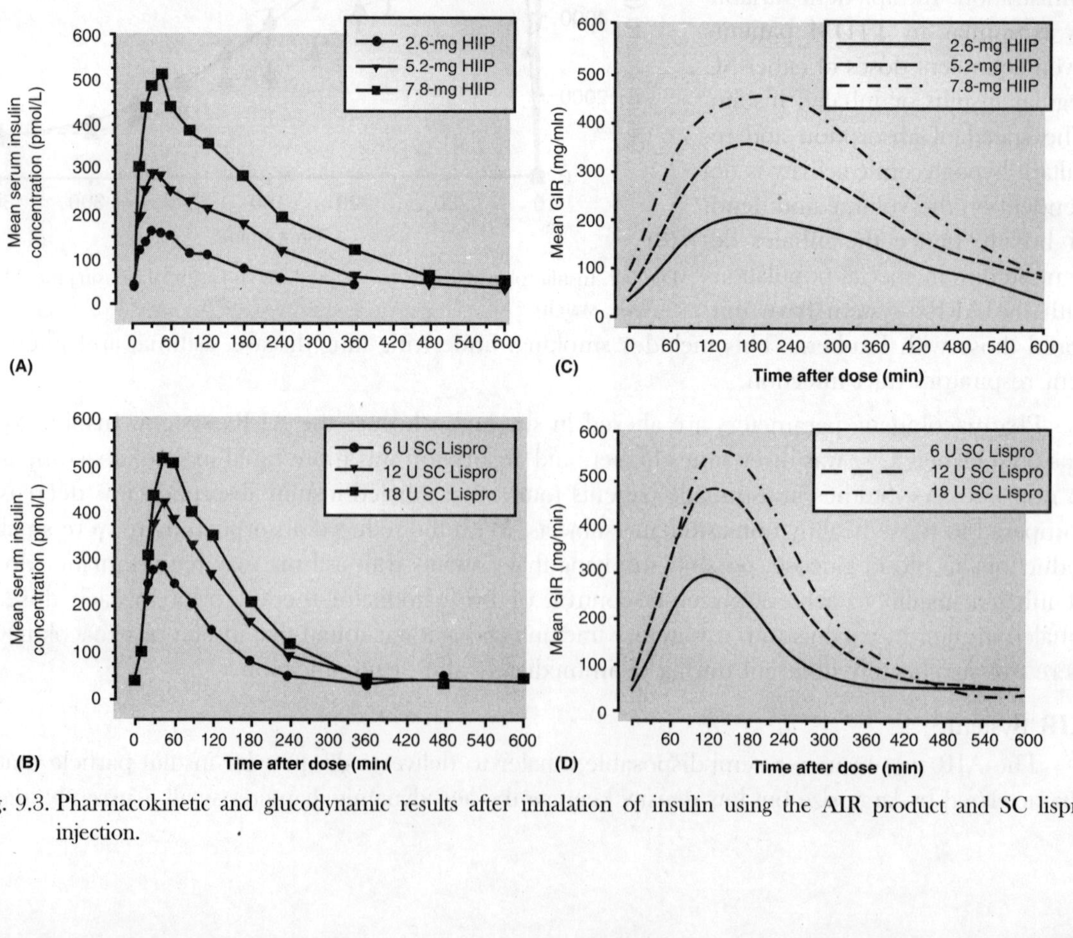

Fig. 9.3. Pharmacokinetic and glucodynamic results after inhalation of insulin using the AIR product and SC lispro injection.

3 inhalations of 2 unit AIR capsules are equivalent to one inhalation of a 6-unit capsule. One study evaluated the pharmacokinetics and glucodynamics of the AIR system in patient with COPD. The mean AUC and metabolic effect were reduced in patients with COPD compared to healthy subjects. Moreover, the intrasubject variability was greater in patients with chronic bronchitis. More extensive studies are needed in this population to further investigate the safety profile.

Technosphere Technology

The Technosphere/Insulin technology is unique in that it captures and stabilizes peptides in small particles. A small organic molecule, 3,6-bis [N-fumaryl-N-(n-butyl)amino-2,5-diketopiperazine] self-assembles in a mild acidic environment into microspheres of approximately 2 μm. Peptides in the reaction solution, such as insulin, are trapped in the microspheres during self assembly. The particles are then dried and suitable for pulmonary delivery. In the neutral pH of the deep lungs, the microspheres dissolve and facilitate the absorption of the insulin into the systemic circulation. The carrier molecules are excreted as ammonium salts in the urine within hours.

The Technosphere/Insulin is provided in gelatin capsules that contain 50 IU of dry powder regular human insulin. In studies evaluating Technosphere/Insulin technology a specifically developed breath-powered unit dose dry powder inhaler was utilized. The inhaler does not require any external energy and subjects were instructed to take three deep breaths per capsule. Delivery of the drug is dependant on the breath effort of the patients. In a pilot study of five healthy subjects, a euglycemic clamp technique was used to evaluate the biologic efficacy and pharmacokinetic properties of the Technosphere/Insulin technology. Subjects were randomized to one of three groups inhaled 100 IU of Technosphere/Insulin, 10 units of SC regular insulin and 5 units of intravenous (IV) regular insulin. Technosphere/Insulin showed an onset of action similar to that of IV regular insulin (13 vs. 9 min) and much quicker than that of SC regular insulin (121 min).

Insulin concentrations with the Technosphere/Insulin technology returned to baseline after 3 h. In a following dose ranging study, subjects were randomized to receive 25, 50, and 100 IU of Technosphere/ Insulin. This study found that there is a linear dose dependant increase in metabolic activity and systemic insulin uptake. Technosphere/Insulin technology differs from the other inhaled technologies in that the onset is faster, although they have not been directly compared.

Spiros Insulin

The Spiros insulin consists of a dry powder inhaler and blisterdisks. The blisterdisks contain a preparation of 11% micronized insulin crystals blended with lactose monohydrate NF. The blend is filled into blisterdisks at a target fill weight of 9.5 mg per well. This results in 1.05 mg of insulin in each blister well that is equivalent to 30 units. The median aerodynamic particle size of the insulin crystals is 2-3 μm and the median size of the lactose particles is 150 μm. The inhaler is a hand held breath activated device that delivers the dry powder crystals. Inspiratory flow activates a battery-powered twin blade impeller that assisted in deagglomeration of the drug and carrier during inspiration.

In an open label euglycemic glucose clamp in healthy volunteers four doses of the Spiros system's (60, 90, 120, and 150 U) pharmacokinetics and glucodynamics were evaluated against those subjects receiving 2 of the 3 possible doses of SC regular insulin (8, 14, and 20 U). Serum insulin levels following inhalation of insulin peaked approximately 60 min earlier when compared to SC regular insulin (P <0.001). The time to maximum glucose infusion rate was reached approximately 70 min earlier with inhaled insulin than with SC regular insulin. The Spiros system intra- and inter-subject variability were found to be comparable to that of SC regular insulin. Smokers have increased serum insulin levels; however they have reduced glucodynamic effect due to increased insulin resistance.

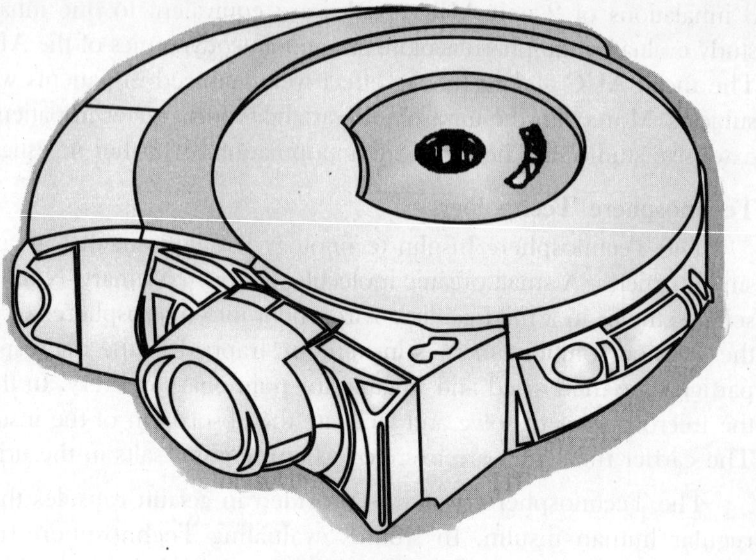

Fig. 9.4. Spiros dry powder inhaler.

PHARMACEUTICAL STUDIES

Exubera System

The efficacy and safety of Exubera has been evaluated in over 4000 individuals with T1DM (4 studies) and T2DM (7 studies).

Type 1 diabetes mellitus

Skyler reported on two trials in T1DM. In a proof-of-concept trial conducted in 72 T1DM patients, a similar reduction in A1C was achieved when randomized to twice daily isophane insulin plus Exubera before meals versus once daily ultralente insulin plus SC regular insulin before meals. Second, 328 patients with T1DM (baseline A1C approximately 8.2%) received premeal Exubera or regular SC insulin in combination with Neutral Protamine Hagedorn (NPH) insulin before breakfast and at bedtime. After 24 weeks, the A1C was 7.7 ± 0.1% in the Exubera group and 7.8 ± 1.2% in the SC insulin group.

The two treatment regimens were comparable for A1C reduction and number of subjects who reached an A1C <7.0%, but Exubera significantly lowered the fasting plasma glucose (FPG) greater than SC insulin (-35 vs. 4 mg/dl [CI -57.50 to -21.56]), respectively. Quattrin et al. randomized 335 T1DM subjects with a baseline A1C of approximately 8.3% to either premeal Exubera plus bedtime ultralente or NPH insulin twice daily (breakfast and evening or bedtime) in combination with two regular insulin doses prior to breakfast and the evening meal for 24 weeks. Exubera and SC insulin reduced the A1C to 7.9 ± 1.1% and 7.7 ± 0.9% (adjusted treatment group difference; 0.16% [95% CI, -0.01 to 0.32]), respectively.

In addition, a comparable numbers of subjects achieved A1C <7% (15.9% and 15.5%), respectively. The FPG in patients receiving Exubera (baseline FPG = 194 mg/dl) was reduced greater than in patients receiving SC insulin (baseline FPG = 203 mg/dl), [-25.2 mg/dl (95% CI -143.4 to -7.0 mg/dl)]. Additionally, Heise et al. examined 47 T1DM subjects given NPH twice daily in combination with pre-meal SC regular insulin or Exubera. A1C and FPG were not significantly different between regimens. Overall, in T1DM, Exubera given pre-meal appears to be equally efficacious, as measured by the A1C and attainment of A1C <7.0%, as SC regular insulin. Exubera has shown in several trials to significantly lower FPG greater than SC insulin, and this may result in the need to adjust the basal insulin, or at a minimum be cognizant of the potential for nocturnal hypoglycemia in a particular patient.

Type 2 diabetes mellitus

Exubera versus or in combination with oral agents

Weiss et al. evaluated the effect of the addition of Exubera or continuation of baseline oral antihyperglycemics (baseline A1C approximately 9.8%) on glycemic control in 68 T2DM subjects on near maximal or maximal therapeutic doses of a sulfonylurea, metformin, or the combination. After 12 weeks, subjects randomized to Exubera three times daily prior to meals versus continued baseline oral therapy had an A1C reduction of 2.3% versus 0.1% (P<0.001). Eleven subjects (34%) taking Exubera versus none continuing oral agents achieved an A1C <7%. DeFronzo et al. randomized 145 T2DM patients with poor glucose control, A1C at baseline approximately 9.5%, despite diet and exercise, to receive premeal Exubera (average dose = 15.3 mg at end of study) or rosiglitazone 4 mg twice daily.

After 12 weeks, the A1C was reduced greater in the Exubera group versus rosiglitazone (-2.3% vs. -1.4%, P < 0.0001), and the odds of reaching an A1C of either <7% or ≤6.5% was significantly higher with Exubera. Changes in FPG were comparable between treatments (-64 and -56 mg/dl) for Exubera and rosiglitazone. Barnett et al. reported on two open-label, parallel, 24-week trials in T2DM subjects randomized to addition of pre-meal Exubera versus addition of metformin in subjects poorly controlled on sulfonylurea monotherapy, or versus addition of a sulfonylurea in poorly controlled subjects on metformin monotherapy. In 423 subjects taking near maximal effective doses failing sulfonylureas, addition of Exubera (average daily dose = 12.1 mg) lowered the A1C greater than addition of metformin (average dose = 1 g twice daily), (-2.06% vs. -1.83%, p = 0.014), respectively.

The high A1C arm (baseline A1C ≥ 9.5%) saw reductions of 2.17% versus 1.79%, p = 0.002, with Exubera and metformin, respectively. A1C ≤ 7.0% was achieved by 54 (25%) Exubera subjects and 45 (23%) of metformin subjects (P = NS). 476 subjects failing metformin (>1.5 grams/day for ≥2 months) were given Exubera (average daily dose = 13.2 mg) or glibenclamide (glyburide in United States, at average daily dose = 7.6 mg). A1C was reduced (2.03% and 1.88%, P = NS), with Exubera and glibenclamide, respectively, but in the high A1C arm, reductions were greater with Exubera (-0.37% [95% CI -0.62 to -0.12, P = 0.004]). Odds of achievement of an A1C ≤ 7.0% were significantly better with Exubera if the baseline A1C was >9.5%.

Overall, Exubera proved to be non-inferior to metformin or glibenclamide at an A1C <9.5%, and superior to addition of metformin or sulfonylurea if the baseline A1C is >9.5%. In an open-label, multicenter study (baseline A1C approximately 9.3%) in 309 T2DM, subjects were randomized to one of three groups: (1) continued current oral antihyperglycemics; (2) premeal Exubera as monotherapy (discontinuation of all oral agents); or (3) current oral agent therapy plus Exubera. After 12 weeks, A1C reductions in the three groups were 0.2%, 1.4%, and 1.7%, respectively, with both Exubera monotherapy and Exubera addition to oral agents significantly lowering A1C versus continued oral agents. Odds of attainment of an A1C <7.0% was significantly higher with both Exubera monotherapy (17%) and addition of Exubera to oral agents (32%) versus continued oral agents (1%). The mean dose of Exubera was higher with monotherapy (26.4 mg) than when oral agents are continued (13.1 mg).

Exubera versus SC insulin regimens

In a small, three month trial, 26 insulin-treated T2DM not taking oral agents were randomized to continue their conventional insulin regimen (2-3 injections/day) or to start premeal Exubera with a single bedtime injection of ultralente insulin. In the Exubera group, A1C declined by 0.71% (baseline A1C 8.67%), and remained unchanged in the conventional insulin regimen group. Subjects were on 14.6 mg Exubera and 35.7 units of ultralente, compared to 19 units of regular insulin and 51 units of long-acting insulin at baseline.

Thus, it appears Exubera may have the potential to lower the basal insulin requirements, despite being prandial insulin. Hollander et al. randomized 298 T2DM subjects, poorly controlled (baseline A1C = ~8.5%) on at least two daily injections of insulin, to receive premeal Exubera plus bedtime ultralente or at least two SC injections of mixed regular/NPH insulin for 6 months. The reduction in A1C in the Exubera (-0.7%) and mixed regular/NPH insulin (-0.8%) groups were similar. An A1C ≤ 7% was achieved in 67 (47%) subjects receiving Exubera versus 46 subjects (32%) on the mixed regular/NPH regimen (adjusted odds ratio 2.27 [95% CI 1.24–2.14]). The FPG was significantly lower with Exubera, but did not show a significant difference in 2-h post-prandial glucose readings versus the mixed regular/NPH group.

Glycemic control

In all studies in T1DM and T2DM, Exubera as proven to be as effective as regular insulin in improving glycemic control. In addition, Exubera appears to lower the FPG when compared to regular insulin, which could result in the need to reduce the basal insulin dose and also supports the potential use of Exubera as monotherapy. Though Exubera can be used in monotherapy, combination with oral agents appears to be advantageous in that you can attain glycemic control with a lower dose of inhaled insulin. It should be noted that the majority of trials were started prior to the introduction of the basal insulin glargine or detemir, and intermediate-acting or ultralente acted as the basal insulin in these trials. The choice of basal insulin can be important, and most clinicians currently would recommend the use of insulin glargine or detemir. Insulin glargine and detemir have less risk of hypoglycemia and/or nocturnal hypoglycemia when compared to NPH, but trials do not support superior efficacy. Thus, if cost is paramount, NPH could be substituted, as lente and ultralente insulins have been discontinued by the manufacturer.

Dosing criteria

Guidelines for initial premeal Exubera dosing are based upon the patient's body weight or the appropriate cross-over dose when switching from regular SC insulin. Based upon clinical trials in which patients consumed three meals per day, an initial meal-time dose can also be calculated by: (body weight in kg) × (0.05 mg/kg), where tenths of a milligram are rounded down to the nearest whole milligram.

These are only initial premeal doses, and further individualization based on glycemic control, timing in regards to meal, content, and size of meal are appropriate. For example, a patient with an A1C of 10% may start at this dose, but is likely to be controlled on a higher dose, whereas if the patient's starting A1 C was 7.0%, post-prandial or fasting hypoglycemia may be of more concern if incorrectly dosed, base on the meal.

It should be intuitive that T2DM patients are more insulin resistant, thus higher doses of Exubera are likely to be required in patients with T2DM versus those with T1DM. Also, Exubera is dosed in milligrams, and each milligram increase is equal to 2-3 units injected subcutaneously. Clinically, this is usually a large dosage increase for T1DM, and must be balanced with the current blood glucose readings.

Exubera is available in a Combination Pack 12 (90, 1-mg blister packs and 90, 3-mg blister packs, plus 2 insulin release units) and a Combination Pack 15 (180, 1-mg blister packs and 90, 3-mg blister packs, plus 2 insulin release units) for subjects who may use 1 2 or 1 5 mg/day, respectively.

Blister packs in the foil can be used, if stored at room temperature, until the expiration date, but once the foil overlay has been opened they are good for 3 months. The insulin release units within the Exubera device should be replaced every 2 weeks so as to lower the risk of insulin powder blockage where the compressed air disperses the insulin. This could lower the amount of insulin delivered from each blister pack. In addition, it is currently recommended that the device be replaced after 1 year of use.

AERx System

Clinical data is limited, but in a 12 week study in T2DM, three AERx insulin inhalations in conjunction with bedtime NPH (baseline A1C 8.5%) produced a similar reduction in A1C (-0.69 ±0.77%) as fast-acting human insulin given SC prior to each meal (-0.77 ± 0.77%). No significant difference between groups was seen for FPG or post-prandial glucose readings. The amount of inhaled insulin versus SC insulin was explored, which is important, as the AERx is dosed in "units" of insulin. The mean meal-related insulin ratio was 0.85 units of SC insulin for each unit of AERx insulin.

AIR System

Clincal data is limited, but in a 12 week study in T1DM subjects who were randomized to receive prandial coverage with SC insulin or AIR system inhaled insulin in conjunction with glargine insulin, the reduction in A1C with AIR system inhaled insulin was equivalent to SC injected insulin.

Technosphere

Clinical data is limited, but in 227 poorly controlled T2DM subjects given prandial Technosphere/Insulin at doses of 28, 42, and 56 units for 12 weeks in conjunction with insulin glargine, A1C declined by 0.67%, 0.70%, and 0.78%, respectively.

PRECAUTIONS

Pulmonary Parameters

Forced expiratory volume in 1 second (FEV_1) and diffusing capacity of carbon monoxide (DL_{CO}) are parameters utilized to evaluate lung function in studies testing inhaled insulin products. FEV1 measures the volume of air exhaled during the first second of a forced expiratory maneuver. Values are reported as a percentage of predicted normal values and are used to measure airway obstructions or changes in airway diameter. DL_{CO} reveals oxygen- diffusing capacity influenced by lung surface area, speed of blood flow and thickness of the air-blood barrier. A reduction in DL_{CO} is a hallmark finding in patients with emphysema and interstitial lung disease. Lung function can be affected by various factors.

Lung diseases such as COPD, interstitial lung disease, or asthma and patient specific factors such as age, smoking, ethnicity, weight or height can have a marked impact on lung function. Patients with diabetes have an unexplained decrease in lung function compared to those subjects without diabetes. In one study, the forced vital capacity (FVC) and FEV_1 were approximately 8% lower in diabetic individuals compared to non-diabetics. This difference is similar to the difference seen between smokers and a subject who has never smoked.

Inhaled System

The preponderance of the safety data comes from studies evaluating the Exubera system while there are only a few studies assessing the other systems.

Exubera system

In patients using the Exubera system, cough was reported more frequently (T1DM = 29.5% and T2DM = 21.9%) compared to controls (10%). The cough was generally mild, rarely productive and accounted for 1.2% discontinuation in clinical trials. Furthermore, the cough occurred within seconds to minutes of dosing and was not associated with declines in lung function or bronchoconstriction. All reports published regarding pulmonary function with inhaled insulin, regardless of the system, have shown a small decline from baseline and was comparable to SC insulin delivery. In studies investigating Exubera, patients with T1DM who received inhaled insulin experienced a statistically significant decline in DL_{CO} and a non-significant decrease in FEV_1 compared to those patients randomized to SC insulin.

Although there was a statistically significant decrease in DL_{CO} there were no clinical manifestations and was deemed not clinically significant. This decline occurred within 2 weeks after insulin initiation and did not progress any further. Moreover, patients who did not receive inhaled insulin had a small decline in lung function over time. Studies with Exubera, AERx, and AIR systems agree that changes in lung function are not clinically significant and reversible upon treatment discontinuation.

Thoracic high resolution computed tomography scans after 2 years of Exubera have shown no significant signals for concern. Additionally, the rate of pulmonary fibrosis, two cases with Exubera (0.40 cases/10,000 subject-months exposure) compared to one case in an oral agent comparator group (0.50 cases/10,000 subject-months exposure), is not increased.

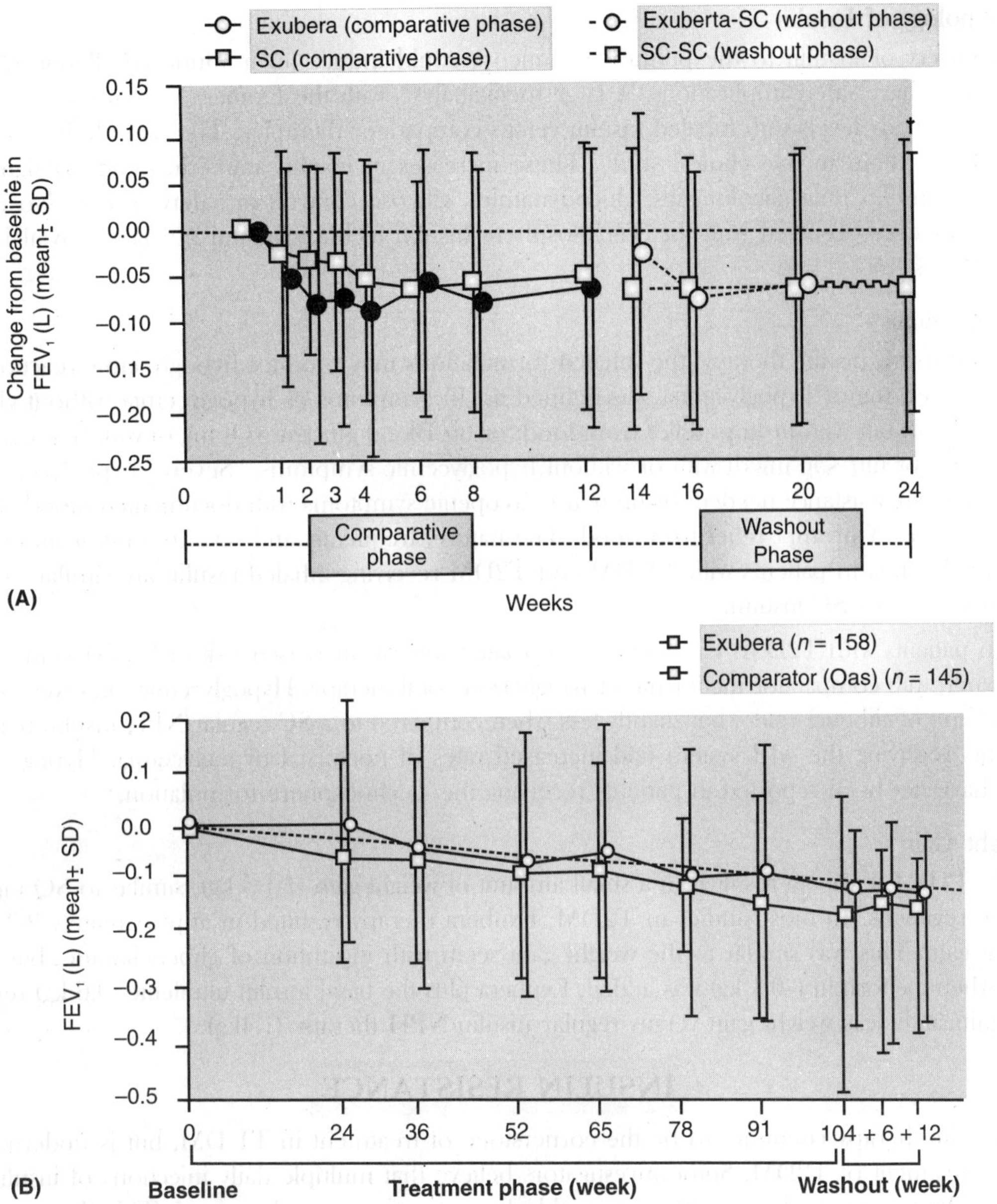

Fig. 9.5. FEV_1 with Exubera inhaled insulin therapy after (A) 12 weeks and (B) 2 years.

Treatment

Prescribing information for the Exubera system recommends all patients have baseline spirometry testing, a follow-up 6 months after initiation of treatment, and yearly follow-up thereafter. Exubera is not recommended for patients with baseline FEV_1 or DL_{CO} < 70% predicted. In patients who experience a ≥20 % decrease in FEV_1, tests should be repeated. If this finding is confirmed, then treatment should be discontinued. Treatment should also be stopped in the rare case of acute bronchospasm.

Immunology of Insulin

Delivery of insulin to the pulmonary mucosa could induce an immune or allergic reaction which may have safety implications. A large meta-analysis with the Exubera system showed higher insulin antibody levels with inhaled insulin versus comparator therapies. The same holds true with the AERx system in one clinical study. These increases in insulin antibodies had no significant effect on insulin pharmacokinetics, glucodynamics, glucose control, or safety profile. No insulin antibodies were observed with the Technosphere/Insulin technology and no data is available for the AIR system.

Hypoglycemia

As with SC insulin therapy, the inhaled formulations may produce hypoglycemic reactions. In most studies, minor hypoglycemia was defined as: (i) symptoms of hypoglycemia without glucose documented, but with prompt relief from food; or (ii) Blood glucose <60 mg/dl with hypoglycemic symptoms; or (iii) <50 mg/dl with or without hypoglycemic symptoms. "Severe" hypoglycemia was defined as: (i) assistance needed; or (ii) neuroglycopenic symptoms with documented blood glucose <50 mg/dl, or if prompt relief from food. The nature, frequency and severity (minor and major) of hypoglycemia in patients with T1 DM and T2DM receiving inhaled insulin are similar to those patients receiving SC insulin.

In patients with T2DM, Exubera was associated with an increased risk of hypoglycemia when compared with comparator metformin or rosiglitazone oral therapy. Hypoglycemic rates were similar to addition of glibenclamide, but slightly less when compared to a SC regular/NPH insulin regimen. Patients receiving the AIR system had increased rates of nocturnal hypoglycemia. Hypoglycemic rates have not been reported in patients receiving the Technosphere formulation.

Weight Gain

In T1DM, Exubera resulted in a small amount of weight gain (1-1.5 kg), similar to SQ injected insulin regimens. In most studies in T2DM, Exubera therapy resulted in approximately 2-3 kg of weight gain. This was similar to the weight gain seem with institution of glibenclamide, but more than when metformin (-0.1 kg) was added. Exubera plus the basal insulin ultralente (0.6 kg) resulted in significantly less weight gain versus regular insulin/NPH therapy (1.4kg).

INSULIN RESISTANCE

Insulin therapy continues to be the cornerstone of treatment in T1 DM, but is underutilized in the treatment of T2DM. Some investigators believe that multiple daily injections of insulin are disliked by patients and are partly responsible for patients not reaching goal. This demonstrated "*resistance*" to the implementation or administration of insulin therapy has been well documented. Often the reasons cited by clinicians to delay insulin therapy include: fear of patient loss from practice, lack of time to implement, and clinical inertia, or the "*wait and see*" approach.

Patient's reasons for delaying insulin therapy include: the frequent asymptomatic course of diabetes, patient refusal to follow directions, and importantly, injections and hypoglycemia. Thus, inhaled insulin may help to overcome some of these concerns. In a study of theoretical treatment choices in inadequately controlled T2DM individuals, subjects were more likely to choose insulin therapy when the insulin therapy choices included inhaled insulin therapy. Rosenstock et al. reported on patient satisfaction in the pooled-analysis of three, 12-week studies in subjects with T1DM and T2DM who were randomized to receive Exubera with ultralente or 2-3 injections of a SC split/ mixed insulin regimen, followed by a 1 year open-label extension.

Using the Patient Satisfaction with Insulin Therapy questionnaire, overall satisfaction and ease of use were significantly higher with Exubera versus SC insulin, and social comfort tended (P = 0.11) to be higher with Exubera. Eighty-five percent of the patients treated with Exubera during the 12-week period chose to continue Exubera during the 1-year extension, whereas only 21% treated with SC insulin during the 12-week study chose to remain on the SC regimen. Most (75%) who were treated with SC insulin during the 12-week study and chose to stay in the 1-year extension were switched to Exubera.

CONCLUDING REMARKS

Inhalation of insulin is a novel, non-injectable method for delivering insulin into the human body. Many patients do not like injections, and would prefer alternative methods of insulin delivery. Inhaled insulin therapy offers additional options for clinicians to improve glycemic control in patients with diabetes. The dry-power system of Exubera is the most extensively studied and the only Food and Drug Administration approved inhaled insulin. Onset and peak effect are similar to rapid-acting insulins such as lispro, aspart, or glulisine, but Exubera has a duration of action similar to regular SC insulin.

Underlying lung disease, smoking, and asthma can all alter the dose of insulin absorbed, and inhaled insulin should be avoided in these populations. Exubera is dosed as milligrams, and this is distinctly different than any other insulin product. It has been shown to be effective in reducing A1C levels in T1DM and T2DM. As with other insulin, a risk of hypoglycemia is present, and Exubera has similar rates of hypoglycemia as SC regular insulin. Safety data is complete through 2 years with Exubera, and small changes in FEV1 and DL_{CO} have been observed, but are reversible upon discontinuation. Other inhalation systems such as AERx, AIR, and Technosphere are in development, and the area of alternative delivery of peptides continues to be an active area of research.

10

Diabetes Insipides

Diuretics have appropriately maintained a stable place in the management of *hypertension* and *heart failure* (HF) because of their proven efficacy and low cost. The aldosterone antagonists spironolactone and its analog eplerenone are diuretics but have added actions that improve myocardial function and are an important part of our armamentarium for the management of patients with HF.

SYMPTOMS

1. Edema associated with corticosteroids, estrogen, or vasodilator therapy.
2. Edema due to renal dysfunction, or ascites due to cirrhosis.
3. Hypertension.
4. *Spironolactone* has been shown to have an important role in the management of class III and IV HF.
5. Management and relief of symptoms of HF: dyspnea, orthopnea, paroxysmal nocturnal dyspnea, and edema.

Note. Edema of the legs presumed to be caused by HF, edema resulting from obstruction of venous return, and dependent edema caused by lack of muscle pump action are some of the commonest reasons for diuretic abuse.

Preventive Measures

1. *Pulmonary embolism* with shortness of breath: Edema caused by cor pulmonale should not be treated aggressively with diuretics. Correct the hypoxemia, and then try to accomplish a very mild diuresis over several weeks.
2. *Ascites* with impending hepatic coma.
3. *Edema* with acute renal failure.

Table 10.1 Generic and trade names of diuretics

Generic name	*Trade name*	*Tablets (mg)*	*Usual maintenance (mg daily)*
Group I: Thiazides			
Chlorothiazide	Diuril, Saluric	250, 500	500-1000
Hydrochlorothiazide	HydroDiuril, Hydrosaluric, Esidrix, Esidrex, Oretic, Direma	25, 50, 100	12.5-25
Bendrofluazide	Aprinox, Berkozide, Centyl, Neo-NaClex	2.5, 5	2.5-5
Bendroflumethiazide	Naturetin	2.5, 5, 10	2.5-10
Benzthiazide	Aquatag, Exna, Hydrex	50	50-100
Cyclothiazide	Anhydron	2	2
Hydroflumethiazide	Diucardin, Hydrenox, Saluron	50	50
Chlorthalidone	Hygroton	25, 50, 100	25-50
Methylclothiazide	Enduron, Aquatensen, Diutensen-R	2.5, 5	2.5-5
Polythiazide	Renese, Nephril	1, 2, 4	0.5-4
Trichlormethiazide	Naqua, Metahydrin	2, 4	2-4
Cyclopenthiazide	Navidrex, Navidrix	0.5	0.5-1
Metolazone	Zaroxolyn, Metenix	2.5, 5, 10	2.5-5
Quinethazone	Aquamox, Hydromox	50	50-100
Indapamide	Lozol, Natrilix, Lozide (C)	2.5	2.5
Group II: Loop diuretics			
Furosemide, Frusemide (UK)	Lasix, Dryptal, Frusetic, Frusid	20, 40, 80, 500	40-120
Ethacrynic acid	Edecrin	25, 50	50-150
Bumetanide	Burinex, Bumex	0.5, 1, 5	1-2
Piretanide	Arlix	6 (capsule)	6-12
Torsemide	Demadex	5, 10, 20, 100	5-20
Group III: K^+-sparing diuretics			
Spironolactone	Aldactone	25, 50 (UK), 100	25-100
Triamterene	Dyrenium, Dytac	50, 100	50-100
Amiloride	Midamor	5	5-10
Group IV			
Thiazide + K^+-sparing	Aldactazide, Dyazide, Moduretic, Moduret		
Frusemide + K^+-sparing	Frumil, Frusene, Lasoride		
Group V			
Acetazolamide	Diamox	250	–

4. Obstructive and restrictive *cardiomyopathy.*
5. *Cardiac tamponade.* When the jugular venous pressure (JVP) is grossly raised (>7 cm H_2O) and the patient is not responding to conventional therapy for HF, before giving diuretics to such patients, consideration should be given to a diagnosis of cardiac tamponade or constrictive pericarditis.
6. Tight mitral *stenosis* or aortic stenosis.

Monitor, intensive diuretic therapy as follows: If any of the following six-point checklist occurs, discontinue diuretics for 24 h and then recommence at approximately half the dose.

1. More than 2 kg weight loss per day is associated with symptoms (1 kg of water loss = 140-150 mEq (mmol) of Na^+ loss in the presence of normal serum Na^+).
2. Systolic blood pressure is <95 mmHg or orthostatic hypotension is present.
3. Electrolytes:
   (a) Blood urea: >7.0 mmol/L from baseline.
   (b) Serum chloride (Cl^-): <94 mEq (mmol)/L.
   (c) Serum sodium (Na^+): <128 mEq (mmol)/L.
   (d) Serum potassium (K^+): <3 mEq (mmol)/L.
   (e) CO_2: >32 mEq (mmol)/L.
   (f) Uric acid: >10 mg/dL (588 mmol/L).
   (g) JVP is <1 cm if previously raised. The Frank-Starling compensatory mechanism is lost if diuresis is too excessive and filling pressures fall below a critical point, thereby causing cardiac output and tissue perfusion to fall.
4. Arrhythmias develop or worsen.
5. The 24-h urinary Na^+ excretion is >150 mEq (mmol).

   For the management of moderate to severe HF with bilateral edema and pulmonary crepitations, the goal should be

6. Weight loss, a little more than 1 kg/d for 3 d, and then 0.5 kg/d for 7 d, with a minimum of 4 kg to a maximum of 10 kg in 10 d.

Note. A 24-h urinary Na^+ excretion < 20 mEq (mmol) indicates inadequate diuretic therapy; 24-h urinary excretion > 100 mEq (mmol), with no weight loss, requires reduction of Na^+ intake. This estimation is, however, seldom required.

PROCESS OF LOOP DIURETICS

The rapid onset of action of loop diuretics, together with their potency in the presence of normal and abnormal renal function and their venodilator effect, renders them more effective than thiazides for the management of acute and chronic HF and life-threatening pulmonary edema.

Physiological Approach

Loop diuretics inhibit the $Na^+/K^+/Cl^-$ transport system of the luminal membrane in the thick ascending loop of Henle, and thus they block Cl^- reabsorption at the site where approximately 40% of filtered Na^+ is normally reabsorbed. Loop diuretics, through their action on Na^+-Cl^- cotransport, inhibit Ca^{2+}, K^+, and Mg^{2+} reabsorption in the loop where some 25% of filtered K^+, 25% of Ca^{2+}, and 65% of Mg^{2+} are normally reabsorbed.

Drug name: *Furosemide*; *Frusemide*

Trade name: Lasix

Supplied: 20, 40, 80, 500 mg

Dosage: 20, 40, or 80 mg each morning until desired effect achieved; maintenance 20–40 mg daily or every second day

Dosage

Patients with severe HF may require between 80 and 160 mg furosemide daily, rarely 240 mg daily for a few days and then a lower maintenance dose. In such patients, it is preferable to give the total dose of furosemide as *one dose each morning*. If a second dose is necessary, this should be given before 2 PM to avoid bothersome nocturia.

Hypokalemia is more common with twice-daily dosage. Also, if the patient was formerly resistant to 80 mg, the renal tubule may be resistant to the 80 mg given later in the day. If a dose of furosemide >60 mg per day is predicted to be necessary for several weeks, then it is advisable to add a K^+-sparing diuretic or angiotensin-converting enzyme (ACE) inhibitor. This will increase diuresis by inhibiting aldosterone and at the same time will conserve K^+. The addition of spironolactone to the HF regimen of ACE inhibitor and furosemide has been shown to improve survival.

Intravenous (IV) dosage. Ampules are available in 10 mg/mL, 20 mg/2 mL, 40 mg/ 4 mL, and 250 mg/25 mL. The IV dose is given slowly (20 mg/min), and if renal failure is present, it should not exceed 4 mg/min (to prevent ototoxicity).

Clinical Actions

Furosemide inhibits Na^+ and Cl^- reabsorption from the ascending limb of the loop of Henle with, in addition, weak effects in the proximal tubule and the cortical diluting segment. The drug is excreted by the proximal tubule. Because of the site and potency of action, loop diuretics are much more effective than thiazides when the glomerular filtration rate (GFR) is markedly reduced. Loop diuretics remain effective even at GFRs as low as 10 mL/min. If a diuretic is required in a patient with a serum creatinine level > 2.3 mg/dL (203 μ/mol/L), it is reasonable to choose furosemide.

Furosemide is also used in preference to thiazide as maintenance therapy in patients with moderate to severe or recurrent HF, that is, in patients in whom further episodes may be predicted because of the extent of cardiac disease. IV furosemide has a *venodilator* effect, and, when it is given to patients with pulmonary edema, relief may appear in 5-10 min.

Intravenous Route

This route is indicated in emergency, life-threatening situations, such as the following:

1. Pulmonary edema or interstitial edema resulting from left ventricular failure.
2. Severe HF; with poor oral absorption.
3. Hypertensive crisis.
4. Hypercalcemia and hyperkalemia.

In addition to the mechanism of action of loop diuretics outlined earlier, furosemide causes venodilation. This action involves prostaglandins and can be inhibited by nonsteroidal antiinflammatory drugs (NSAIDs). Furosemide has a half-life of 1.5 h and a duration of action of 4-6 h. Diuresis commences some 15-20 min after IV administration, but relief of shortness of breath may be apparent within 10 min because of an increase in systemic venous capacitance, reduced cardiac preload, and a decrease in left atrial pressure. After oral administration, diuresis peaks in 60-90 min.

Symptoms

1. In women of child-bearing potential, except in life-threatening situations, in which IV furosemide may be absolutely necessary. Furosemide has caused fetal abnormalities in animal studies.
2. Hepatic failure.
3. Hypokalemia or electrolyte depletion, hyponatremia, or hypotension.
4. Hypersensitivity to furosemide or sulfonamides.

Alarming Symptoms

1. Monitor electrolytes, blood urea, creatinine, complete blood counts, and uric acid, especially when the dose exceeds 60 mg daily.
2. Commence with a minimum dose of 20-40 mg, especially in the elderly.

Abnormalities

Hypokalemia, dehydration, anemia, leukopenia, thrombocytopenia, rare agranulocytosis, and thrombophlebitis have been noted, but aplastic anemia seems to be more common with thiazides than with furosemide. Hypotension, hyperuricemia and precipitation of gout, hypocalcemia, and precipitation of nonketotic hyperosmolar diabetic coma may occur.

Biochemical Activities

1. Use carefully in the presence of renal dysfunction when combined with *cephalosporin* or *aminoglycoside antibiotics*, because increased nephrotoxicity has been noted.
2. Care should be taken when loop diuretics or thiazides are given to *lithium-treated* patients. The decreased Na^+ reabsorption in the proximal tubules causes an increased reabsorption of lithium and may cause *lithium toxicity.* Patients receiving concomitant chloral hydrate may experience hot flushes, sweating, and tachycardia. Prostaglandin inhibitors indomethacin and other NSAIDs antagonize the actions of loop diuretics as well as thiazides.

3. The effects of tubocurarine may be increased.

Drug name:	*Ethacrynic acid*
Trade name:	Edecrin
Supplied:	50 mg
Dosage:	Oral dose of 50–150 mg daily or in an emergency IV 50 mg diluted with 50 mL 5% dextrose/water given slowly

This is a potent loop diuretic, similar in action to furosemide. The drug may cause slightly more Cl^- loss than furosemide. Ethacrynic acid has slightly more side effects (although there is less increase in serum uric acid concentration), and it is therefore reserved for patients who are resistant to the effects of furosemide. Ethacrynic acid is not a sulfonamide and exhibits no cross-sensitivity with the thiazides. Ethacrynic acid is therefore useful in patients allergic to sulfonamides. The anticoagulant effect of warfarin is increased by ethacrynic acid.

Drug name:	*Bumetanide*
Trade names:	Burinex, Bumex
Supplied:	Tablets: 0.5, 1, and 5 mg
	Ampules: 2, 4, 10 mL; 500 mg/mL
Dosage:	Oral: 0.5-1 mg daily increased if required to 2-4 mg; 5 mg in oliguria
	IV: 1–3 mg over 1–2 min, repeated after 20 min

Bumetanide is as effective as furosemide, and it has a similar site of action in the medullary diluting segment. The drug is excreted along with its metabolites in the urine and is believed to cause less Mg^{2+} loss than furosemide during long-term administration. Bumetanide and furosemide have similar pharmacokinetic characteristics. Bumetanide is more potent than furosemide: 1 mg bumetanide = 40 mg furosemide.

The drug is absorbed more rapidly in patients with HF and has a bioavailability twice that of furosemide. Bumetanide is more nephrotoxic but appears to be less ototoxic than furosemide, so it is prudent not to use the drug with aminoglycosides or other potentially nephrotoxic drugs. Indications are as listed for furosemide. The drug is not approved in the United States for hypertension.

Drug name:	*Torsemide*; *torasemide* (UK)
Trade names:	Demadex, Torem (UK)
Supplied:	Tablets: 5, 10, 20, 100 mg
	Ampules: 50 mg
Dosage:	Oral: 2.5-40 mg once daily; HF dose 20-200 mg
	IV: 5-50 mg

Toresemide given orally achieves high bioavailability, and a 10-mg dose produces diuresis equivalent to 40 mg furosemide.

POTASSIUM IN ACTION

Increased renin release from the juxtaglomerular cells is caused by several conditions:

1. Reduction in renal blood flow from HF, blood loss, hypotension, or ischemia.
2. Na^+ diuresis.
3. Beta-adrenergic stimulation.

Renin converts liver angiotensinogen to angiotensin I. Angiotensin II and adrenocorticotrophic hormone stimulate adrenal aldosterone production. The "*aldosterone antagonists*" are very weak diuretics. Aldosterone handles approximately 2% of filtered Na^+ at the distal tubule, so only a small diuresis is achieved.

Diuretics that block aldosterone or act at the same site distal to the macula densa cause a small amount of Na^+ excretion and prevent exchange of K^+. Only spironolactone and potassium canrenoate antagonize aldosterone. Spironolactone interferes with the effect of aldosterone to increase the rate of Na^+-K^+ exchange at the basolateral surface. Amiloride and triamterene interact with lumen membrane transporters to prevent urinary Na^+ entry into the cytoplasm. They are direct inhibitors of K^+ secretion. K^+-sparing diuretics play a vital role in conserving K^+ and Mg^{2+} in patients treated with thiazides or loop diuretics.

The addition of a K^+-sparing drug to low-dose thiazide therapy results in a reduced risk of cardiac arrest. The four available K^+-sparing diuretics are *amiloride* (Midamor), *spironolactone* (Aldactone), *triamterene* (Dyrenium, Dytac), and *potassium canrenoate* (Spiroctan-M). These weak diuretics are very important in the following situations:

1. *Spironolactone* 25 mg added to an ACE inhibitor in patients with HF causes a more complete block of aldosterone production than is achieved with ACE inhibition and reduces mortality and morbidity in these patients. The salutary effects are related not only to Na^+ loss but also to a decrease in cardiac fibrosis, retardation of endothelial dysfunction, and increased nitric oxide production caused by inhibition of aldosterone. Aldosterone, a compensatory "good" hormone, appears to have fibrogenic and other deleterious properties. Tissue collagen turnover and fibrosis appear to be important facets of HF, and spironolactone may attenuate deleterious structural remodeling.
2. When added to thiazides or loop diuretics, diuresis is greatly augmented. The serum K^+ often remains within the normal range.
3. Clinical situations in which secondary aldosteronism is involved. The "aldosterone antagonists" are first-line diuretics in conditions associated with secondary aldosteronism:
   (a) Cirrhosis with ascites.
   (b) Nephrotic syndrome.
   (c) Chronic *recurrent HF*. (They can be extremely effective and beneficial when added to loop or to thiazide diuretics and ACE inhibitors.)
   (d) Cyclical edema.
   (e) Renovascular hypertension.

Spironolactone is also of value in the diagnosis of primary aldosteronism and in treatment of Bartter' s syndrome. The following are suggested guidelines for the diuretic management of the patient with marked ascites caused by cirrhosis. Ensure that

1. Hepatic encephalopathy is not present.
2. The patient can sign his or her name and constructional apraxia is not present.
3. The patient can tolerate a 60-80-g protein diet for 1 wk without the precipitation of encephalopathy.
4. Jaundice is either not present or not increasing.

If these four assessments are passed for more than 7 d, then commence spironolactone 25-100 mg twice daily. Some 2-8 wk later, if there is no encephalopathy, add 25 mg of hydrochlorothiazide (HCTZ) every second day and then if needed daily, or furosemide 40 mg daily. It may be necessary to wait 2-3 mo to achieve a 75% reduction in the ascites.

Symptoms

The following are contraindications and warnings for the use of K^+-sparing diuretics or combinations with thiazides or furosemide:

1. Acute or chronic renal failure. These drugs should be avoided if there is any evidence of renal failure, in particular a serum creatinine level > 1.3 mg/dL (115 μmol/L) in patients aged <70 yr, and > 1.0 mg/dL (88 μmol/L) for those > 70 yr, or urea > 7 mmol/L. These ground rules should be broken only if the patient is under strict observation in hospital and an order is written to discontinue the medications if the serum K^+ level is >5 mEq (mmol)/L.
2. Do not use in conjunction with K^+ supplements or ACE inhibitors or in patients who have metabolic acidosis because these diuretics may themselves increase acidosis by retaining H^+ along with K^+.
3. Do not use triamterene combined with indomethacin or other NSAIDs; acute renal failure may be precipitated. Triamterene is relatively insoluble and may precipitate as *renal calculi.* This agent is contraindicated in patients who have had a renal stone.

Spironolactone is a competitive inhibitor of aldosterone.

Drug name: Spironolactone

Trade names: Aldactone, Spiroctan (UK)

Dosage: 25-100 mg daily in single or two divided doses

Beneficial Activities

1. The drug appears to decrease cardiac fibrosis and endothelial dysfunction and to increase nitric oxide bioactivity. These actions appear to explain the beneficial effects of the drug in patients with HF, as shown in the Randomized Aldactone Evaluation Study (RALES).
2. Spironolactone is metabolized in the liver, whereas amiloride is excreted by the kidney, so if renal dysfunction supervenes, the risk of hyperkalemia is less with spironolactone.
3. Spironolactone does not cause aplastic anemia or have any other serious hematologic effects. Amiloride and triamterene have been associated with aplastic anemia (although rarely).

4. The drug does not cause megaloblastic anemia, seen rarely with triamterene.
5. Spironolactone has a positive inotropic effect independent of and additive to that of digitalis. Stroke volume is increased.

Harmful Activities

1. *Gynecomastia.* This depends on the dose and its duration. Keep the maintenance dose at less than 50 mg/d. Eplerenone is as effective as spironolactone. Most important, the drug does not cause gynecomastia and will replace spironolactone.
2. *Caution.* Potential human metabolic products are carcinogenic in rodents and tumorigenic in rats.

   Drug name: *Eplerenone*

   Trade name: Inspra

   Dosage: 12.5–25 mg once daily, max. 50 mg daily

   Eplerenone is an important addition to our therapeutic armamentarium for the management of HF and hypertension.

Biochemical Approach

It is now well established that thiazide diuretics cause significant K^+ loss, which increases the incidence of arrhythmias, cardiac arrest, and cardiac mortality. Conservation or replacement of K^+ is therefore essential and has been proved to decrease the risk of cardiac arrest. K^+-sparing diuretics are useful and can prevent the use of gastric-irritating potassium chloride (KCl) preparations in most patients if these drugs are used with careful restrictions.

1. *Dyazide.* Tablets (capsules, US) of 50 mg triamterene and 25 mg HCTZ.
2. *Dytide*: Tablets of triamterene 50 mg and 25 mg benzthiazide. *Dosage:* one tablet each morning or every second day; for mild to moderate hypertension, a dose of one tablet daily for maintenance; *contraindicated* in patients who have had a *renal stone* or renal failure.
3. *Aldactazide.* Tablets of 25 mg spironolactone and 25 mg HCTZ.
4. *Aldactide.* Tablets of 25 mg spironolactone and 25 mg hydroflumethiazide; tablets of 50 mg spironolactone and 50 mg hydroflumethiazide.
5. Moduretic (moduret). Tablets of 50 mg HCTZ and 5 mg amiloride hydrochloride; Moduret 25 or nonproprietary in the United Kingdom, co-amilozide = HCTZ 25 mg, amiloride 2.5 mg; *Dosage:* 1 tablet each morning; not advisable to use more than 1 tablet daily.
6. *Maxzide.* Tablets of 75 mg triamterene and 50 mg of HCTZ. *Dosage:* half to 1 tablet each morning.
7. *Frumil* (lasoride): Tablets of 40 mg frusemide and 5 mg amiloride, given 1 to 2 tablets daily.
8. *Frusene.* Tablets of 40 mg frusemide and 50 mg triamterene.

Alarming Symptoms

1. K^+-sparing diuretics must not be given concomitantly with K^+ supplements or ACE inhibitors except under strict supervision; severe hyperkalemia may result.

2. Elderly diabetic patients may develop hyporeninemic hypoaldosteronism and may therefore retain K^+ despite a normal serum creatinine level.
3. If the patient develops gynecomastia during spironolactone therapy, then triamterene or amiloride can replace the spironolactone because these two agents do not cause gynecomastia. They are devoid of the hormonal effects of spironolactone. Triamterene, however, is slightly insoluble and can precipitate as renal calculi.
4. K^+-sparing diuretics may rarely produce mild metabolic acidosis.

OTHER DRUGS CAUSING DIURESIS

Drug name: *Metolazone*
Trade names: Zaroxolyn, Metenix (UK)
Supplied: 2.5, 5, 10 mg
Dosage: 2.5-5 mg once daily; rarely 10 mg

Metolazone has a prolonged action of up to 24 h. The drug acts in both the proximal convoluted tubule and the distal nephron, similar to thiazide. Both thiazides and metolazone have secondary effects in the proximal tubule that are not usually manifest because the proximally rejected ions are ordinarily reabsorbed in the loop.

Thus, combinations of metolazone and loop diuretics are very effective in the management of intractable HF. Sequential nephron blockade is a proven concept. Thiazides become ineffective when the GFR falls to <30 mL/min, whereas loop diuretics and metolazone retain effectiveness. The combination of metolazone and a loop diuretic is very potent and useful, but K^+ loss is often pronounced.

Drug name: *Acetazolamide*
Trade name: Diamox
Supplied: 250 mg
Dosage: 250 mg three times daily, max. 4 d; treatment can be repeated once or twice in a month.

This drug is a carbonic anhydrase inhibitor. It causes excretion of HCO_3^-, retention of Cl^-, and, consequently, metabolic acidosis and hyperchloremia. Acetazolamide is a very weak diuretic, and the action is lost after 4 d. It is of importance only in the management of hypochloremic metabolic alkalosis in the presence of a normal serum K^+ level. The typical case is one of refractory HF in a patient taking furosemide and K^+-sparing diuretics.

The electrolyte picture shows $Cl^- < 92$ mEq (mmol)/L, CO_2 >30 mEq (mmol)/ L, and K^+ 3.5-5 mEq (mmol)/L. In such cases, acetazolamide, added to the spironolactone with the furosemide dose discontinued or halved, results in continued diuresis and correction of the normokalemic hypochloremic metabolic alkalosis.

Acetazolamide is contraindicated in patients with renal failure, renal calculi, metabolic acidosis, and severe cirrhosis.

HYPOKALEMIA

To physicians caring for cardiac and hypertensive patients, diuretics are the commonest cause of hypokalemia. The incidence of hypokalemia is about 5-30% with HCTZ and 5-20% with loop diuretics, but as high as 50-100% with chlorthalidone. The author concurs with these observations and ceased to use chlorthalidone in 1974. Nephrologists may use the drug successfully in patients with mild renal impairment because K^+ is usually retained in that situation. Thus, the nephrologist may recommend this drug, inappropriately, for general use. Mild hypokalemia of 3-3.5 mEq (mmol)/L is of concern and must be corrected if the patient is taking digoxin or has arrhythmias, cardiac disease, or weakness.

If metabolic acidosis is present, a serum K^+ level < 3.5 mEq (mmol)/L constitutes a definite total body K^+ deficit. Minor decreases in serum K^+ concentration to 3 mEq (mmol)/L may be corrected in most patients by asking them to ingest foods that are rich in K^+.

Table 10.2 K^+-Enriched foods

Food	*Amount*	*K^+ (mEq)*
Orange juice	Half cup	6
Milk (skim-powdered)	Half cup	27
Milk (whole-powdered)	Half cup	20
Melon (honeydew)	Quarter	13
Banana	One	10
Tomato	One	6
Celery	One	5
Spinach	Half cup	8
Potato (baked)	Half	13
Beans	Half cup	10
Strawberries	Half cup	3
Meats, shellfish, and avocado	All contain increased K^+	

It is a relatively useless exercise to tell the patient to take an extra glass of orange juice, as is commonly done (6 oz = 8.4 mEq K^+). Note that salt substitutes, such as Co-salt, Nosalt, and other brands, also aid in increasing serum K^+ levels. However, salt substitutes contain KCl and may therefore cause gastric irritation in some patients. It is important not to give KCl mixtures along with K^+-sparing diuretics or an ACE inhibitor and then continue with enriched diets and salt substitutes without knowledge of renal function, because this occasionally causes hyperkalemia. Patients taking thiazide diuretics should be allowed to continue for at least 1-2 mo and the electrolytes reassessed.

Depending on the dose, hypokalemia occurs in 30-50% of patients. If, however, patients are instructed to follow a diet containing K^+-rich foods, such as those outlined in Table 10.2, the incidence of hypokalemia can be reduced. Patients showing even mild hypokalemia should be given KCl supplements or preferably taken off thiazides and given a thiazide K^+-sparing diuretic.

As an alternative, patients may be started on Moduretic or Dyazide or Frumil if renal function is normal. If the serum K^+ concentration is <2.5 mEq (mmol)/L, IV K^+ is given. In the range of 2.5-3.5 mEq (mmol)/L, oral K^+ is usually sufficient, but it must always be in the form of Cl^- except in renal tubular acidosis, in which citrates and bicarbonates can be used.

Diuretics cause significant hypokalemia, but only rarely do they cause a significant fall in total body K^+. Compare extracellular 65 mEq (mmol) with 4000 mEq (mmol) intracellular total K^+. Thus, the importance of determining the serum K^+ level is emphasized, because it is abnormalities at this level that dictate alterations of the K^+ gradient across the myocardial cell membrane and that can result in severe electrical changes and cardiac arrhythmias.

Fluctuations of the serum K^+ concentration are often exaggerated by *acidosis*, causing hyperkalemia, and by *alkalosis*, causing hypokalemia. It is necessary to watch for the occurrence of the two conditions because they can be altered within minutes (e.g., metabolic acidosis during seizures, diabetic ketoacidosis, cardiac arrest, or respirator hyperventilation causing respiratory alkalosis and perhaps triggering ventricular tachycardia or ventricular fibrillation). A low serum K^+ level *reduces ventricular fibrillation threshold* and therefore increases the potential for sudden death.

Not to forget

1. Hypokalemia produced by catecholamines is mediated by $beta_2$-adrenoceptors. The increase in catecholamines that occurs during acute myocardial infarction can cause a significant decrease in serum K^+ concentration. Catecholamine-induced hypokalemia may be prevented by $beta_2$-blockade.
2. The $beta_2$-stimulants salbutamol, terbutaline, and pirbuterol may precipitate hypokalemia that is transient but perhaps important.

Effect of Potassium Chloride

1. *Mixtures.* Patients dislike the taste of these costly mixtures. A dose of 20 mEq (mmol)/ L is given twice daily, and it is usually adequate, along with a K^+-rich diet. Patients in whom the serum K^+ level consistently falls to <3 mEq (mmol)/L, despite this regimen, and who are taking necessary doses of diuretics may require as much as 40 mEq KCl three times daily. It is preferable in these patients to add a K^+-retaining diuretic rather than using KCl.
2. The *effervescent* potassium preparations contain very little KCl and are not recommended.
3. KCl tablets, capsules, or slow-release wax matrix are not completely satisfactory because there is a significant incidence of ulceration of the gastrointestinal tract, including perforations. Controlled-release preparations, K-Dur 20 (US) and K-Contin (UK), are safer than wax matrix KCl. Nonwax matrix preparations include Micro-K and K-Dur 20. The dispersion and slow-release characteristics of these preparations are believed to minimize contact between erosive K^+ and the mucosal lining, but caution is required.

   Diuretic-K^+ combinations are not recommended.

Sodium versus Potassium

The patient who has edema, HF, or hypertension must be on a restricted-Na^+ diet. There is a definite case, therefore, for salt substitutes in which K^+ takes the place of Na^+. There are many such products on the market–Nosalt, Cosalt, Morton's salt substitute, and Featherweight-K all have a reasonable taste. It is important, however, to recognize that the occasional patient may develop gastric discomfort. Thiazide and loop diuretics cause K+ and Mg^{2+} losses. Some patients with severe K^+ deficiency may require supplemental Mg^{2+} to achieve correction of the K^+ and Mg^{2+} deficiency. The Mg^{2+} deficiency often continues undetected. Importantly, K^+-sparing diuretics are also Mg^{2+} sparing.

Intravenous Potassium Chloride

Care should be taken with IV KCl because death from *iatrogenic hyperkalemia* is not uncommon.

Table 10.3 Potassium supplements

	Ingredients	K^+ *mEq (mmol)*	Cl^- *mEq (mmol)*
Liquids			
Kay Ciel; Kay-Cee-L 1 mmol/mL	KCl	20	20
Potassium chloride 10%	KCl	20	20
K-Lor (paquettes)	KCl	20	20
K-Lyte Cl	KCl	25	25
Kaochlor 10% (or sugar free)	KCl	20	20
Klorvess 10%	KCl	20	20
Kolyum	KCl	20	3
Kaon Elixir	K gluconate	20	–
Kaon Cl 20%	KCl	40	40
K-Lyte (effervescent)	KH_2CO_3	25	–
Potassium triplex	not KCl	15	–
Potassium sandoz	KCl	12	8
Rum K	KCl	20	20
Tables/capsules: slow release			
K-Long	KCl	6	6
Kalium durules	KCl	10	10
Kaon	K gluconate	5	–
Leo K	KCl	8	8
Nu-K	K	8	8
Sando K	K	12	8
Slow-K	KCl	8	8
K-Dur 20	KCl	20	20
Micro-K	KCl	8	8
Micro-K-10	KCl	10	10

1. Use IV KCl only when necessary, that is, when $K^+ < 2.5$ mEq (mmol)/L.
2. Ensure an adequate urine output and that renal failure is absent.
3. Do not give along with K^+-sparing diuretics, captopril, or enalapril.
4. Correct metabolic acidosis or alkalosis.

Not to forget

1. In the presence of *metabolic alkalosis* with a pH > 7.5 and CO_2 > 30 mEq (mmol)/L, smaller amounts of KCL are required.
2. *Metabolic acidosis,* with a pH < 7.3 and serum K^+ level < 3 mEq (mmol)/L, means a major K+ deficit and therefore the need for correction over several days.

Dilute KCl as much as possible: 40-60 mEq (mmol)/L. In noncardiac patients, dilute KCl in *normal saline,* especially if severe hypokalemia is present. In life-threatening situations, dilute in saline for all patients.

Limitations

1. If aggressive therapy is not required and there is a need to limit Na^+ load, as in the presence of HF or poor cardiac contractility in patients with recurrent or past HF.
2. If on the first day of therapy the KCl was diluted in saline.

Precautions

1. The container must not usually contain >40 mEq (mmol) KCl.
2. The rate should not exceed 10 mEq (mmol)/h, except in severe hypokalemia <2.5 mEq (mmol)/L or when symptoms or arrhythmias are present.
3. If the rate must exceed 10 mEq (mmol)/h, an electrocardiographic monitor is necessary with observations every 15 min. The maximum rate of 30 mEq (mmol)/h is rarely necessary.

ANNOTATION

1. The loop diuretic *torsemide* (torasemide, UK) is completely absorbed and appears to have higher reliability and predictability compared with furosemide. The variability of absorption is 9 for torsemide and 30 for furosemide.
2. Torsemide has been shown in a small study to cause a dramatic decrease in hospitalization and reduced mortality in patients with class III and IV HF compared with furosemide. The simple expedient of the choice of a diuretic can result in a decrease in hospitalization for HF of >33%.
3. Food further reduces absorption of furosemide, and not many physicians recognize that this well-known drug must be taken on an empty stomach.

Aldosterone antagonists hold the key. The therapeutic value of aldosterone antagonists must not be underestimated in the management of HF of all grades. The drug should not be reserved for class IV HF.

Index